National Center for Construction Edu

Project Supervision

Annotated Instructor's Guide

This information is general in nature and intended for training purposes only. Actual performance of activities described in this manual requires compliance with all applicable operating, service, maintenance, and safety procedures under the direction of qualified personnel. References in this manual to patented or proprietary devices do not constitute a recommendation of their use.

Copyright © 2003 by the National Center for Construction Education and Research (NCCER) and published by Pearson Education, Inc., Upper Saddle River, New Jersey 07458. All rights reserved. Printed in the United States of America. This publication is protected by Copyright and permission should be obtained from the NCCER prior to any prohibited reproduction, storage in a retrieval system, or transmission in any form or by any means, electronic, mechanical, photocopying, recording, or likewise. For information regarding permission(s), write to: NCCER, Curriculum Revision and Development Department, POB 141104, Gainesville, FL 32614-1104.

10 9 8 7
ISBN 0-13-103596-7

PREFACE

Field supervisors play a major role in every construction company and every construction project. They are the frontline managers on the job, directly supervising workers and other field supervisors. They are both the engine and the anchor of the construction team, driving it toward effectiveness and efficiency, and stabilizing it with consistency and good judgment. In essence, their skills and leadership largely determine whether the job is built on time and within budget.

To fill this enormous role, field supervisors need more than experience in the field. They also need management skills in problem solving, planning, estimating, safety supervision, scheduling, controlling costs and resources, and, perhaps most important, managing people. These are skills most easily acquired through education.

Project Supervision *provides the basis for that education. It is a comprehensive, competency-based program that gives both veteran and new field managers a step-by-step approach to honing their natural abilities, developing essential skills, and generally improving their performance as leaders.*

The program consists of a Participant Manual and an Instructor's Guide. The Participant Manual contains the substance of the course — the information every supervisor needs to master. The manual also contains exercises and self-check evaluations that permit the instructor to evaluate participants' learning while allowing participants to monitor their own progress. For the convenience of the instructor and those participants who study the program on their own, an Instructor's Guide has been prepared as a companion to the Participant Manual. The Instructor's Guide contains information necessary for organizing and teaching the program, including recommendations on the number of classroom hours for each section of each module, key questions for class discussion, and the answers to the participant activities and self-check exercises.

To gain the most from this program, the participant should attend formal classes taught by a trained and experienced instructor. However, the structure of the program also allows a participant to study and complete the program independently by using the materials in the Participant Manual and the Instructor's Guide.

A note to participants: not all the information and examples in the program will exactly match the type of work you do. No two construction companies operate in the same way, and you may find that some points in the manual do not apply to your company or your job. Nonetheless, all the information in this program is important and should be learned, because it reflects common practices in the industry as a whole and may be helpful to you in the future. Details of field supervision may vary from company to company and project to project, but the overall process does not. All field supervisors share the same general responsibilities.

Contents

MT201-01	Orientation to the Job	1.1
MT202-01	Human Relations and Problem Solving	2.1
MT203-01	Safety	3.1
MT204-01	Quality Control	4.1
MT205-01	Contract and Construction Documents	5.1
MT206-01	Document Control and Estimating	6.1
MT207-01	Planning and Scheduling	7.1
MT208-01	Resource Control and Cost Awareness	8.1

Project Supervisor

Module MT201-01
Orientation to the Job

Orientation to the Job
Instructor's Guide

Module MT201

MODULE OVERVIEW

This module introduces the project supervisor trainee to the role of a supervisor. This module will enable the trainee to manage people, meet project schedules, stay within the budget, and maintain safety on the job site.

PREREQUISITES

There are no prerequisites for this module.

LEARNING OBJECTIVES

Upon completion of this module, the trainee will be able to:

1. Explain the scope and purpose of the *Project Supervision* program.
2. Understand the role of a construction supervisor.
3. Explain the history, trends, and economic conditions affecting the construction industry.
4. Outline the progress of a successful construction project from initial development through completion.
5. Identify the milestones in the growth of a construction company and the reasons for a formal and informal organizational development.
6. Explain the functions of management.
7. Explain the purpose and content of a satisfactory job description.
8. Discuss company policies and procedures.

PERFORMANCE OBJECTIVES

This is a knowledge-based module – there is no performance profile examination.

NCCER STANDARDIZED TRAINING PROGRAM

The National Center for Construction Education and Research (NCCER) provides a standardized national program of accredited craft training. Key features of the program include instructor certification, competency-based training, and performance testing. The program provides trainees, instructors, and companies with a standard form of recognition through a National Craft Training Registry. The program is described in full in the Guidelines for Accreditation, published by the NCCER. For more information on standardized craft training, contact the NCCER by writing us at P.O. Box 141104, Gainesville, FL 32614-1104; calling 352-334-0911; or e-mailing info@nccer.org. More information may be found at our Web site at www.nccer.org.

HOW TO USE THIS ANNOTATED INSTRUCTOR'S GUIDE

Each page presents two sections of information. The larger section displays each page exactly as it appears in the Trainee Module. The narrow column ties suggested trainee and instructor actions to each page and provides icons to call your attention to material, safety, audiovisual, or testing requirements. The bottom of each page includes space for your notes.

 If you see the Teaching Tip icon, that means there is a teaching tip associated with this section. Also refer to the suggested teaching tips at the end of the module.

PREPARATION

Before teaching this module, you should review the Module Outline, Learning Objectives, and the Materials and Equipment List. Be sure to allow ample time to prepare your own training or lesson plan and gather all required equipment and materials.

MATERIALS AND EQUIPMENT LIST

Materials:

Transparencies

Markers/chalk

Module Examinations*

Sample drawings, specifications, bid documents, quantity take-off sheets, organizational charts, and purchase orders**

Equipment:

Overhead projector and screen

Whiteboard/chalkboard

*Located in the Test Booklet packaged with this Annotated Instructor's Guide.
**If available on loan from your workplace or other resource.

ADDITIONAL RESOURCES

This module is intended to present thorough resources for task training. The following reference works are suggested for both instructors and trainees interested in further study. These are optional materials for continued education rather than for task training.

Construction Contracting, 1994. Richard H. Clough and Glenn A. Sears. New York: John Wiley & Sons.

Professional Construction Management: Including Contracting, 1991. Donald S. Barrie and Boyd C. Paulson (Contributor). New York: McGraw-Hill Higher Education.

Construction Management, 1997. Daniel W. Halpin and Ronald W. Woodhead. New York: John Wiley & Sons.

Construction Operations Manual of Policies and Procedures, 2000. Andrew Civitello Jr. New York: McGraw-Hill.

TEACHING TIME FOR THIS MODULE

An outline for use in developing your lesson plan is presented below. Note that each Roman numeral in the outline equates to one session of instruction. Each session has a suggested time of 2 1/2 hours. This includes 10 minutes at the beginning of each session for administrative tasks and one 10-minute break during the session. Approximately 5 hours are suggested to cover *Orientation to the Job*.

Topic	Planned Time
Session I. Introduction	
A. Introduction	_____
B. The Role of a Supervisor	_____
C. History and Roots of Construction	_____
D. The Construction Industry	_____
1. Business Failures	_____
2. Business Concepts	_____
3. The Union Shop	_____
4. The Non-Union Shop	_____
5. Training	_____
E. Phases of a Construction Project	_____
1. Development Phase	_____
2. Design (Planning) Phase	_____
3. Construction Phase	_____
4. Construction Flow	_____
5. Bidding Phase	_____
6. Pre-Construction Phase	_____
Session II. Phases of a Construction Project	
A. Phases of a Construction Project	_____
1. Construction Phase	_____
2. Closeout Phase	_____
B. Construction Organization	_____
1. The Growth of an Organization	_____
2. Formal Organization	_____
3. Informal Organization	_____
4. Span of Control	_____
5. Authority and Responsibility	_____
C. Management Functions	_____
1. Planning	_____
2. Organizing	_____
3. Staffing	_____
4. Directing	_____
5. Controlling	_____
D. Employment Requirements	_____
1. Employing Site Personnel	_____
2. New Employee Orientation	_____
3. Coaching and Mentoring	_____
4. Training	_____
E. Policies and Procedures	_____

F. Summary
   1. Summerize Module
   2. Answer Questions
G. Module Examination
   1. Trainees must score 70% or higher to receive recognition from the NCCER.
   2. Record the testing results on Craft Training Report Form 200 and submit the results to the Training Program Sponsor.

Project Supervision – Module MT201

Orientation to the Job

Instructor's Notes:

ACKNOWLEDGMENTS

The NCCER wishes to acknowledge the dedication and expertise of Phil Copare, the original author and mentor for this module on leadership development.

Philip B. Copare, MBA

President, Construction Services Enterprise

Education and Safety Consultant

Zellwood, FL

We would also like to thank the following reviewers for contributing their time and expertise to this endeavor:

J.R. Blair

Tri-City Electrical Contractors
An Encompass Company

Mike Cornelius

Tri-City Electrical Contractors
An Encompass Company

Dan Faulkner

Wolverine Building Group

David Goodloe

Clemson University

Kevin Kett

The Haskell Company

Danny Parmenter

The Haskell Company

Course Map

This course map shows all of the modules of the *Project Supervision* curriculum. The suggested training order begins at the bottom and proceeds up. Skill levels increase as you advance on the course map. The local Training Program Sponsor may adjust the training order.

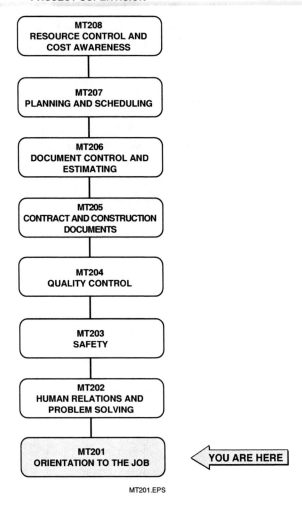

Instructor's Notes:

NATIONAL CENTER FOR CONSTRUCTION EDUCATION AND RESEARCH

MODULE MT201

TABLE OF CONTENTS

1.0.0	**INTRODUCTION**	1.1
2.0.0	**THE ROLE OF A SUPERVISOR**	1.1
3.0.0	**HISTORY AND ROOTS OF CONSTRUCTION**	1.2
4.0.0	**THE CONSTRUCTION INDUSTRY**	1.3
4.1.0	Business Failures	1.3
4.2.0	Business Concepts	1.4
4.2.1	*The Union Shop*	1.4
4.2.2	*The Non-Union Shop*	1.4
4.3.0	Training	1.4
5.0.0	**PHASES OF A CONSTRUCTION PROJECT**	1.5
5.1.0	Development Phase	1.5
5.2.0	Design (Planning) Phase	1.7
5.3.0	Construction Phase	1.7
5.4.0	Construction Flow	1.7
5.4.1	*Bidding Phase*	1.9
5.4.2	*Pre-Construction Phase*	1.9
5.4.3	*Construction Phase*	1.11
5.4.4	*Closeout Phase*	1.11
6.0.0	**CONSTRUCTION ORGANIZATION**	1.11
6.1.0	The Growth of an Organization	1.11
6.2.0	Formal Organization	1.12
6.3.0	Informal Organization	1.12
6.4.0	Span of Control	1.12
6.5.0	Authority and Responsibility	1.13
7.0.0	**MANAGEMENT FUNCTIONS**	1.13
7.1.0	Planning	1.13
7.2.0	Organizing	1.13
7.3.0	Staffing	1.14
7.4.0	Directing	1.14
7.5.0	Controlling	1.14
8.0.0	**EMPLOYMENT REQUIREMENTS**	1.14
8.1.0	Employing Site Personnel	1.14
8.2.0	New Employee Orientation	1.14
8.3.0	Coaching and Mentoring	1.15
8.4.0	Training	1.15
9.0.0	**POLICIES AND PROCEDURES**	1.16
	SUMMARY	1.17
	GLOSSARY	1.19

LIST OF FIGURES

Figure 1	•	Shift in Use of Time From Crew Member to Supervisor......1.2
Figure 2	•	Project Flow Diagram....................................1.6
Figure 3	•	Construction Flow.......................................1.9
Figure 4	•	Example of a Job Description.........................1.15
Figure 5	•	Sample Safety Policy and Procedure 1.16

Instructor's Notes:

NATIONAL CENTER FOR CONSTRUCTION EDUCATION AND RESEARCH

MODULE MT201

Orientation to the Job

Ensure that you have all the necessary materials to teach the course. Check the Materials and Equipment list at the front of the module. Prepare for teaching Session I by reading Sections 1.0.0–5.4.2.

OBJECTIVES

Upon the completion of this module, you will be able to do the following:

1. Explain the scope and purpose of the *Project Supervision* program.
2. Understand the role of a construction supervisor.
3. Explain the history, trends, and economic conditions affecting the construction industry.
4. Outline the progress of a successful construction project from initial development through completion.
5. Identify the milestones in the growth of a construction company and the reasons for a formal and informal organizational development.
6. Explain the functions of management.
7. Explain the purpose and content of a satisfactory job description.
8. Discuss company policies and procedures.

SECTION 1

1.0.0 INTRODUCTION

Congratulations! Whether you are participating in this course because of a company promotion, company-ordered training, or self-interest, you are taking the next step in advancing your career. The role of *supervisor* is an important position and knowing what is involved is the first step in preparing yourself for the job.

The duties of a supervisor are varied, challenging, plentiful, and rewarding. Becoming a supervisor and performing a supervisor's duties brings up a lot of questions for those new to supervision. What is the role of the supervisor? As a supervisor, where do I fit into the management structure? Do I need to know every company policy, procedure, or job-site function? Do I have to know a lot about the construction industry? These questions and more are all answered in this module.

SECTION 2

2.0.0 THE ROLE OF A SUPERVISOR

A supervisor has been authorized by the company to use the workforce, equipment, and materials to complete a particular job or task. In general, a supervisor is responsible for completing work to the satisfaction of all code requirements, drawings, specifications, and contractual agreements while staying within budget.

Ask trainees to explain the different types of supervisors and their functions within their respective companies.

Show Transparency 1-1, (Course Objectives).

Assign reading of Module MT201 Sections 1.0.0–5.4.2

Copyright © 2003 National Center for Construction Education and Research, Gainesville, FL 32614-1104. All rights reserved. No part of this work may be reproduced in any form or by any means, including photocopying, without written permission of the publisher.

Ask trainees to give examples of how labor, materials, and technology have changed in the construction industry.

Show Transparency 1-2 (Figure 1).

In this manual, the term *supervisor* includes both frontline and second-line field supervisors. Traditionally, we define frontline supervisor (also referred to as a foreman or lead) as one who supervises one or more craftworkers. A second-line supervisor, or superintendent, is one who supervises one or more frontline supervisors.

Today, there is a serious need for new, trained supervisors. Advancing workers from the ranks to supervisory positions based only on their length of time in a position or with a company is no longer acceptable. Even the deeply entrenched belief that new supervisors learn more through experience than they can in the classroom is being discarded. Contractors now realize that they waste thousands of dollars on low productivity, rework, and other costly mistakes whenever they leave supervisors to learn the job on their own. If new supervisors are to be successful, these new skills must be learned quickly. Formal training is the fastest and most effective way to acquire these skills. Because of this, your company has chosen you to attend this program.

As employees move from crew member to supervisor, they may find themselves spending more hours on supervisory tasks and less time on technical work. *Figure 1* shows this shift in the use of time.

This shift from technical to supervisory tasks presents problems. New supervisors are generally promoted on the basis of their personal ability to get the work done, get along with people, meet project schedules, and stay within budget. Suddenly, the new supervisor is responsible for harnessing these abilities in several people.

SECTION 3

3.0.0 HISTORY AND ROOTS OF CONSTRUCTION

Today's construction industry provides more than buildings where people live, work, shop, worship, and learn. It also builds highways, bridges, airports, and tunnels that enable goods and people to move freely about the country. It builds reservoirs, dams, power stations, irrigation systems, and sewage and flood control networks that provide water and power and protect public health. Without the construction industry, our lives would be considerably different and less comfortable.

The construction industry is the largest industry in the United States — larger than the steel and automobile industries combined. In the mid-1990s, the total value of new construction in the United States was over $500 billion, and the industry is expected to keep growing. With this growth comes the need for more and more trained construction personnel. However, before we look to the future needs of the construction industry, let's look back to the past.

Construction predates the earliest written records and can be traced back to man's need to build shelters against the elements and human and animal enemies. Shelter building expanded into village and city building. Early craftworkers

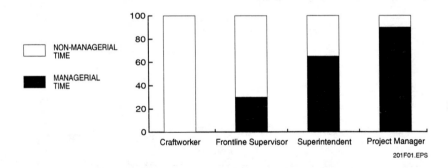

Figure 1 • Shift in Use of Time From Crew Member to Supervisor

Instructor's Notes:

had to develop from scratch the technologies for building everything from roads to great monuments in honor of gods and ancestors. As societies grew and changed, the skilled builder was there to meet mankind's ever-changing needs.

One thing has remained the same, however. Like the builders of the Great Pyramids of Egypt and the aqueducts of the Roman Empire, today's contractors rely on three elements to do their job:

1. Available materials
2. Labor
3. Know-how (technology)

These three elements are as much a part of the World Trade Center in New York and the Sears Tower in Chicago as they are of Stonehenge or the ancient Incan temples found in South America. They are the roots of construction today, as they were yesterday. These elements function just as well in providing skyscrapers that make the best use of scarce land in the modern city as they did in providing sprawling palaces for the kings and queens of ancient Europe and Asia.

What has changed are the characteristics of materials and technology. Today's materials are tougher and more reliable than any known before. Today's technology has solved construction problems undreamed of a hundred years ago. Who in the 1800s could have imagined building cities in outer space? We are now tackling even bigger technological challenges as engineers, architects, and contractors search for new ways to satisfy society's new needs.

Labor, too, has changed. Today's craftspeople are better trained, more open to new ideas, and more interested than their fathers and grandfathers in having a say in project decisions. Construction supervisors explore new materials and technologies, new avenues of human relations, new means of enhancing productivity, and new techniques of effective management.

SECTION 4

4.0.0 THE CONSTRUCTION INDUSTRY

The construction industry is huge and is on a path of steady growth. Such growth demands even greater sophistication and knowledge about business, management, and best practices. Supervisors must be lifelong learners, ready to acquire new skills and meet new challenges through continuous training and development.

4.1.0 Business Failures

The construction industry employs a large portion of the country's workforce, makes a major contribution to the gross national product, and, unfortunately, has one of the highest annual business failure rates in the economy. Studies show that the number of failures in the construction industry is much higher than it should be. Owners know the techniques of construction, but they have not developed adequate business management skills.

Most business failures occur within the first three years of operation. Some of the reasons for failure are:

- Limited working capital
- Failing to qualify for loans
- Too many projects starting at the same time
- Borrowing money from relatives and friends
- Company officers taking too large of a salary
- Purchase or lease of unnecessary, expensive vehicles
- Inaccurate bookkeeping records
- Poor project estimating due to:
  — Not knowing the cost of material
  — Underestimating labor costs
  — Not understanding overhead or general conditions
  — Low profit margins

Many companies are successful and continue to prosper despite the rate of failure in the construction industry. Owners of successful companies:

- Recognize potential causes for failure
- Continue to learn through educational seminars

Ask trainees to discuss their experiences with businesses that failed and what these companies could have done to succeed.

Ask trainees what training they have received to prepare them as supervisors.

- Develop company policies and procedures
- Get involved in their communities
- Share their time and knowledge with others through associations like Associated Builders and Contractors (ABC) and Associated General Contractors (AGC).

4.2.0 Business Concepts

Construction firms operate under one of two business "shop" concepts: **union shop** or **non-union shop.** National and state laws govern both kinds of business operations. Because of legal responsibilities, the supervisor must understand these two concepts and their advantages and disadvantages.

The following industry guidelines have been endorsed by the Associated Builders and Contractors:

- Provide opportunities for both union and non-union firms and employees in the construction industry.
- Give management control over the selection of craftworkers and construction techniques.
- Give management the freedom to promote and pay employees according to their skills, achievements, and desire.
- Stress performance, pride, economy, and efficiency within each organization.
- Promote teamwork, encourage training, and recognize employees as individuals.

Your employer is meeting these objectives and guidelines by supporting national and local associations through membership.

4.2.1 The Union Shop

In a union shop, the contractor agrees to abide by a collective bargaining agreement that contains work rules, wages, and hiring practices that both the contractor and employee must follow. Under law, if a majority of a firm's employees wish to take part in a collective bargaining agreement through a union, the employer is legally obliged to bargain in good faith with that union.

4.2.2 The Non-Union Shop

In a non-union shop, management and employees agree on wages, hours, and working conditions without collective bargaining. Sometimes a non-union shop works alone; other times it may work with other firms that are not bound by agreement to work only with union firms.

4.3.0 Training

Current construction industry efforts to train craftspeople are not enough. In almost all industries except construction, employers hire and train their own employees. They select employees for their general ability to do the task, and then provide any training needed to help the employees meet future job requirements. Employees remain with the same employer for comparatively long periods, which allows them to receive a fair amount of training that increases the quality of the entire workforce.

In construction, there are different arrangements for employment depending on the project. Sometimes, employees are hired for short-term projects and therefore are not with the same employer long enough to receive adequate training. Consequently, both the employee and the industry suffer.

In response to the need for well-trained personnel, the National Center for Construction Education and Research offers management and safety programs. These programs include NCCER's Construction Management Academies. The academies have delivered quality management education for more than ten years and have graduated over 3,000 students. The academies are held at the finest construction education institutions in the country.

NCCER offers leadership programs for crew leaders, project supervisors, and project managers, and has partnered with Clemson University to offer a Master's Degree in Construction Science and Management.

Despite the success of these and other programs, the need for continuing craft training is growing beyond the industry's ability to keep up. The ABC's report, *2001 Construction Industry Economic Overview,* states: "The most important asset to the continued strength of the U.S. construction economy is the skilled construction craftworker. We estimate that some 250,000 new craftworkers are needed yearly just to fill the demand caused by attrition and retirement."

Instructor's Notes:

The U.S. Department of Labor's Bureau of Labor Statistics report, *Employment by Major Industry Division, 1988, 1998, and Projected 2008*, estimates that: "Construction industry employment from 1998-2008 will increase 9 percent. Construction manager/supervisor positions are expected to increase 12.4 percent from 1998-2008. Employment of construction managers is expected to increase about as fast as the average for all occupations through 2008, as the level and complexity of construction activity continues to grow." Unless training efforts increase significantly, a severe shortage will result.

The construction industry is experiencing a shortage of trained supervisors in addition to the shortage of skilled craftworkers. According to the Business Roundtable on Supervisory Training: "The lack of proper training of these supervisors has contributed to the continued rise of construction costs. Their inability to plan work, communicate with workers, and direct work activities adequately is judged to be an important factor in the declining cost effectiveness of the construction industry."

Two major challenges face the construction industry today. They are:

- Developing adequate programs to provide trained craftworkers to meet future growth needs
- Providing supervisory training to ensure adequate job-site leadership throughout the industry

If the future demand for craftworkers and supervisory personnel is so great and the consequences of not meeting those demands are so catastrophic, why aren't more contractors training their own personnel? In fact, many large contracting companies *have* developed their own in-house programs.

In the past, many contractors avoided training their own employees because they believed that training costs would prevent them from keeping their bids competitive. They were afraid of losing trained employees to competitors. Because they had limited experience with formal training programs, they were afraid of trying an unproven program or simply did not understand the value of improved productivity that a well-trained workforce brings. These attitudes are changing as the construction industry gets more competitive and increases its investment in a new workforce.

Your participation in this program proves that your company recognizes the need for training and is committed to meeting that need. You are part of a growing sector that knows of the pressing needs of today and is working to ensure prosperity and growth for tomorrow.

SECTION 5

5.0.0 PHASES OF A CONSTRUCTION PROJECT

The three phases of a construction project, known as the *project flow,* are:

- Development
- Design (Planning)
- Construction

To give you an idea of the steps involved in each of the three phases, a sample project flow diagram is shown in *Figure 2*. Flow diagrams are important tools for management and communication on all types of projects, and the field supervisor should be familiar with them.

5.1.0 Development Phase

A building project begins with someone recognizing a need for a new facility and being willing and able to finance building this new facility. This owner might be a company in need of more space or more modern quarters, a government agency fulfilling a community need, or an individual investor. The **development phase** is the process of shaping that need into a workable plan for construction.

The development phase includes:

- Undertaking land research
- Performing feasibility studies
- Producing conceptual drawings

Conceptual drawings are developed by architect/engineers to define the project and to provide the owner with sketches of room layouts, estimates of elevations, and suggestions on construction materials.

Show Transparency 3A-C (Figure 2).

Transparency 3 is divided into three parts: 3A, 3B, and 3C. Each transparency is layered to illustrate each phase individually as well as how the phases flow together.

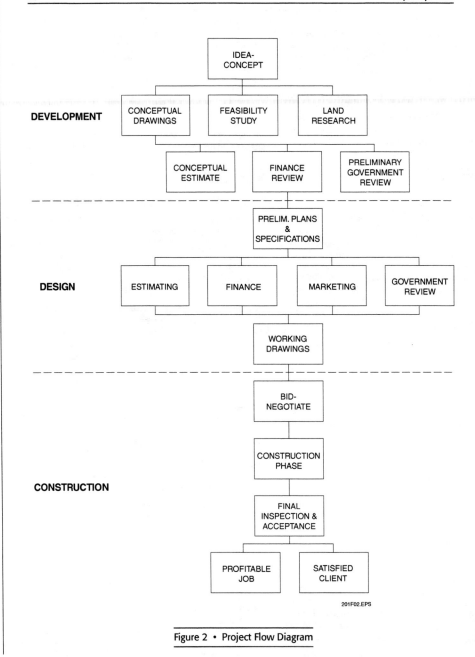

Figure 2 • Project Flow Diagram

Instructor's Notes:

Next, a project budget is developed, taking into account all anticipated costs or expenses. Here the owner analyzes both the cost of the project and the potential return on investment. If the anticipated value of the project balances favorably against anticipated cost, the project is considered financially feasible, and the owner seeks financing from lending institutions.

At the same time, the architect/engineers and/or the owners begin preliminary reviews of the project with government agencies. These reviews ensure the project meets all applicable zoning laws, building restrictions, landscape requirements and environmental concerns.

5.2.0 Design (Planning) Phase

In the design phase, the architect/engineer starts work on the preliminary drawings and specifications by bringing in other design professionals (structural, mechanical, electrical, and others) to perform the calculations, analyze all technical data, and determine all project details.

This work is translated into drawings and specifications that provide general contractors, subcontractors, suppliers, and vendors with information needed to install the hundreds of parts in a project. Bid documents are produced for the owner to gather bids from interested contractors. With the contract documents complete, the owner selects a method to follow to choose contractors. The owner may choose to negotiate the project with several contractors or to select a contractor through competitive bidding. During the design phase:

- Cost estimates are refined
- Regulatory agencies requirements are met
- A construction loan is secured

If the project is to be marketed—for example, if it involves building condominiums, offices, shopping centers or other facilities that will be occupied by people other than the owner—then a marketing program is also developed at this stage.

A common model of design and construction is the design-build model. The owner works directly with a design-build contractor, who takes responsibility for the complete project.

5.3.0 Construction Phase

A general contractor traditionally organizes the construction phase. General contractors:

- Employ their own workers to do certain parts of the project
- Engage the services of specialty contractors to do other parts of the project, such as:
  - Mechanical components
  - Electrical wiring
  - Elevator installation

In some cases, a construction manager (an administrator who acts on the owner's behalf but does none of the actual construction) organizes construction. Whether a general contractor or a construction manager heads the project, that person is responsible for managing all trades involved in the project.

As construction nears completion, the architect/engineers, the owner, and the government agencies start their final inspections and acceptance of the completed work.

Final inspection procedures can be completed quickly and easily if:

- The local codes have been met
- Architect/engineers have inspected the project regularly
- Work has been performed satisfactorily
- The project has been managed properly throughout to the satisfaction of the owner and mutual profit of everyone

If the inspection reveals faulty workmanship, poor use of materials, and violations of codes, the inspection and acceptance procedures can result in a dissatisfied client and a loss of profits for everyone involved.

5.4.0 Construction Flow

Just as a project can be divided into phases—development, design, and construction—so each of the phases themselves can be further divided into detailed steps. For the field supervisor, construction is the primary phase of any project, so we will concentrate on the steps involved in the construction phase—steps that we refer to as the *construction flow*. *Figure 3* shows the construction flow of a typical project.

Show samples of drawings, specifications, and bid documents and explain their purpose in the design (planning) phase.

Show Transparency 4A-D (Figure 3).

Transparency 4 is divided in 4 parts: 4A, 4B, 4C, and 4D. These can be layered to illustrate each phase individually as well as how the phases flow together.

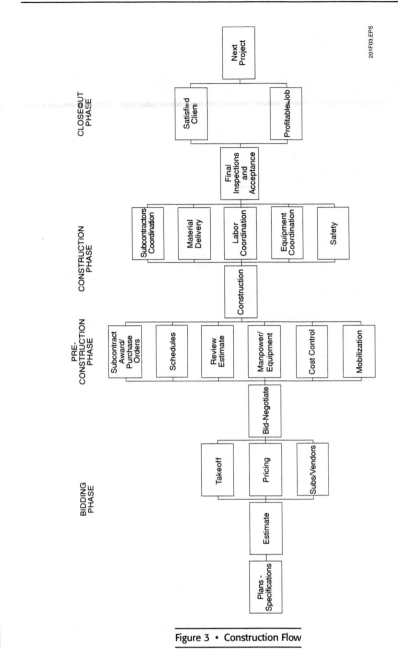

Figure 3 • Construction Flow

Instructor's Notes:

The flow includes four construction phases:

1. Bidding
2. **Pre-construction**
   - Selecting subcontractors
   - Issuing purchase orders
   - Developing a schedule
   - Reviewing the estimate
   - Identifying staffing resources and equipment requirements
   - Establishing a cost control system
   - **Mobilizing**
3. Construction
4. Closeout

5.4.1 Bidding Phase

The bidding phase starts when an owner or architect requests a general contractor or specialized contractor to submit a bid on all or part of a construction project. The contractor reviews the documents to determine the extent of the work and the requirements of the project. If the project meets the needs of the contractor, then an estimate of what it will cost to do the work is prepared.

Usually, the contractor assigns an in-house estimator to prepare the estimate. The estimator studies the project documents and prepares a **quantity survey (takeoff)** — a list of the various materials and parts necessary to build the project. Using this list, the contractor then contacts specialized subcontractors and suppliers to arrive at an estimated cost for each item on the takeoff. The total cost of the items on the takeoff, including materials, parts, labor, and subcontracts, is added to a company's estimated overhead cost and the target gross profit. The sum becomes the bottom line of the bid the company submits to the owner or the architect.

In a competitive bid situation, the company that submits the lowest bid usually is given the work. If the project is being negotiated, the owner or architect may not award the project solely on the basis of the lowest bid but may consider other factors, such as the bidders' qualifications, financial strength, personnel, and schedule. In any case, the bidding phase ends when the project contract is awarded.

5.4.2 Pre-Construction Phase

The pre-construction phase is often the most important part of the construction flow. During this phase, subcontractors are selected, **purchase orders** are issued, construction schedules are developed, and estimates are reviewed in depth. The project team holds discussions on staffing, equipment, and cost control, and then mobilization begins.

STEP 1
Selecting Subcontractors

Just as the general contractor obtains a contract by submitting a bid to the project owner, a subcontractor obtains work in the same way from the general contractor. Using a description of the work to be done, the subcontractor develops an estimate based on a takeoff and on prices obtained from suppliers, vendors, and the company's cost reports. For example, a mechanical contractor calculates a bid by determining the cost of the work the mechanical crew will perform and adding the bids from contractors needed to support the crew's work, such as sheet metal contractors, insulation contractors, control contractors, and equipment suppliers.

When the general contractor receives bids from subcontractors, the bids are reviewed carefully, making certain they are complete. Items usually reviewed include:

- Scope of work
- Cost
- Staffing resources
- Schedule

The general contractor then awards the contract.

STEP 2
Issuing Purchase Orders

Purchase orders (often called simply "POs") refer to the list of approved materials for the job. In subcontracting, the PO may also contain materials and labor. Contractors issue POs only after reviewing all project specifications and making certain that suppliers being considered can meet the requirements of the specifications.

Refer to Teaching Tip at the end of this module.

Show an example of a project schedule, and then discuss how the scheduling effort is prioritized.

STEP 3
Developing a Schedule

The project schedule should be developed with cooperation and input from:

- Members of the company's project team
- Major subcontractors
- Key suppliers
- Others who can contribute to a realistic schedule

High priority throughout the scheduling effort should be:

- Long lead items
- Critical operations
- Sequencing
- Duration of activities
- Equipment deliveries and installation
- Staffing resources

STEP 4
Reviewing the Estimate

The **pre-construction review** of the estimate is not a search for errors but a review to be sure all costs have been identified. Many times, errors both in the contractor's favor and against it are uncovered during the review. Even if no errors are found, a thorough project team review ensures that everyone on the project has a clear understanding of the project and the costs involved.

Many contractors add an additional step that requires the construction team to prepare a construction estimate to compare against the original bid. This helps identify any lapses or doubling up of items that were in the original estimate.

STEP 5
Identifying Staffing Resources and Equipment Requirements

Labor and construction equipment requirements for all construction phases should be identified during review of the estimate and preparation of the schedule.

Determining the number of craftworkers from the various trades required at each stage of work gives the project team a basis for:

- Determining the availability of personnel
- Identifying possible worker shortages and surpluses
- Controlling labor cash flow

Reviewing equipment requirements ensures that cranes, lifts, generators, and other construction equipment is available when needed, used wisely and efficiently on site, and removed from the job site when no longer required.

STEP 6
Establishing a Cost Control System

Every step of the project, from beginning to end, is based on cost management. This makes a cost control system essential to all aspects of the proper management of any project.

- The owner considers cost when developing the project concept.
- The architect/engineer considers cost throughout the design phase.
- Cost is the basis for accepting and rejecting bids.
- Cost plays a central role in whether the project is profitable for those involved.

During the pre-construction meeting, a review of the cost system, the extent of control, and the need for control must be established and understood.

STEP 7
Mobilizing

Mobilization includes:

- Moving all items required to start construction to the job site
- Coordinating job-site arrangements with the other trades on the job
- Organizing all job-site details in preparation for construction
- Setting up office and storage trailers
- Locating temporary facilities
- Establishing storage areas
- Assigning specific responsibilities to members of the project team

Instructor's Notes:

5.4.3 Construction Phase

The construction phase is the actual building of the project. At this point, the field supervisor is involved not only in the physical details of the project but with subcontractors, material deliveries, manpower assignments, equipment scheduling, and safety. In all of these areas, the supervisor's primary role is planning and coordination.

Planning is "knowing what is to be done, when it is to be done, how it is to be done, and who is to do it." It allows the supervisor to minimize risks and eliminate surprises in advance instead of merely reacting to problems and emergencies as they pop up.

Coordination is planning applied to the interaction of all related trades on the job, and many times, it holds the key to the success of a project. For example, the general contractor's supervisor must coordinate the electrical, mechanical, and reinforcing steel subcontractors to ensure their tasks are completed prior to the installation of concrete structures. Otherwise, concrete work can become an obstacle to these subcontractors, causing delays and added costs.

Material deliveries are another important area for planning and coordination. The supervisor must know delivery dates, quantities expected, unloading and storage procedures, and proper procedures for documenting deliveries. For equipment, too, the supervisor must be fully informed and must look ahead. The supervisor needs to know what construction equipment is needed on the job site when, what types of equipment are available, how to operate the equipment most effectively, and how to develop and execute an equipment layout plan.

Many points about the planning and coordination of subcontractors, materials, and equipment affect crew composition as well. In fact, personnel coordination is among the supervisor's most important responsibilities. For each step of the project, the supervisor must develop a list of the crafts required, the number of craftworkers required for each trade, and an estimate of how long the task will take to complete.

Coordination and planning are not one-time efforts at the beginning of the construction phase. They are continuous processes that demand the supervisor's attention right through to the last day of the job if the company's resources are to be used profitably.

5.4.4 Closeout Phase

Closeout involves final inspections and acceptance of the work by the owner's technical representatives, state and local building authorities, and others. If all parties have performed according to the contract documents, closeout can be completed in a timely manner. If the parties were negligent in their responsibilities, however, closeout can become a lengthy and expensive process that drains profits and leaves everyone involved dissatisfied.

To avoid the latter situation, the field supervisor must plan for closeout from the very first day of the job by insisting on quality workmanship, adherence to drawings and specifications, and the meeting of all schedules. Here again, planning and coordination are the key to ensuring success.

SECTION 6

6.0.0 CONSTRUCTION ORGANIZATION

Effective organization is the basis of any successful construction business. Organization provides the structure for assigning authority and ensuring that management's plans are carried out; it provides the foundation for the operation of the business. As a result, organization and management are inseparable.

For a small construction firm, the organizational structure is simple. The owner of the firm usually serves as chief executive officer, project manager, estimator, superintendent, purchasing agent, negotiator, record keeper, scheduler, and cost control officer. In other words, the owner personally directs the operation of the business, hiring only those people needed to do the actual construction work.

6.1.0 The Growth of an Organization

As a company grows, however, it expands past the ability of one person to wear all the management hats, and the need increases for additional people and a structure for organizing their individual efforts. A full-time estimator

Ensure that you have all the necessary materials to teach the course. Check the Materials and Equipment list at the front of the module. Prepare for teaching Session II by reading sections 5.4.3-9.0.0.

Assign reading of Module MT201 Sections 5.4.3-9.0.0.

Refer to the end of the module for a teaching tip.

prepares the estimate instead of the owner. Rather than one superintendent, the organization employs several, and instead of reporting directly to the owner, the superintendents now report to a project manager or a general superintendent.

Throughout all this growth, the function of management remains the same: to direct and control the company's people and resources toward the profitable completion of every project. What changes is the organizational structure needed to maintain direction and control. In a small company, organization is simple and informal; in a larger company, it is more complex and formal. This change is necessary if growing numbers of people are to act as a true team, working effectively and efficiently.

6.2.0 Formal Organization

Regardless of the size of a company, a formal organizational structure is important. It ensures that tasks are completed and company goals are reached. It also contributes to employee morale by letting each employee know who the supervisor is and how the work the employee does fits into the scheme of the company's plans. Nothing does more to create a solid company work force.

Generally, an organizational structure develops in two segments. First, horizontal segments are designed to accomplish specific tasks. Good examples of horizontal segments are separate departments for estimating, purchasing, and accounting. Second, vertical segments, or lines of authority, are developed to establish clear channels of communication and a chain of command.

A company's organizational structure is usually presented in a diagram called an **organizational chart.** An organizational chart shows the horizontal segments, the vertical segments, and the lines of authority among them. By clearly showing each individual's duties and area of authority, a well-prepared organizational chart also displays areas where authority and responsibility of different individuals overlap, and areas in which no one is assigned authority or responsibility.

6.3.0 Informal Organization

In addition to the formal lines of authority and responsibility shown in the organizational chart, there are informal channels of communication and contact within every company. Informal channels among company personnel are the result of accident, necessity, and social relationships, and they exist at all levels of management, supervision, and the workforce.

Like the formal company organization, the informal organization grows out of a need to accomplish company goals. For example, a project manager may require immediate verification that a bill was paid. Instead of going through the formal channels outlined on the organizational chart, the manager simply picks up the phone, calls the accounts payable department directly, gets the answer immediately and, at the same time, establishes a new channel of communication—one that might be useful again in the future.

Another good example of informal company organization is the "grapevine." In most companies, this network based on friendship and casual contact is able to spread information through an office or a project faster than any company memo. Many times, damaging rumors can be stopped and incorrect information can be corrected more quickly through the grapevine than through the formal channels of communications.

A company's informal organization is an important supervisory tool, and supervisors must recognize its existence and learn to use it to their advantage. Coffee breaks, company social functions, and daily interaction on and off the job are all important opportunities for exploring the informal organization.

6.4.0 Span of Control

Supervisors should not be responsible for supervising more employees than they can effectively monitor and control. There is no set number of employees a supervisor can effectively manage, but generally between four and eight employees is about right. However, much depends on specific circumstances. A supervisor might be able to supervise a crew of twelve skilled workers who are performing similar

Instructor's Notes:
___
___
___
___
___
___

tasks in a small area. The supervisor would probably lose control if expected to supervise three crews of four workers each if the crews were performing widely differing types of work over a large area.

The supervisor's **span of control** is the maximum number of workers a supervisor can control in a given situation.

Several factors determine a supervisor's span of control on a given job. They are:

- The ability and personality of the supervisor
- The quality of communications between supervisor and workers
- The type of work being done
- The skill levels of the workers and the supervisor
- The duties other than supervision for which the supervisor is responsible

Any one of these factors can contribute to a supervisor's losing effective control of the crew. However, often the reason for poor supervision lies in trying to supervise more work and more workers than can be managed properly. In other words, failure comes from a supervisor trying to work beyond his or her span of control.

6.5.0 Authority and Responsibility

As an organization grows, the manager must assign some duties to others in order to have time to devote to managing. **Delegation** is the act of assigning duties to others. In delegating activities, the supervisor assigns not just the task but the authority and responsibility to complete that task.

Having authority means having the power to act or make decisions. Having responsibility means being accountable for the outcome of an assignment. In every act of delegation, authority and responsibility must be carefully defined and properly balanced. The supervisor must also keep in mind that, regardless of to whom a task is delegated, ultimate authority and responsibility for that task still belongs to the supervisor. The supervisor is the one accountable to the company for completion of tasks under his or her supervision.

SECTION 7

7.0.0 MANAGEMENT FUNCTIONS

Regardless of the size of the company, every supervisor is responsible to some extent for five management functions, namely:

- Planning
- Organizing
- Staffing
- Directing
- Controlling resources and people

7.1.0 Planning

Planning is determining a course of action that attains a specific objective. In construction, the goal is the completion of work on time and within budget. Planning:

- Focuses the supervisor's attention
- Prepares the supervisor to meet uncertainties
- Minimizes costly errors

A successful plan is a logical sequence that includes:

- Establishing a goal
- Identifying what must be done to attain the goal
- Determining how to ascertain that the goal has been accomplished

7.2.0 Organizing

Organizing involves gathering and coordinating resources in order to fulfill a plan. It includes assigning the various tasks outlined in the plan to individuals and crews and then giving them the authority to carry them out.

The way a supervisor organizes a project depends largely on the way the company is organized. Therefore, understanding how a construction company grows and how that growth affects the company's organizational structure is an important first step in understanding the supervisor's organizational responsibilities.

Ask trainees for examples of effective span of control in their company. Discuss what steps a supervisor should take to prevent loss of control of the crew.

Refer to the end of the module for a teaching tip.

Classroom

Discuss the elements of a job description, and then ask trainees to write their own job description as supervisors within their respective companies.

7.3.0 Staffing

Staffing involves:

- Acquiring competent workers
- Placing them into specific positions
- Keeping those positions filled

In many companies, staffing crews is the responsibility of the supervisor, and a job description of each position to be filled is an important tool to help the supervisor perform this function.

7.4.0 Directing

Directing involves guiding and supervising employees toward accomplishing the company's objectives. This includes making assignments and seeing that they are carried out. People cannot be forced to do their best work; they must be motivated to do it. Therefore, it is the supervisor's job to learn and employ correct motivational skills. The most important of these skills are solid leadership and good communications.

7.5.0 Controlling

Control is the means of measuring performance and correcting any apparent change from the project plan. The best control techniques establish standard units for measuring progress and identifying changes before they occur.

SECTION 8

8.0.0 EMPLOYMENT REQUIREMENTS

Within each company, a supervisor's role will vary. Many supervisors are responsible for employing site personnel, new employee orientation, and training.

8.1.0 Employing Site Personnel

Knowing how to screen, employ, and train personnel is essential. Having the right mix of personnel on the crew and knowing how to motivate and manage them can make or break a supervisor's success.

The supervisor should know all relevant job descriptions before attempting to employ personnel. A job description gives each employee a clear picture of what is expected, making the transition into the project team quicker and easier. It also gives the supervisor a standard for evaluating an employee's performance and assigning duties, as well as determining a new employee's need for training.

A job description should be brief yet detailed enough to leave no misunderstandings regarding the duties and responsibilities of the position. It should also contain all the information necessary to evaluate the employee's performance later.

The essential elements of a job description are:

- Job title
- General description of the position
- The supervisor to whom the position reports
- Specific duties and responsibilities
- Other requirements, such as required tools/equipment for the job

The job description shown in *Figure 4* contains all of these elements.

8.2.0 New Employee Orientation

A supervisor should always take the time to orient a new employee as soon as possible after hiring. As part of the orientation procedure, the supervisor should:

- Introduce the employee to supervisors and fellow workers.
- Explain the employee's duties and the performance standards by which the employee will be evaluated.
- Set rules for working hours.
- Explain special procedures, such as parking restrictions at the job site.
- Review the company safety policy with the employee.
- Review the company's benefits program and pay procedures with the employee.
- Explain what to expect during a typical week.
- Describe the type of physical and/or mental work typically required.

Instructor's Notes:

```
SAMPLE JOB DESCRIPTION

POSITION: Frontline Construction Supervisor – Field

GENERAL SUMMARY:
First line of supervision on a construction crew installing underground and underwater cathodic protection systems.

REPORTS TO: Field Superintendent

DUTIES AND RESPONSIBILITIES:
• Oversee crew
• Provide instruction and training in construction tasks as needed
• Make sure proper materials and tools are on the site to accomplish tasks
• Keep project on schedule
• Enforce safety regulations
• Work within budget provided

KNOWLEDGE, SKILLS, AND EXPERIENCE REQUIRED:
• Extensive travel throughout the eastern United States, home base in Atlanta
• Ability to operate a backhoe and trencher
• Valid commercial driver's license with no DUI violations
• Ability to work under deadlines with the knowledge and ability to foresee problem areas and develop a plan of action to solve the situation
```

Figure 4 • Example of a Job Description

Have trainees discuss the training programs in place at their companies.

8.3.0 Coaching and Mentoring

Employees will need a great deal of **coaching** and **mentoring** at every stage of their employment, especially at the beginning. Coaching is guiding and advising or correcting employees along the way. Mentoring is a method of developing employees by using the more experienced workers to share, teach, guide, and advise the less experienced. Effective mentoring can improve team morale and reduce turnover.

In today's fast-paced and ever-changing work environment, supervisors must provide direction and dedicate the necessary time for employee development. Most managers view coaching and mentoring as important only for their problem employees, but it is just as necessary for your best workers. Every employee thrives on recognition. Effective managers know this and work daily to provide feedback about job responsibilities, expectations, goals and work processes.

8.4.0 Training

Many companies provide a period of training for new employees as part of their continuing training program for all employees. In most cases, the supervisor does not develop or design a training program, but does implement it. Supervisors should become involved in the training effort, suggest changes, and even try their hand at being an instructor. All of these steps on the supervisor's part help to ensure a good training program that will keep the company's work force prepared, flexible, and up to date on new techniques and procedures.

Have trainees discuss some of the policies and procedures that work well in their workplace.

SECTION 9

9.0.0 POLICIES AND PROCEDURES

Most companies have policies and procedures. They may be formal or informal, written or verbal, company-wide or project-specific. Supervisors might not be involved in developing policies and procedures but should be familiar with them since they are designed to help everyone in the company do a good job.

Policies are general statements that guide front-line supervisors in making decisions.

Procedures are detailed methods for implementing policies in everyday activities. They are statements of specific action.

To avoid misunderstandings, formal policies and procedures should be in writing. Written policies and procedures are best because they leave no doubt about what is expected of everyone in the company. They firmly establish the

---

SAMPLE POLICY AND PROCEDURE

Workplace Safety Policy:

Your safety is the constant concern of this company. Every precaution has been taken to provide a safe workplace.

Common sense and personal interest in safety are still the greatest guarantees of your safety at work, on the road, and at home. We take your safety seriously and any willful or habitual violation of safety rules will be considered cause for dismissal. We are sincerely concerned for the health and well-being of each member of the team.

Workplace Safety Procedure:

To ensure your safety and that of your co-workers, please observe and obey the following rules and guidelines:

- Observe and practice the safety procedures established for the job.
- In case of sickness or injury, no matter how slight, report at once to your supervisor. In no case should an employee treat his or her own or someone else's injuries or attempt to remove foreign particles from the eye.
- In case of injury resulting in possible fracture to legs, back, neck, or any accident resulting in an unconscious condition or a severe head injury, the employee is not to be moved until medical attention has been given by authorized personnel.
- Do not wear loose clothing or jewelry around machinery. It may catch on moving equipment and cause a serious injury.
- Never distract the attention of another employee, since this might cause an injury. If necessary to get the attention of another employee, wait until it can be done safely.
- Where required, you must wear protective equipment, such as goggles, safety glasses, masks, gloves, or hairnets.
- Safety equipment such as restraints, pull backs, and two-hand devices are designed for your protection. Be sure such equipment is adjusted for you.
- Pile materials, skids, bins, boxes, or other equipment so as not to block aisles, exits, fire fighting equipment, electric lighting or power panel, valves, etc. FIRE DOORS AND AISLES MUST BE KEPT CLEAR.
- Keep your work area clean.
- Use compressed air only for the job for which it is intended. Do not clean your clothes with it and do not fool with it.

Figure 5 • Sample Safety Policy and Procedure

company's position on various matters, and they stand as a visible barrier against employees' repeating the kinds of mistakes that gave rise to the policies and procedures in the first place.

The sample policy and accompanying procedures on workplace safety in *Figure 5* illustrate how policies and procedures can be written to document company guidelines.

In addition to formal (written) policies and procedures, many companies also have informal ones — ways of doing things that have become common practice over time, but are not written down. Small companies, generally, have more informal policies and procedures. Informal policies allow for considerable individual freedom of action. As a company grows, however, it needs to formalize its policies and procedures to make sure it treats all its employees consistently and fairly.

A typical example is a company's policy regarding company vehicles. A small company may not have a written policy regarding the use of company vehicles because it is not difficult to control the use of a small company fleet. However, as the company grows, its fleet grows, too. It soon becomes necessary to establish a formal written policy to control the use of this company resource and to ensure the safety of its growing number of employees.

Policies and procedures are usually company-wide and apply to everyone. However, it may be necessary to establish policies and procedures that apply to a single construction project and that differ from the company's normal practices. Job-specific policies and procedures may address government regulations, labor agreements, or special features of the contract documents and may include such items as special working hours and guidelines for unloading and storing materials.

Copies of company policies should be available to every supervisor, and every supervisor should read them carefully. Supervisors new to a company often make the mistake of creating their own policies or following those of their previous employers. This creates confusion on the job and weakens the supervisor's ability to lead.

All employees should realize that following company policies is a condition of employment and failure to do so leads to termination. A supervisor should set the example in this regard.

SUMMARY

The construction industry is the largest industry in America. It is affected by changes in economic conditions, business trends, and technology more than any other industry. The industry has recovered from the recession of the late-1980s; set a new record of over $500 billion in new construction work in the mid-1990s, and ended December 2000 at a $811.5 billion seasonally adjusted annual rate.

In order for the industry to continue to grow, it must continue to train both craftworkers and supervisors. Training programs such as this one are meeting the needs of supervisors as they move through the ranks from craftworker to management.

The supervisor should understand the tasks involved in completing a project, particularly the various phases — development, design, and construction — that must be coordinated in any successful construction project.

The success of any company or project depends on effective organization. Many companies have very simple organizations, others very complex ones. In either case, the employees should be aware of their authority and responsibility within the organization. Having clear, complete job descriptions and knowing the company's policies and procedures helps employees understand what is expected of them on the job.

Instructor's Notes:
_____
_____
_____
_____
_____
_____

NATIONAL CENTER FOR CONSTRUCTION EDUCATION AND RESEARCH

GLOSSARY

Trade Terms Introduced in This Module

Closeout: The proceedure which includes final inspections and acceptance of the work by the owner's technical representatives, state and local building authorities, and others.

Coaching: Guiding and advising or correcting employees along the way.

Delegation: The act of assigning duties to others. The supervisor assigns not just the task but the authority and responsibility to complete that task.

Mentoring: A method of developing employees by using the more experienced workers to share, teach, guide, and advise the less experienced.

Mobilization: Moving all items required to start construction to the job site, coordinating job site arrangements with the other trades on the job, and organizing all job site details in preparation for construction.

Non-union shop: In a non-union shop, management and employees agree on wages, hours, and working conditions without collective bargaining.

Organizational chart: Shows the horizontal segments, the vertical segments, and the lines of authority across a company or department.

Pre-construction phase: During this phase, subcontractors are selected, purchase orders are issued, construction schedules are developed, and estimates are reviewed in depth.

Pre-construction review: A review done before construction begins to be sure all costs have been identified.

Project flow: The three phases of a construction project: Development, Design (Planning), and Construction.

Purchase order (PO): Refers to the list of approved materials for the job. In subcontracting, the PO may also contain materials and labor.

Quantity survey (takeoff): The estimator studies the project documents and prepares a list of the various materials and parts necessary to build the project.

Span of control: The maximum number of workers a supervisor can control in a given situation.

Union shop: The contractor agrees to abide by a collective bargaining agreement that contains work rules, wages, and hiring practices that both the contractor and employee must follow.

Have trainees read the Glossary for unfamiliar terms. Answer trainee questions.

Have trainees prepare for the Module Examination.

Administer the Module Examination. Be sure to record the results of the Exam on Craft Training Report Form 200 and submit the results to the Training Program Sponsor.

ORIENTATION TO THE JOB — INSTRUCTOR'S GUIDE MODULE MT201

MOD MT201-01—TEACHING TIPS

The following are suggested activities or instructional methods to help you teach the material in this AIG.

Section 5.4.1   **Bidding Phase**

Show an example of a quantity survey (takeoff) and explain the purpose of this document in the bidding phase.

Show an example of a purchase order (PO), then explain how it is issued and by whom.

Section 6.2.0   **Formal Organization**

Show an example of a formal organizational chart. Discuss the advantages and disadvantages of informal channels of communication in a company and how they work as a supervisory tool.

Section 7.0.0   **Management Functions**

Discuss the five management functions a supervisor carries out. Ask trainees for examples of how each function is performed in their company. Does the current method work? If yes, how does it work? If no, how would the trainee change it so the method does work?

CONTREN™ LEARNING SERIES — USER UPDATES

The NCCER makes every effort to keep these textbooks up-to-date and free of technical errors. We appreciate your help in this process. If you have an idea for improving this textbook, or if you find an error, a typographical mistake, or an inaccuracy in NCCER's Contren™ textbooks, please write us, using this form or a photocopy. Be sure to include the exact module number, page number, a detailed description, and the correction, if applicable. Your input will be brought to the attention of the Technical Review Committee. Thank you for your assistance.

Instructors – If you found that additional materials were necessary in order to teach this module effectively, please let us know so that we may include them in the Equipment/Materials list in the Instructor's Guide.

Write: Curriculum Revision and Development Department
National Center for Construction Education and Research
P.O. Box 141104, Gainesville, FL 32614-1104

Fax: 352-334-0932

E-mail: curriculum@nccer.org

Craft _____ Module Name _____

Copyright Date _____ Module Number _____ Page Number(s) _____

Description _____

_____

_____

_____

_____

(Optional) Correction _____

_____

_____

_____

(Optional) Your Name and Address _____

_____

_____

_____

Project Supervisor

Module MT202-01

Human Relations and Problem Solving

Human Relations and Problem Solving
Instructor's Guide

Module MT202

MODULE OVERVIEW

This module introduces the project supervisor trainee to human relations and problem solving. The module will enable the trainee to enhance their communications with employees when acting in a leadership role. The project supervisor will apply new skills to delegating tasks, conducting interviews, motivating employees, managing and resolving conflict, solving problems, and making decisions.

PREREQUISITES

There are no prerequisites for the module; however, prior to training with this module, it is recommended that the trainee complete the following module:
Project Supervision, Module MT201

LEARNING OBJECTIVES

Upon completion of this module, the trainee will be able to:

1. State how a supervisor's performance is evaluated. Identify the challenges that the transition into supervision brings.
2. List the resources, techniques, and characteristics a successful leader uses to get the job done.
3. State the advantages of using various approaches to lead people effectively.
4. Communicate effectively.
5. Complete a task analysis.
6. Name nine essential elements on a job orientation checklist.
7. List the six steps for on-the-job training.
8. Explain the nine steps for conducting a performance appraisal.
9. Identify the root causes of performance problems and how to handle conflict.
10. Explain how moving up the management ladder affects a supervisor's ownership of time.
11. Construct a "To Do List" for on-the-job use.
12. Explain the nature of managerial decision-making and problem solving.
13. Determine the relationship between problem solving and decision-making.
14. Identify environmental influences on decision-making.

PERFORMANCE OBJECTIVES

This is a knowledge-based module – there is no performance profile examination.

NCCER STANDARDIZED CRAFT TRAINING PROGRAM

The National Center for Construction Education and Research (NCCER) provides a standardized national program of accredited craft training. Key features of the program include instructor certification, competency-based training, and performance testing. The program provides trainees, instructors, and companies with a standard form of recognition through a National Craft Training Registry. The program is described in full in the Guidelines for Accreditation, published by the NCCER. For more information on standardized craft training, contact the NCCER by writing us at P.O. Box 141104, Gainesville, FL 32614-1104; calling 352-334-0911; or e-mailing info@nccer.org. More information may be found at our Web site at www.nccer.org.

HOW TO USE THIS ANNOTATED INSTRUCTOR'S GUIDE

Each page presents two sections of information. The larger section displays each page exactly as it appears in the Trainee Module. The narrow column ties suggested trainee and instructor actions to each page and provides icons to call your attention to material, safety, audiovisual, or testing requirements. The bottom of each page includes space for your notes.

If you see the Teaching Tip icon, that means there is a teaching tip associated with this section. Also refer to the suggested teaching tips at the end of the module.

PREPARATION

Before teaching this module, you should review the Module Outline, Learning Objectives, and the Materials and Equipment List. Be sure to allow ample time to prepare your own training or lesson plan and gather all required equipment and materials.

MATERIALS AND EQUIPMENT LIST

Materials:

Transparencies

Markers/chalk

Module Examinations*

Sample task analysis forms**

Equipment:

Overhead projector and screen

Whiteboard/chalkboard

*Located in the Test Booklet packaged with this Annotated Instructor's Guide.
**If available on loan from your workplace or other resource.

ADDITIONAL RESOURCES

This module is intended to present thorough resources for task training. The following reference works are suggested for both instructors and trainees interested in further study. These are optional materials for continued education rather than for task training.

Construction Management, 1997. Daniel W. Halpin and Ronald W. Woodhead. New York: John Wiley & Sons.

Construction Operations Manual of Policies and Procedures, 2000. Andrew Civitello Jr., New York: McGraw-Hill.

TEACHING TIME FOR THIS MODULE

An outline for use in developing your lesson plan is presented below. Note that each Roman numeral in the outline equates to one session of instruction. Each session has a suggested time of 2 1/2 hours. This includes 10 minutes at the beginning of each session for administrative tasks and one 10-minute break during the session. Approximately 20 hours are suggested to cover *Human Relations and Problem Solving*.

Topic	Planned Time

Session I. Transition to Construction Management
- A. Introduction _____
- B. Transition to Management _____
    1. Participant Activity _____
    2. Psychological Barriers _____
    3. Social Barriers _____
    4. Time Management and Delegation _____
    5. Participant Activity _____

Session II. Communication
- A. Leadership _____
    1. Attributes of a Good Leader _____
    2. Functions of a Leader on the Job site _____
    3. Styles of Leadership _____
    4. Participant Activity _____
- B. Communication _____
    1. Components of the communication Process _____
    2. Mastering the Communication Process _____
    3. Participant Activities _____

Session III. Employee Relations and Task Training
- A. Employee Relations _____
    1. Task Analysis _____
    2. Interviewing _____
    3. Participant Activity _____
    4. Selecting an Employee _____
    5. Employee Orientation _____
    6. Performance Evaluations _____
- B. Performance, Assessment, and Task Training _____
    1. Training Is Your Responsibility _____
    2. Value of Training _____
    3. Types of Training _____

Session IV. Performance and Assessment
- A. Task Training, Continued _____
    1. Effectiveness and Efficiency _____
    2. Identification of Performance Discrepancies _____
- B. The Employee Response Model _____
    1. Participant Activity _____

Session V. Motivation, Mentoring, and Confrontation
- A. Motivation, Mentoring, and Confrontation
    1. Employee Motivators
    2. Mentoring
    3. Confrontation
    4. The Employee Responsibility Model
- B. Conflict Resolution
    1. Managing Conflict

Session VI. Problem Solving and Decision Making
- A. Problem Solving and Decision Making
    1. Organizational vs. Operational Decision
    2. Decision Making
    3. Participant Activity
    4. Influence of Schedules and Budget

Session VII. Problem Solving and Decision Making, Continued
- A. Problem Solving and Decision Making
    1. Steps in Decision making and Problem Solving
    2. Problem Solving for Non-Routine Decision Making

Session VIII. Problem Solving and Decision Making, Continued
- A. Problem Solving and Decision Making
    1. Implementing a Decision
    2. Participant Activities
    3. Evaluating Decisions
    4. Participant Activity
- B. Summary
    1. Summarize module
    2. Answer review questions
- C. Module Examination
    1. Trainees must score 70% or higher to receive recognition form the NCCER.
    2. Record the testing results on Craft Training Report Form 200 and submit the results to the Training Program Sponsor.

 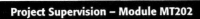

Human Relations and Problem Solving

Instructor's Notes:

Instructor's Notes:

ACKNOWLEDGMENTS

The NCCER wishes to acknowledge the dedication and expertise of Tom Tansey, the original author and mentor for this module on human relations.

M. Thomas Tansey, CLA, Master Trainer
CEO, Tansey and Associates, Inc.
Management Consultant
Tucson, AZ

We would also like to thank the following reviewers for contributing their time and expertise to this endeavor:

J.R. Blair
Tri-City Electrical Contractors
An Encompass Company

Mike Cornelius
Tri-City Electrical Contractors
An Encompass Company

Dan Faulkner
Wolverine Building Group

Kevin Kett
The Haskell Company

David Goodloe
Clemson University

Danny Parmenter
The Haskell Company

Course Map

This course map shows all of the modules of the *Project Supervision* curriculum. The suggested training order begins at the bottom and proceeds up. Skill levels increase as you advance on the course map. The local Training Program Sponsor may adjust the training order.

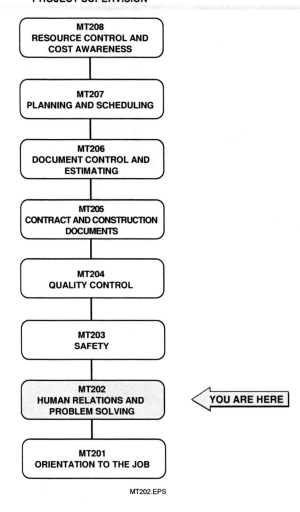

Instructor's Notes:

MODULE MT202

TABLE OF CONTENTS

1.0.0	**INTRODUCTION**	2.1
2.0.0	**TRANSITION TO MANAGEMENT**	2.2
2.1.0	Psychological Barriers	2.2
2.1.1	*The Comfort Zone*	2.2
2.1.2	*Supervising Former Co-Workers*	2.3
2.2.0	Social Barriers	2.3
2.3.0	Time Management and Delegation	2.3
2.3.1	*Establishing Priorities*	2.3
2.3.2	*Delegating Tasks*	2.4
3.0.0	**LEADERSHIP**	2.6
3.1.0	Attributes of a Good Leader	2.6
3.1.1	*Consistency*	2.6
3.1.2	*Concern for Employee Growth*	2.6
3.1.3	*Willingness to Train*	2.6
3.1.4	*Willingness to Coach or Mentor*	2.6
3.1.5	*Excellent Communication Skills*	2.6
3.2.0	Function of a Leader on the Job Site	2.6
3.3.0	Styles of Leadership	2.7
4.0.0	**COMMUNICATION**	2.7
4.1.0	Components of the Communication Process	2.8
4.1.1	*The Sender*	2.8
4.1.2	*The Encoder*	2.8
4.1.3	*The Transmitter*	2.8
4.1.4	*The Receiver*	2.8
4.1.5	*The Decoder*	2.8
4.1.6	*Noise*	2.8
4.2.0	Mastering the Communication Process	2.8
4.2.1	*Active Listening*	2.8
4.2.2	*Giving Clear Instructions*	2.10
5.0.0	**EMPLOYEE RELATIONS**	2.12
5.1.0	Task Analysis	2.12
5.2.0	Interviewing	2.12
5.3.0	Selecting an Employee	2.13
5.4.0	Employee Orientation	2.13
5.5.0	Performance Evaluations	2.13

6.0.0	**PERFORMANCE, ASSESSMENT, AND TASK TRAINING**	2.14
6.1.0	Training Is Your Responsibility	2.15
6.2.0	Value of Training	2.15
6.3.0	Types of Training	2.15
6.4.0	Effectiveness and Efficiency	2.16
6.4.1	*Identification of Performance Discrepancies*	2.16
6.4.2	*The Employee Response Model*	2.16
6.4.3	*Diagnosis of Employee Response*	2.17
6.4.4	*Constructive Solutions*	2.18
7.0.0	**MOTIVATION, MENTORING, AND CONFRONTATION**	2.20
7.1.0	Employee Motivators	2.20
7.1.1	*Recognition and Praise*	2.21
7.1.2	*Accomplishment*	2.21
7.1.3	*Opportunity for Advancement*	2.21
7.1.4	*Job Importance*	2.21
7.1.5	*Change*	2.21
7.1.6	*Personal Growth*	2.21
7.1.7	*Rewards*	2.21
7.2.0	Mentoring	2.23
7.3.0	Confrontation	2.23
7.4.0	The Employee Responsibility Model	2.23
8.0.0	**CONFLICT RESOLUTION**	2.25
8.1.0	Managing Conflict	2.25
8.1.1	*Factors Affecting Conflict Management*	2.25
8.1.2	*Strategies for Dealing With Conflict*	2.26
9.0.0	**PROBLEM SOLVING AND DECISION MAKING**	2.26
9.1.0	Organizational vs. Operational Decisions	2.27
9.2.0	Decision Making	2.28
9.2.1	*Choice*	2.28
9.2.2	*Alternatives*	2.28
9.2.3	*Goals*	2.28
9.2.4	*Consequences*	2.28
9.2.5	*Variables*	2.28
9.2.6	*Limits to Rational Decision Making*	2.29
9.3.0	Influence of Schedules and Budget	2.29
9.3.1	*Schedules*	2.30
9.3.2	*Budget*	2.30
9.4.0	Steps in Decision Making and Problem Solving	2.30
9.4.1	*Types of Decisions*	2.30
9.4.2	*Authority for Decision Making*	2.31
9.4.3	*Group Involvement in Decision Making*	2.31

Instructor's Notes:

9.4.4	*Problem Solving for Non-Routine Decision Making*	2.33
9.4.5	*Implementing a Decision*	2.36
9.5.0	*Evaluating Decisions*	2.37
	SUMMARY	2.37
	REVIEW QUESTION	2.39
	GLOSSARY	2.42

LIST OF FIGURES

Figure 1	•	Delegation Checklist	2.5
Figure 2	•	Employee Orientation Checklist	2.13
Figure 3	•	Performance Evaluation Checklist	2.14
Figure 4	•	Employee De-Motivator Checklist	2.22
Figure 5	•	Eleven Commandments of Successful Confrontation	2.24
Figure 6	•	Employee Responsibility Model Checklist	2.25
Figure 7	•	Routine and Non-Routine Decision Process	2.31
Figure 8	•	Problem-Solving Process	2.34
Figure 9	•	Comparing Total Cost of Alternatives	2.36

Instructor's Notes:

NATIONAL CENTER FOR CONSTRUCTION EDUCATION AND RESEARCH

MODULE MT202

Human Relations and Problem Solving

Ensure that you have all the necessary materials to teach the course. Check the Materials and Equipment list at the front of the module. Prepare for teaching Session I by reading Sections 1.0.0-2.3.2.

Show Transparency 1 (Course Objectives).

Homework

Assign reading of Module MT202 Sections 1.0.0-2.3.2.

Objectives

Upon the completion of this module, you will be able to do the following:

1. State how a supervisor's performance is evaluated. Identify the challenges that the transition into supervision brings.
2. List the resources, techniques, and characteristics a successful leader uses to get the job done.
3. State the advantages of using various approaches to lead people effectively.
4. Communicate effectively.
5. Complete a task analysis.
6. Name nine essential elements on a job orientation checklist.
7. List the six steps for on-the-job training.
8. Explain the nine steps for conducting a performance appraisal.
9. Identify the root causes of performance problems and how to handle conflict.
10. Explain how moving up the management ladder affects a supervisor's ownership of time.
11. Construct a "To Do List" for on-the-job use.
12. Explain the nature of managerial decision making and problem solving.
13. Determine the relationship between problem solving and decision making.
14. Identify environmental influences on decision making.
15. Cite the advantages and disadvantages of group involvement in decision making.
16. Distinguish between routine (programmed) and non-routine (non-programmed) decisions.
17. Describe the critical role of implementation and evaluation on future actions.

SECTION 1

1.0.0 INTRODUCTION

It takes a thorough understanding of human relations to be successful in the world of work today. Flexibility, quality, teamwork, diversity, ethics, productivity, and the need to balance the demands of family and work are all part of human relations.

Supervisors must acquire knowledge and skills for communicating productively and respectfully in the workplace with subordinates, peers, and superiors. Understanding how to communicate as a leader is very important.

An A/E/C Job Bank article recently reported: "More than a decade ago, many of the nation's leading trade associations like the Associated General Contractors (AGC) began warning its members of an impending labor shortage."

The article goes on to state: "The Center for Creative Leadership in San Diego commented in a recent survey that firms which offered employee development, good communication, ethics, and other positive human factors enjoyed better retention rates and 20% higher profits."

Copyright © 2003 National Center for Construction Education and Research, Gainesville, FL 32614-1104. All rights reserved. No part of this work may be reproduced in any form or by any means, including photocopying, without written permission of the publisher.

Ask trainees to give examples of times when they were required to think outside the box.

Show Transparency 2 (Participant Activity).

Good communication and employee development are key factors in employee satisfaction and career planning. This module is geared towards helping an individual achieve both of these goals. You will learn about human relations on the job site. You may find that many of the decision-making and confrontation techniques are beneficial when applied off the job site as well.

SECTION 2

2.0.0 TRANSITION TO MANAGEMENT

The initial step into management, the step from craftworker to supervisor, is among the most challenging career changes. The supervisor's job is tough, and it demands new management skills, people-handling skills, and leadership skills. Supervisors must develop the ability to think beyond the task at hand. In the transition, evaluation criteria also change. As a craftworker, you are judged by job performance, but as a supervisor, you must take responsibility for productivity and profitability on the job.

As a supervisor you must learn to rely on your crew to get the job done. You will need to develop the ability to manage and direct the work of others without doing it yourself. Your ability to **think outside the box** will be critical to your success as a manager. Thinking outside the box means looking at situations with a new "eye" that does not depend on how you have always done it or what seems logical. Thinking outside the box demands that you see things from another perspective, using creativity and limitless thinking to approach your work.

PARTICIPANT ACTIVITY

List at least two "people" (human relations) problems that you have experienced or might experience as a member of management. Keep each statement to five words or less.

In class discussion, analyze the problems you listed and gather ideas for handling them properly.

PARTICIPANT ACTIVITY

2.1.0 Psychological Barriers

There are two primary psychological barriers to success in management: failing to expand your **comfort zone** and not overcoming your **self-perceptions** when supervising former co-workers.

2.1.1 The Comfort Zone

The first psychological barrier to the success of new supervisors lies in their comfort zone. A comfort zone is the area of activity in which a person feels most secure and confident. For new supervisors, the comfort zone represents the tendency to do as much of the construction work as they themselves can.

This is natural. Supervisors have usually risen through the ranks and received promotions on the basis of demonstrated technical skills. They become managers because they are among the best in their craft, able to complete tasks better, quicker, and smarter than most. In the new role of supervisor, however, these master craftworkers find themselves at the bottom of a new status ladder. In management, they are the apprentices.

This step can be a shock. Gone are the instructions from the boss outlining daily tasks. Gone is the confidence in the ability to do the job well. Gone is the praise received for excellent work. In place of these things there is often a confusing responsibility and sense of not really knowing what to do. The supervisor is now being judged on the basis of the work accomplished—or not accomplished—by those being supervised.

In addition, new skills must be learned. Planning, directing, controlling, following up, reporting customer relations, conflict resolution, and motivational techniques all have to be learned and applied. It's natural to take refuge in an old familiar comfort zone by doing as much of the construction work as possible, instead of delegating those tasks. Unfortunately, this is a major step backward. It is, at best, a temporary reprieve from facing the tasks of a manager. At worst, it can lead to failure as a supervisor, since a large portion of the supervisor's responsibility is to delegate, follow up, and report on task progress.

Instructor's Notes:

Everyone has a comfort zone. The secret to success in management is not to eliminate your comfort zone but to create a new one that includes planning, delegating, follow-up, handling customers, and motivating team members.

2.1.2 Supervising Former Co-Workers

The second common psychological barrier to a supervisor's success lies in supervising former co-workers. When assigning work (especially unpopular work) to a former co-worker, the supervisor often thinks:

- "I wonder if Bob will be upset with me?"
- "I hope Mary doesn't think I'm big-timing her."
- "I hope everyone's forgotten what I used to say about how dumb management acts."
- "How do I justify this when everyone knows that I used to goof off, too?"
- "Joe's my best buddy. I know he'll do right by me."
- "Am I acting 'uppity'? Is that why the team doesn't joke with me since I got promoted?"

These are all examples of the supervisor's self-perception, and they have an impact on the treatment of employees that they used to work with on a peer level. Unless you can overcome doubts in your own self-perception, you will not be able to be an effective supervisor.

Above all, as a supervisor you must remember that the job is the boss, that doing what has to be done to complete the project is more important than your perception of what team members think of you. In your role as supervisor, you must learn new responsibilities and model the appropriate behavior to get the job done.

2.2.0 Social Barriers

Socially, there are several barriers to successful management. Among the most common are:
- Socializing with employees off the job
- Having relatives working on the job
- Family connections
- Commuting

Most humans need a social life. Most people need friends and need to feel they are a part of a group. Separating professional life from personal life is a very hard task.

In the days as a craftworker, a supervisor may have gone fishing, hunting, or bowling with the people who now must be supervised. Often, these people are members of the same church group, and their spouses and children may be friends. A supervisor may be in the same carpool as the employees. Some of them might even be relatives. Now, as a manager, off-the-job friendships can have complicated consequences.

A manager should never attempt to dictate the choice of family friends. Professional and social lives should be separated and careful consideration should be given to the effect drinking, hunting, and fishing with employees might have when the time comes to confront one of them over poor performance, absenteeism, or tardiness.

If supervisors can separate social life from professional life, then they should continue on- and off-the-job relationships. If the two cannot be successfully separated, then they should seek new friends and other forms of social life.

2.3.0 Time Management and Delegation

Time cannot be saved. It cannot be borrowed, stolen, or bought. We all have the same number of hours in a day, in a week, in a year. How we learn to balance the many priorities that we have in life will be the mark of our success. Organizing our personal life and work through balance and efficiency leads to happy, productive workers. If things are out of balance, work and home life will suffer. If supervisors are to have the time to complete all the tasks that they are being paid to accomplish, then they have to have a way to control time. They have to have a strategy for spending time wisely. Two strategies for spending time wisely are establishing priorities and delegating specific tasks.

2.3.1 Establishing Priorities

Some tasks and responsibilities require urgency. Some are important but not crucial. Others are immediate but not urgent, and others are little more than busy work. Finally, there are those tasks that are simply a waste of time. The supervisor must learn to assign each task a priority based on its urgency and importance, and then use that list of priorities as a guideline for scheduling time and effort.

Ask trainees to give examples of how they faced the challenge of separating professional and social lives.

Ask trainees to describe the difference between immediate but not urgent and important but not crucial tasks.

Show Transparency 3 (Participant Activity).

Sticking to priorities is important. Sometimes we have to focus on our priorities for a long time before we get the desired results. This can be difficult without a continuous process of prioritizing and revisiting or revising goals. Learning to set goals and prioritize tasks is an important skill.

Urgent tasks are those that must be done immediately; they are top-priority tasks. For example, your boss demands that you complete a report by 11:00 a.m. Assigning a priority to such a task is rarely a problem.

Time management begins to become important with tasks that are crucial but not immediate. These tasks can be postponed a little while. However, once postponed they run the risk of being ignored until it is too late. Most of the supervisor's responsibilities fall into this category, and because they are so important, the supervisor must learn to assign reasonable deadlines for them as soon as they pop up.

Tasks that are immediate but not important are those activities that can be assigned a low priority if examined objectively. Ordinarily, they are the tasks that we agree to do because someone is standing in front of us or waiting on the phone for a reply. Again, the supervisor should schedule these tasks as soon as they are known. However, they should be scheduled around more important or pressing tasks.

Tasks that are busy work are probably worth doing but aren't important or urgent. A supervisor might do these tasks because they provide a "quick fix" feeling of accomplishment and an excuse for not handling more urgent or immediate tasks. At times, doing these tasks may provide a relief from other types of work. Alternating tasks can actually make you more efficient.

In the last category are tasks that are a waste of time, tasks that are often blamed when tasks that are more important are not completed on time. Supervisors who spend time on such tasks rarely survive in construction management.

2.3.2 Delegating Tasks

No supervisor can do all of the tasks on the prioritized list. Certain tasks must be delegated. If this is not done, the supervisor ends up racing from job to job "putting out brush fires," cultivating inexperienced job supervisors, destroying morale, and, most importantly, losing job profits.

Delegating is assigning responsibility and authority to do a task. *Responsibility* is the obligation to complete the task. *Authority* is the power to carry out the tasks.

Delegation has several advantages, particularly:

- It gives the supervisor time to complete very important tasks.
- It allows work to get done when the supervisor is not on site.
- It provides supervisory training.
- It enhances the project team's effectiveness and efficiency.

Supervisors should delegate whenever they find that they are unable to keep up with everything on the prioritized list of tasks, beginning with the items of lowest priority on that list. Failing to delegate is a sign that the supervisor cannot distinguish between crucial and trivial responsibilities.

Figure 1 presents a checklist that will help a supervisor distinguish between the crucial *few* that the supervisor must personally accomplish, and the trivial *many*. It will also help determine what tasks to delegate and to whom.

PARTICIPANT ACTIVITY

Discuss the following:

1. Does moving up the management ladder increase or decrease the amount of time you own?
2. List five steps to time management.
3. What are the three steps to reducing blocked time?
4. Name four tips to control time usage.

PARTICIPANT ACTIVITY

Instructor's Notes:

SAMPLE DELEGATION CHECKLIST

A. List five major tasks or responsibilities:

1. _____
2. _____
3. _____
4. _____
5. _____

B. Break down each of the above major tasks into five subtasks:

1	2	3	4	5
___	___	___	___	___
___	___	___	___	___
___	___	___	___	___
___	___	___	___	___
___	___	___	___	___

Utilize the following code to differentiate among the subtasks:

M.E.: The supervisor is absolutely responsible and accountable for this subtask.
S.O.: Someone else could do this subtask with proper guidance, instruction, or training.
D.O.: Delegate out immediately.

Figure 1 • Delegation Checklist

Once the supervisor has used the checklist to break down tasks for delegation and then selected someone to delegate them to, the necessary responsibility and authority must be assigned for getting those tasks done. This involves the following steps:

1. Give a clear statement of what the employee is to do.
2. Explain to the employee exactly how far to go—the extent of the authority.
3. Establish a definite deadline.
4. Explain how you are going to follow up on the assignment with inspections and checklists.
5. Outline the importance of the task in relation to the individual's other assignments – what priority the assignment has in relation to other duties.
6. Explain why you are delegating the task to the employee selected.
7. Let everyone involved in the task know that the person chosen has the authority.
8. Manage the work that you have delegated.

Using the delegation checklist and following the steps listed above ensures that the tasks the supervisor delegates will be completed properly and on schedule.

Ask trainees to give an example of a situation where they used a checklist.

Show Transparency 4 (Figure 1).

Ensure that you have all the necessary materials to teach the course. Check the Materials and Equipment list at the front of the module. Prepare for teaching Session II by reading Sections 3.0.0-4.2.2.

Assign reading of Module MT202 Sections 3.0.0-4.2.2.

Ask trainees to list other traits a good leader possesses.

SECTION 3

3.0.0 LEADERSHIP

There are few "born" leaders. *Leadership* is the process of getting a job done through other people. An effective leader is the one who is able to do the job daily, week after week, month after month, regardless of the project.

Many new supervisors are surprised to find out that not everyone they work with likes good leaders. The supervisor might not be the most highly skilled craftworker and might not set the job schedule, but an effective leader *does* understand employees and is able to integrate their goals to the project goals.

The selection and training of supervisors is one of the construction industry's major concerns, yet there is no established set of attributes or traits that objectively identifies an "ideal" supervisor. The reason is simple. The qualities a supervisor needs for success on one project might be insufficient to do even a mediocre job on another. Consequently, it is impossible to define the "perfect" supervisor.

Nevertheless, there are at least five recognizable traits that good managers possess and routinely demonstrate during the execution of their daily tasks. They are:

- **Consistency**
- Concern for employee growth
- Willingness to train
- Willingness to mentor
- Excellent communication skills

3.1.0 Attributes of a Good Leader

Good leaders exhibit many qualities that earn them respect from their employees. These characteristics must be learned and communicated through words and actions. A leader's consistency, concern for employee growth, and willingness to coach, mentor, and train are essential for success.

3.1.1 Consistency

Consistency is the ability to deal with all situations according to a standard set of rules. When working under a consistent leader, employees know what is expected of them in all cases, and they know the consequences of their actions if they neglect their responsibilities.

This is not to say that leaders treat everyone alike or do not vary their approach in motivating people. On the contrary, effective leaders try various approaches to problems because they realize that people and situations are complex and varied. Still, the variations all fall within the same set of rules and values and do not violate consistency.

3.1.2 Concern for Employee Growth

Solid leaders are actively concerned with the long-term growth of their employees. Employees are expected to make mistakes, but are not punished for their errors. Instead, ways are found to help them learn from their mistakes and not repeat them. At the same time, the leader also learns from the situation.

3.1.3 Willingness to Train

The best supervisors take an active role in employee training. They encourage their employees to learn and willingly give of their time and energy to teach them. They do this because they realize that well-trained employees make the supervisor's job easier and the team more efficient.

3.1.4 Willingness to Coach or Mentor

Good leaders take the time to coach or mentor their employees so that they might do their jobs more efficiently. This is part of the interest good leaders have in training, and their understanding that an employee who knows how to do the job well is a benefit to the team and the company.

3.1.5 Excellent Communication Skills

Few successful leaders are poor communicators. An effective leader must be equally comfortable talking or listening to subordinates, managers, and other supervisors.

3.2.0 Function of a Leader on the Job Site

Although the functions of a leader will vary from project to project, there are certain functions that are common to virtually all job sites:

Instructor's Notes:

- Planning
- Organizing
- Staffing
- Following-up
- Reporting
- Training
- Coaching and mentoring
- Resolving conflict
- Handling customers

The degree of responsibility for each of these functions will vary, as will the procedures and techniques employed.

3.3.0 Styles of Leadership

Leadership styles can be divided into three major categories: directive, involved, and uninvolved. The **directive style** is characterized by little if any involvement of employees. High levels of communication between supervisors and their teams characterize the **involved style.** The **uninvolved style** is the other extreme, characterized by employees making decisions with little input by the supervisor. Additional characteristics of each style follow:

Directive
— Expects employees to work
— Does not seek input from employees
— Solves problems alone
— Praises and criticizes on a personal basis

Involved
— Discusses problems with employees
— Listens actively
— Explains and instructs
— Coaches and counsels employees

Uninvolved
— Believes no supervision is best
— Lets the employees decide
— Seldom confronts poor performance
— Praises without merit

A mixture of the above is the preferred style. Each situation and employee requires a slightly different style.

PARTICIPANT ACTIVITY

1. List some resources a leader uses to get the job done.
2. Name five qualities of a good leader.
3. List techniques that a leader uses to get the job done.
4. Identify an important quality a leader uses to confront problems.
5. Why do you think a leader uses different approaches to guide people?
6. List the nine functions of a leader on a job site.
7. Name the three basic styles of leadership.

PARTICIPANT ACTIVITY

SECTION 4

4.0.0 COMMUNICATION

Communication is the transfer of ideas, thoughts, feelings, or facts between people. Communication is successful if the communicator, such as the speaker or writer, transfers information in such a way that the receiver understands it as it is intended.

Although, like breathing, communication is a function of being alive, we are not born with successful communication skills. We learn them, just as we learn the skills of our craft. As with any other learned skills, it takes time and practice to master the art of communication.

Why must a supervisor be an effective communicator? Research indicates that almost 66 percent of all errors on the job are due to communication problems. This is remarkable, and it puts a great burden on the supervisor because the supervisor is the hub of the company's communication system. The supervisor is responsible for relaying job-site problems and concerns to the other members of the project team and for communicating company goals, policies, and procedures to the tradesmen.

Ask trainees to choose which function they would find easier and why, and which function would be a challenge and why. Discuss how leadership styles affect the function of a leader.

Show Transparency 5 (Participant Activity) then show Transparency 1 (1 of 3) (Course Objectives).

Ask trainees to give examples of how they changed their method of communication after decoding an individual's listening style.

Who is in a better position than the supervisor to control this two-way communication between the job site and the company? The supervisor is closest to the job and has a finger on the pulse of the project.

The first step in being a good communicator is to be open, honest, and complete in your communications. If you and your crew are behind schedule, make sure you let upper management know so adjustments can be made. If you have a disagreement with the general contractor, the architect, a vendor, or another contractor, tell your boss about it. If there are morale problems among the crews, let the office know. Hiding unpleasant news or omitting important facts from your reports cripples the project team's efforts and sets the stage for ugly surprises later on.

4.1.0 Components of the Communication Process

The communication process is similar to a movie. For example, a director wants to communicate or send an idea to the world. The director uses cameras, film, specific music, and special lights as tools to encode or translate the idea into a movie. The movie is transmitted by a projector, screen, and speakers. Then, audiences around the world receive the movie when they see and hear it. Each person who watches and decodes the movie has a unique experience with it. This experience has to do with a number of factors, many of which are related to the six parts of the communication process: the sender, the encoder, the transmitter, the receiver, the decoder, and noise or interference.

4.1.1 The Sender

The *sender* is the person who wants to communicate a message or who has thoughts, feelings, ideas, instructions, or other information to be shared with someone else. If you have an employee, a peer, or a boss that you have a hard time communicating with, identify that individual's **decoding,** or listening, style and use that information to change your method of communication.

4.1.2 The Encoder

The mind and the brain are the machines that we use to translate (**encode**) thoughts, perceptions, ideas, feelings, and facts into a form that others can understand. That form is usually language.

4.1.3 The Transmitter

The body transmits the encoded message through *verbal* or *nonverbal* channels, which can include words, hand gestures, and facial expressions. Therefore, it is important to listen to what is being said and to take in the emotional as well as the verbal content. It is equally important to notice body posture, expressions, eye contact, and other nonverbal cues.

4.1.4 The Receiver

The person receiving the sender's message does so by using the senses—sight, sound, taste, smell, and feel—to accept the message.

4.1.5 The Decoder

The receiver's mind and brain translate (decode) the message sensed into understandable information. Essentially, the sender's message is looked at through one's own "mind's eye."

4.1.6 Noise

Noise is any interference between the sender and the receiver. It can be external or internal. External noise is exactly what the term implies: the distracting sounds of carpenters hammering, street traffic passing, air compressors running, other people talking, telephones ringing, and so on. Internal noise occurs inside the sender's or the receiver's mind and includes thoughts about personal problems, personality conflicts, and random thoughts, like what to have for lunch.

4.2.0 Mastering the Communication Process

To master the complexities of the communication process, a supervisor must learn two essential skills: listening actively and giving clear instructions.

4.2.1 Active Listening

Active listening is the basis for all other management functions. Negotiating, problem solving, disciplining employees, planning, and goal setting all require mastery of listening skills.

Despite its importance, listening is the communication skill misused most and taught least. The proof lies in the fact that the average American retains only about 25 percent of what is heard. Retaining what you hear is the purpose of active listening, and it involves much more than simply hearing the words said by the

Instructor's Notes:

sender. It requires listening for ideas and feelings that the sender is trying to communicate.

The active listener (receiver) must pay attention to the sender's facial expressions, tone of voice, gestures, and body language. Feedback must be continually provided to the sender to let the sender know that the receiver is listening and trying to understand. This is done through the processes of physical attendance and reflection.

Physical attendance is the act of using body language to communicate that you are listening. Among the body language signals you can use to show the sender your physical attendance are:

- Maintaining eye contact
- Leaning forward
- Using the sender's name when responding
- Nodding your head when in agreement
- Keeping your arms uncrossed

Reflection is the receiver's act of processing the information the sender is transmitting. It involves rephrasing the message in the receiver's own words and defining the message in terms of facts and feelings. Reflection is not judgment. The active listener who practices reflection does not judge, question, or argue with the sender; rather, an attempt is made to understand the sender's message.

Reflection is the active listener's way of comprehending sender's message and understanding the essence of what the sender is trying to say. Often, reflection shows that what the sender says is not what is meant.

Show Transparency 6 (Participant Activity).

Discuss possible answers to the Participant Activity. If the trainees get stuck, some suggested answers are located at the end of this module.

PARTICIPANT ACTIVITY

To the right of each statement, write the real message the speaker (one of your team members) is trying to communicate. This exercise shows how reflection helps the active listener understand the emotion of the situation. In a confrontation, reflection helps to clear away strong emotions and expose the real source of the problem.

"This is the most stupid, mixed-up job I ever worked on." _____

"It's impossible, but I'll have it done for you by Thursday afternoon." _____

"The laborer you assigned to work with the electricians is really sharp." _____

"I guess the other guys must be too busy to do the job you gave me." _____

"Maybe you'd better talk more with your people about this." _____

"Oh, I thought I was working as fast as expected." _____

"How come I always get the grunt work?" _____

"What's wrong with kids these days? Don't they want to work?" _____

"That architect doesn't know what he is doing." _____

"Sorry I'm late. Things have been going real bad at home." _____

PARTICIPANT ACTIVITY

Ask trainees which of their presentation styles needs strengthening and which style is their strongest.

Have the trainees complete the Participant Activity.

4.2.2 Giving Clear Instructions

People decode messages the way they need to. Unfortunately, the way employees decode messages is not always the way the supervisor sends them. To make sure that a message is encoded in a way that makes it easy for the receiver to decode it accurately, the supervisor can use the technique known as the *six-pack*.

The six-pack describes the decoding styles of most listeners. Once the supervisor learns to identify the type of listener who is to receive the instructions, the message can be encoded to fit that listener's decoding system. The result: clearly understood instruction.

The six types of listeners (and listening styles) described by the six-pack are the:

1. Laid-back listener
2. Imager
3. Sensitive listener
4. Terminator
5. Encloser
6. Numberer

The Laid-Back Listener – The laid-back listeners hear only positive messages. On the job, they hear praise but seem to block out criticism and confrontation. This style of decoding can be extremely frustrating to a sender trying to transmit negative messages that must be heard. In transmitting to a laid-back listener, use words with a positive tone—words like *new, easier, faster, quicker, smoother, smarter,* or *improved*. For example: "Jack, this is an easier way to do that job."

The Imager – Imagers usually are more impressed with the presentation of the message than with its content. On the job, an imager will usually receive your message if it is direct, smooth, and easy flowing. All the "i's" must be dotted and all "t's" crossed. Using image (picture) phrases also helps, such as *in my view*, the *way it looks to me*, or *as I see it*.

The Sensitive Listener – The sensitive listener looks for the reasons behind your message. Often the message is missed because the listener is concentrating on discovering the whys-and-wherefores of what you are saying. To communicate clearly with a sensitive listener, state the reason for your message first. For example, "The reason for this change order is…" When you begin your message this way, the sensitive listener knows what needs to be known and can concentrate on listening to the message.

The Terminator – The terminator is interested only in the bottom line of the message. This individual often jumps to conclusions and literally races the sender to the end of the message. The most successful presentation method for communicating to a terminator is to state the bottom line of the message first. Use words and phrases such as *outcome, goal, end-result,* or *bottom line*.

PARTICIPANT ACTIVITY

1. Read the six-pack definitions.
2. Print your initials next to the decoding style that best describes your listening style.
3. Print your immediate supervisor's initials next to the style that best describes his or her listening style.
4. Select the decoding style of an employee who usually hears what you say. Place that person's initials next to that style.
5. Select the decoding style of an employee who seldom, if ever, hears you. Place the initials of that individual next to that style.

The decoding style of the employee who usually hears you represents your strongest presentation style. The decoding style of the employee who seldom hears you represents your weakest presentation style.

As a rule, do not change your strongest presentation style. However, you should strengthen your weakest presentation style.

PARTICIPANT ACTIVITY

Instructor's Notes:
_____
_____
_____
_____
_____
_____

The Encloser – The encloser listens for a verbal road map. Unless the message contains signposts indicating where the message is leading, attention will quickly stray. The most successful presentation method for dealing with an encloser is to explain your message step-by-step, with all "what-ifs" included.

The Numberer – The numberer listens for numbers in your message, such as rates, percentages, times, or volumes. If the message does not include the necessary figures, the numberer will insert them. In communicating to this type of listener, cite quantitative relationships. For example, "I need 200 bricks laid in 2 hours;" or "Have this done by 10 a.m."

SECTION 5

5.0.0 EMPLOYEE RELATIONS

The development and use of human resources is every supervisor's prime responsibility. Technical knowledge is important, but all the technical skill in the world will not help a supervisor succeed if the human resources are mishandled.

In order to begin the process of building a staff, the supervisor must know something about recruiting and hiring. This skill will become an important part of your position as you move into management. Knowing how to recruit, select, hire, and train staff is the key to your success as a manager. Understanding company policy as well as legal requirements and regulations at the company, state, and federal levels are also part of a supervisor's role.

The first principle of human resource management is that the supervisor must know what the employees are getting paid for. This does not mean that the supervisor has to be an electrician, a pipe fitter, a bricklayer, or a glazier. The supervisor needs to know the tasks these people do and has to be able to tell if they are doing them properly.

5.1.0 Task Analysis

Every position on a job should have a **task analysis** completed by the responsible supervisor. A time study isn't necessary, but the supervisor should be able to list the major activities involved in each position.

Start with your best employee, such as your best carpenter. What does this individual do well? If you think about the job, you will be able to identify the tasks and skills involved. Among them are probably:

- Reading blueprints
- Hanging doors
- Installing locks
- Building stairways
- Squaring foundations
- Installing windows
- Working with other trades
- Working alone with minimal instructions

A supervisor who makes hiring decisions can use a task analysis as a basis for interviewing potential employees. Ask the applicants to rate their own abilities based on each of the tasks they will be expected to perform. This pays off in two ways. First, it guides you, the interviewer, through the interview process. Second, it lets the applicant know what to expect on the job before starting.

5.2.0 Interviewing

Hiring the right person for every job is critical to your success as well as your company's profitability. Good hiring procedures help avoid turnover and absenteeism that are expensive, frustrating, and hurt employee morale. As a supervisor, you need to be aware of the legal aspects of employment and supervision. Training programs help managers to deal with such issues as sexual harassment, affirmative action, and the Americans with Disabilities Act (ADA). Company policies often contain helpful information about expectations for complying with the law.

Since the interview is the first opportunity that you will have to speak one-on-one with an applicant, avoid asking questions dealing with the subjects of race, creed, national origin, sex, age, marital status, and disabilities. Asking such questions is illegal.

Formalizing the application process begins with your company creating an application form to be used by everyone who applies to work for the company.

Ensure that you have all the necessary materials to teach the course. Check the Materials and Equipment list at the front of the module. Prepare for teaching Session III by reading Sections 5.0.0-6.3.0.

Distribute a sample task analysis sheet and discuss where a supervisor would need one.

Assign reading of Module MT202 Sections 5.0.0-6.3.0.

Ask trainees if their company uses an orientation checklist and what they would add to this checklist, if anything.

Show Transparency 7 (Participant Activity).

PARTICIPANT ACTIVITY

Complete a task analysis for an electrician, a pipe fitter, and a bricklayer by listing below the activities each is responsible for and the skills required.

A good electrician must be able to:
_____
_____

A good pipe fitter must be able to:
_____
_____

A good bricklayer must be able to:
_____
_____

Try this exercise for every position that reports to you.

PARTICIPANT ACTIVITY

The form should at least record the applicant's education and training, job history (including specific dates and places of employment), address, and signature. The completed form should tell you if an applicant has the type of background and experience that you require.

When interviewing, tell the applicant what is expected on the job. Use the task analysis discussed above and the *Employee Orientation Checklist (Figure 2)*.

Ask open-ended questions instead of questions that can be answered with just a word or two. This forces the applicant to reveal information. Among the questions that you might use are:

- How do you feel your training in the Navy will help you on this job?
- What did you learn on your last job that will help you succeed on this job?
- What did you like about your supervisor on your last job?
- Why did you leave your last job?

5.3.0 Selecting an Employee

If there are several applicants for a job, do not compare them to each other in making your decision. Instead, rate each of the applicants against your completed task analysis and select the one that best fits your needs. To do this, rate each applicant on a scale of one to ten.

5.4.0 Employee Orientation

Every new employee must be oriented to the company and the job. Sit down and talk with the new hire. Use the task analysis to explain what you expect. Be sure that you brief the employee about the official and unofficial rules of the job. Establish an open channel for communication to flow. Be willing to mentor and coach employees as they make their way on their career path.

By setting up systems for proactive training, good communication, and mentoring, you will be assuring your employee's success as well as your own.

The Employee Orientation Checklist presented in *Figure 2* will help you and your new employee get off on the right foot.

Instructor's Notes:

Employee Orientation Checklist

Employee: _____

As part of the orientation procedure, I have discussed with the employee the following:

_____ The analysis of the task involved
_____ Daily start and finish times
_____ Attendance, call-in procedure, phone numbers
_____ Lunch/break times
_____ Safety rules and regulations
_____ Job-site rules
_____ Documentation (time cards, etc.)
_____ Personnel policies
_____ Employee handbook
_____ Employee contacts: safety, performance, personal/personality, emergency
_____ Other:

Supervisor: _____
(Signature) _____ Date: _____

Employee: _____
(Signature) _____ Date: _____

Figure 2 • Employee Orientation Checklist

5.5.0 Performance Evaluations

Once new workers are on the job, the supervisor must provide regular appraisals of their performance. In other words, the supervisor has the responsibility to keep the employees informed about their performance.

There are three reasons for a supervisor to complete **performance evaluations** on employees:

- Employees need to know how well they are doing their jobs.
- Good employees need feedback to encourage them to continue their good work habits.
- Craftworkers, apprentices, and laborers who are performing poorly need to be informed in order to correct bad work habits.

Performance evaluations are the supervisor's responsibility. The supervisor is the member of management closest to the ongoing work habits of employees. However, a valid performance rating is much more than a supervisor's opinion. It should be based on observable behavior and rated on production levels, quality of workmanship, safety practices, dependability (as documented by timecards, etc.), and other measurable criteria.

Make each performance evaluation important to you and the employee. Arrange a private meeting, and give the employee your undivided attention when discussing performance. Remember, you are the role model. If you do not show the employee that you consider the performance appraisal to be important by displaying your concern and arranging for privacy and enough time, the employee will not consider the performance appraisal important and will not act on it.

Using a performance evalutaion checklist like the one in *Figure 3* encourages good work habits, provides positive feedback, and delivers constructive criticism. It is not a substitute for day-to-day supervision. Avoid waiting for performance evaluation time to discuss how an individual is doing on the job.

Ask trainees if their company uses an evaluation checklist and what they would add to this checklist, if anything.

Show Transparency 8 (Figure 2).

Show Transparency 9 (Figure 3).

Performance Evaluation Checklist

The Performance Evaluation Checklist will help you prepare and give effective performance evaluations to each employee.

_____ Schedule a time for the appraisal meeting with the employee. Inform the employee of the meeting well ahead of time.

_____ Use task analysis as the foundation for all performance appraisals.

_____ Review each task with the employee and discuss how you rated his/her performance of that task.

_____ Give positive feedback for each task done well.

_____ Tell the employee what can be done to improve performance.

_____ Ask the employee what you can do to help improve this performance.

_____ Ask for a commitment to improve, if necessary.

_____ Follow up the performance evaluation with subsequent regular reviews.

_____ Above all, be fair and consistent with all employees.

Figure 3 • Performance Evaluation Checklist

SECTION 6

6.0.0 PERFORMANCE, ASSESSMENT, AND TASK TRAINING

Employees learn how to do a job one way or another. In fact, people learn whether we want them to or not. They learn what we do not want them to learn just as easily as they learn what we want them to know. Employees want to be able to learn new things on the job. Providing a continuous learning environment is the best way to retain employees.

Supervisors are responsible for what their employees learn on the job and ensuring that their employees develop safe, acceptable work habits and avoid those that are unsafe and improper. Employee training is the only method to assure success in this area, and that puts the supervisor in the role of trainer from the very first day on the job.

Training needs are the basis of many quality and production problems. Indicators of training needs on the job site include low rates of production, rework, excessive overtime rates, and high accident rates.

6.1.0 Training Is Your Responsibility

Training is the supervisor's responsibility. However, it is not necessary for the supervisor to do all the training—only to see that every employee is properly trained. Therefore, if the training needs of a job are more than you can effectively handle, you are free to turn to any good sources of training that are available.

As the supervisor and the major trainer in your company, it is your responsibility to reinforce those skills your employees learn from outside sources. This involves demonstrating concern for employee training by assigning employees to jobs where they can practice newly acquired skills. Providing an opportunity for experienced workers to mentor beginners is an excellent way to build confidence and provide growth opportunities for everyone.

To determine the type of training an employee needs, use the results of the task analysis and performance evaluations discussed earlier as a baseline. Then, put the area of needed training into a logical sequence, starting with the easiest skills and progressing to the most difficult. This sequence becomes the guideline for each employee's training, whether you conduct the training yourself or it occurs elsewhere.

Instructor's Notes:

It allows the employee to gain confidence, avoid frustration, and increase productivity.

As part of every on-the-job training effort:

- Tell the employee what is to be learned.
- Explain how to do the task being taught.
- Demonstrate the skills involved.
- Let the employee try it.
- Give immediate feedback.
- Repeat the sequence until the skill is learned.

You are responsible for the performance of your employees. Whether they are craftworkers, helpers, or apprentices, they are your company's most valuable resource. They use the material resources of your company to provide goods and services to your customers. When your employees perform well, your project functions smoothly and your customers are satisfied, which leads to repeat business.

6.2.0 Value of Training

Meeting the training needs of crew members is one of the most important responsibilities of the supervisor, yet is often overlooked. Training provides great value, and those who benefit from training include:

The client – receives on-time and within-budget projects.

The company – satisfies clients and enjoys continued profitability.

The employee – is provided with a career pathway that leads to a more rewarding career.

You, the supervisor – builds a successful team and creates replacements that will allow your continued growth and promotion within the company.

As a supervisor, you might not lead training; however, you can provide important assistance by assessing your crew members' skills and determining what training they need.

6.3.0 Types of Training

Types of training include:

Orientation – A supervisor should always orient a new crew member (both new hires and those new to the site) as soon as possible after assignment.

As part of the orientation procedure, the supervisor should:

- Describe the type of physical and/or mental work typically required.
- Explain what to expect during a typical week.
- Explain job duties and the performance standards by which the crew member will be evaluated.
- Set rules for working hours and days.
- Review the company's benefit program and pay procedures.
- Introduce the crew member to supervisors and fellow workers.
- Explain safety procedures for both the assigned job tasks and any other site-specific hazard.
- Explain special procedures, such as parking restrictions at the job site.

Task/Apprenticeship – New processes, tools, and materials may require new skills to effectively complete the job. The supervisor's responsibility is to evaluate the skill level of the crew and provide or recommend proper training to improve performance.

Safety – Over the last decade, average profit margins across the industry have fallen to around 6 percent, leaving very little room for unexpected losses. The supervisor is directly responsible for the crew's safety and the company's losses due to accidents. Safety training provides knowledge to crew members so that they may do their jobs in a safe, productive manner. Supervisors must evaluate the hazards facing crew members and either eliminate the hazards or provide training so that the crew members can protect themselves.

Leadership/Management – Job turnover in the construction industry is staggering, with tens of thousands of skilled workers leaving to pursue other careers. Supervisors left with untrained, inexperienced workers struggle to complete projects to the satisfaction of both clients and management. The supervisor can play an important role in retaining valuable, experienced workers by actively identifying potential leaders and recommending appropriate training. The supervisor should remember that it is much easier to be promoted when a trained replacement is already available.

Ask trainees to give specific examples under the five types of training.

Ensure that you have all the necessary materials to teach the course. Check the Materials and Equipment list at the front of the module. Prepare for teaching Session IV by reading Sections 6.4.0-6.4.4.

Ask trainees to give an example of each level of employee response that they have experienced in the workplace.

Assign reading of Module MT202 Sections 6.4.0-6.4.4.

Personal Development – The supervisor should also be aware of special issues that may require training to improve morale and production. These issues include, but are not limited to:

- English as a second language
- Basic reading skills
- Harassment training

6.4.0 Effectiveness and Efficiency

Your success as a supervisor is the result of two factors, effectiveness and efficiency.

Effectiveness is the outcome of meeting your scheduled and assigned workload in a timely manner.

Efficiency is the outcome of meeting your scheduled and assigned workload in a cost-effective manner.

Your job as a supervisor is to establish and maintain conditions that promote effectiveness and efficiency. You will accomplish this by identifying and correcting performance **discrepancies** that impede your efforts to foster and nurture these two factors.

6.4.1 Identification of Performance Discrepancies

One of your craftworkers has missed several scheduled maintenance deadlines in the past week due to being "too busy." A review of your job log indicates that the individual has been failing to keep equipment maintained for the last several weeks and has not alerted you. Aside from this, however, you note that the employee is a good performer in other aspects of their job.

You have a problem. The employee is failing to meet the performance standards for maintenance, yet exhibits model behavior in other areas.

Your task is to correct the problem and ensure that it does not develop again in the future. Your first step in resolving a performance problem is to identify the performance discrepancy. Only after you have determined its root cause can you isolate and finally rectify the performance problem.

6.4.2 The Employee Response Model

There are a variety of causes of poor performance. One is the employee's response or lack of response. The level of response can be classified in the following manner:

Will – Your employee *will* do the job.
Won't – Your employee *won't* do the job.
Can – Your employee *is able* to do the job.
Can't – Your employee *is not able* to do the job.

Isolating and resolving poor performance is your job. If one of your employees is not meeting a performance standard, you must be aware that there can be several reasons.

Can but Won't – The employee knows how to maintain the equipment, but is not doing it. Why is the employee not doing it?

Can't but Will – The employee is trying to maintain the equipment level but doesn't know how it is done. What can you do to help the employee?

Can and Can't – The employee sometimes maintains the equipment, but not always. Ask yourself why the employee is able to complete the job sometimes and is unable to do so at other times.

Can't and Won't – The employee does not know how to maintain the equipment load and does not want to learn.

6.4.3 Diagnosis of Employee Response

You will be more successful as a supervisor if you are able to consistently diagnose your employees' responses. This will allow you to zero in on performance problems more quickly, take steps to correct them, and get on with your workday.

Isolate the root cause. Ask yourself the following questions to identify and isolate the performance discrepancies:

- What causes me to think there is a problem?
- What is the difference between what I want done and what I am getting?
- What makes me say that?
- Why am I dissatisfied?
- What is the root cause of the problem?

Examine the cause behind the problem and develop a means to eliminate it.

Instructor's Notes:

Is the problem a high priority? Solving problems can cost your company a lot of time and money. Before you begin to resolve a performance problem, you should step back and evaluate its significance. Rank problems in order of importance, and strive to resolve those with the highest priority first. You may find that you ignore some discrepancies in order to concentrate on more serious problems.

Missed maintenance is a high priority problem. Ask yourself the following questions:

- Why does this performance discrepancy have a high priority?
- How much does the problem cost? Make sure you look at real costs of the problem. Some possible sources are:
  — Cost of replacement
  — Cost of downtime
  — Equipment unavailable to other jobs
  — Equipment damaged
  — Work not completed
  — Work completed poorly
  — Lost business
  — Extra supervision necessary
  — Customer dissatisfaction
- What would happen if I ignore it?
- How could doing something to resolve it have any profitable outcome?

Now that you have assessed the priority of the problem, it is up to you to determine the cause of the problem.

Is it a lack of skill? Determine whether your employee has the necessary skills to perform the required task. If, for example, the employee has not been trained in maintenance, you cannot expect his or her performance to meet the prescribed standards. Ask yourself the following questions:

- Does the employee know the procedure?
- If the employee's job depended on the task, could the employee do it?
- Does the employee have the necessary skills for the required performance? If so, what are they?

If the employee could not maintain the equipment even if the job depended on it, there is a skill deficiency. To decide how you will proceed, ask yourself the following question:

Did the employee know how to maintain this piece of equipment at one time? Like any other tool, infrequently used skills will quickly dull. Maybe the employee's primary activity has been on another piece of equipment and he or she has not maintained this other type of equipment for years. The employee's skills have simply become rusty.

- How often does the individual maintain this type of equipment? Are the correct procedures being followed?
- Is regular feedback provided on the employee's maintenance performance? How?

If the employee is new to your job site, be sure to frequently provide performance feedback. Do this until both you and your employee are comfortable with the performance. Once the required performance level has been attained, continue to give feedback (positive as well as negative), but less frequently.

6.4.4 Constructive Solutions

There may be an easier way to accomplish the needed performance improvements. Ask yourself the following questions:

- Is there a job aid available?
- How can the employee use written instructions or checklists?
- How can you show the employee how the job should be done?
- Would hands-on training be more effective?

An employee may perform better after the process is presented in a step-by-step fashion as "the way to do it." Often a lack of clear communication will cause performance breakdowns.

Can the employee handle the job? Sometimes a performance discrepancy can be traced to a problem that cannot be resolved.

- Can the employee be trained to maintain the equipment?
- Does the employee have the physical and mental abilities needed to do the job correctly?
- Is the employee overqualified for this job and better qualified for another?

Discuss missed maintenance issues that trainees have encountered in the workplace and its outcome on a project.

Refer to the end of the module for a teaching tip.

Discuss examples of rewarding poor performance that trainees have encountered in the workplace and how they think the practice can be stopped.

Discuss examples of rewarding good performance that trainees have encountered in the workplace and why they think the rewards were effective.

For example, if an employee has difficulty using a tool because of a vision problem, a transfer to a position for which good vision is not a prerequisite may be a solution. If you discover that the employee simply refuses to learn the procedure, then you have a *Can't and Won't* situation and must deal with it accordingly.

If your employee has forgotten how to maintain this particular piece of equipment or has never learned the process, there are solutions:

- If the employee was never trained, a training course will provide the necessary skills.
- If the employee correctly maintained the equipment in the past but has forgotten how, coaching may be necessary.
- If the employee is often responsible for equipment maintenance and is inconsistent in performance, be sure to provide positive feedback when the job is being done correctly.

If the employee can complete the task when required, there is no skill deficiency. Skills training will not correct the problem. You will need to look elsewhere to uncover the real reason for the performance discrepancy.

If the job is consistently done well, is the employee penalized? For example, is that hardworking operator who takes fewer, shorter breaks than average increasing his own workload by being available for more work because of high operating percentage? Employees soon learn the consequences of "good performance" and may avoid it in the future. Ask yourself:

- When the employee maintains the standard level of performance (60-80 percent overall, 100 percent peak), what happens to that individual?
- Is there a penalty involved if the employee does the job right?
- Does the employee associate good performance with penalties?

Remove hidden penalties that inadvertently affect your employee. For example, instead of continually pairing your best operator with your worst assistant, provide a little help. The employee could feel penalized by the recognition. Employees may not like being put "on stage" and therefore sabotage their own performance so that they are no longer the best.

Is the employee rewarded for poor performance? Poor performance may yield rewards for the employee. An employee may omit key steps in maintenance in order to increase the uptime percentage and thereby appear more productive. Neglect may result in unnecessary downtime. However, if the result of such negligence is a reward in the form of praise for a high percentage of uptime, shortcuts in the job will continue.

To find out if your employees are rewarded for poor performance, ask yourself the following questions:

- What happens when the employees do the job their own way instead of according to procedure?
- What prestige, status, or reward does the employee receive for poor performance?
- Does the employee get more attention for poor performance than for excellent performance?
- Are you overlooking vital factors while rewarding performance that doesn't matter?
- Is the employee mentally incapable of doing the work and getting by with the current level of performance?
- Is the employee physically incapable?

Productivity is not the only standard by which your job sites are measured. Your customers continue to call your company because of the superior quality of work and service you provide. If employees try to increase productivity at the expense of your company's standard of quality, you sacrifice your reputation and ultimately your position in the industry.

An employee should not be judged solely on the *quantity* of work completed in a given day. The *quality* of work should be evaluated as it can result in less rework.

Do the employees really care about good performance? How are they affected by poor or exceptional performance? When employees' actions do not affect their status, they will be more likely to do just enough work to get by.

- How important is doing the job right to the employee?
- How are employees rewarded for doing the job right?

Instructor's Notes:
_____
_____
_____
_____
_____
_____

Show Transparency 10 (Participant Activity).

PARTICIPANT ACTIVITY

After isolating the cause of a problem, you may find that there are several different solutions. Think of a problem you have at work and list several solutions:

_____
_____
_____
_____
_____

Now that you have your solutions listed, take a step back and examine them:

1. Have you really listed all the possible solutions?
2. How does each solution work to solve one of the problems that you have identified?
3. What is the cost of each solution?
4. Have you examined hidden costs? What are they?
5. Which solution is most likely to work, most economical, and easiest to execute?
6. Which solution is likely to solve more of the problems while draining fewer resources? Which will provide the "biggest bang for the buck"?
7. Which solution will yield greater results for less effort?
8. Which solution can we test and document most easily?
9. Which solution is most strongly supported?
10. Which solution is closest to company policy?

PARTICIPANT ACTIVITY

- Are there any undesirable consequences for the employee for poor or no performance?
- How can employees take pride in their work?
- How are the employee's personal needs satisfied?

Company employees at all levels need to know how their work affects others. These individuals should be personally responsible for the quality of their work. If, for example, an operator is aware of the problems equipment neglect can cause down the line and knows that downtime will be traced back, less neglect will occur.

Sometimes an employee is prevented from achieving expected standards by encountering performance obstacles. For example, if an employee is unfamiliar with the procedure to follow during maintenance, the same frustration and level of performance will occur every time equipment is shut down for maintenance. Ask yourself:

- What types of obstacles can keep an employee from doing a job right?
- How does the employee know what should be done?
- How does the employee know when to do the job?
- Is the employee receiving conflicting instructions?
- Does the employee need authority, time, or tools, which are not available to do the job?
- Is there a "right way" or "a way we've always done it" that needs to be changed?

Can distraction be reduced by improving lighting, providing a more comfortable work environment, using different colors, or changing a work position? Should visual and audio distractions be reduced? Can competition from within the job be reduced? Some examples would be "brush fires" and less important problems that require immediate solutions.

_____
_____
_____
_____
_____
_____

Ensure that you have all the necessary materials to teach the course. Check the Materials and Equipment list at the front of the module. Prepare for teaching Session V by reading Sections 7.0.0-8.1.2.

Ask trainees how their company measures motivation.

Assign reading of Module MT202 Sections 7.0.0-8.1.2.

SECTION 7

7.0.0 MOTIVATION, MENTORING, AND CONFRONTATION

The need to motivate, mentor, and confront is inevitable in management. An effective leader knows that ignoring or avoiding motivational problems only leads to larger motivational problems later. A good leader also knows the value of mentoring employees and how it helps guide them on the job.

The effective leader outshines the poor leader in dealing with confrontations. An effective leader approaches each confrontation with tact. Tact enables one to find common ground between the needs of the employees and the requirements of the project.

7.1.0 Employee Motivators

The ability to motivate others is a key leadership skill that effective supervisors must possess. **Motivation** refers to behavior set into action because the individual feels a need to perform. It's also the word we use to describe the amount of effort that a person is willing to put forth to accomplish something. For example, a crew member who skips breaks and lunch in an effort to complete a job on time is thought to be highly motivated. A crew member who does the bare minimum or just enough to keep the job is considered unmotivated.

Employee motivation has dimension because it can be measured. Examples of how motivation can be measured include absenteeism levels, employee turnover percentages, and the number of complaints, as well as the quality and quantity of work produced.

Different things motivate different people in different ways. Consequently, there is no "one-size-fits-all" approach to motivating crew members. It is important to recognize that what motivates one crew member may not motivate another. In addition, what works to motivate a crew member once may not motivate that same person again in the future.

Frequently, the needs that motivate individuals are the same as those that create job satisfaction. They include:

- Recognition and praise
- Accomplishment
- Opportunity for advancement
- Job importance
- Change
- Personal growth
- Rewards

A supervisor's ability to satisfy these needs increases the likelihood of high **morale** within a crew. Morale refers to individuals' attitudes toward the tasks that they are expected to perform. High morale means that employees will be motivated to work hard, having a positive attitude about coming to work and doing their jobs.

7.1.1 Recognition and Praise

Recognition and praise refers to the need to have good work appreciated, applauded, and acknowledged by others. This can be accomplished by simply thanking employees for helping out on a project, or it can entail more formal praise, such as an award for "Employee of the Month."

Here are some tips for giving recognition and praise:

- Be available on the job site so that you have the opportunity to witness good work.
- Know good work when you see it, and praise it when you see it.
- Give recognition and praise only when truly deserved; otherwise, it will lose its meaning.
- Acknowledge satisfactory performance, and encourage improvement by showing that you have confidence in the ability of your crew members to do above-average work.

7.1.2 Accomplishment

Accomplishment refers to a worker's need to set challenging goals and achieve them. There is nothing quite like the feeling of achieving a goal, particularly a goal one never expected to accomplish in the first place.

Supervisors can help their crew members obtain a sense of accomplishment by helping them develop performance goals for the year. In addition, supervisors can provide the support and the tools, such as training and coaching, necessary to help their crew members achieve these goals.

Instructor's Notes:

7.1.3 Opportunity for Advancement

Opportunity for advancement refers to an employee's need to gain additional responsibility and develop new skills and abilities. It is important that some employees know that they are not stuck in dead-end jobs. Rather, they want a chance to grow with the company and to be promoted in recognition of excelling in their work.

Effective leaders encourage their crew members to work to their full potential. In addition, they share information and skills with their employees in an effort to help them advance within the organization.

7.1.4 Job Importance

Job importance refers to employees' need to feel that their skills and abilities are valued and make a difference. Employees who feel that they don't make a difference tend to have difficulty justifying why they should get out of bed in the morning and go to work.

Supervisors should attempt to make every crew member feel like an important part of the team, as if the job would be impossible without their help.

7.1.5 Change

Change refers to employees' need to have variety in their work. Change is what keeps things interesting or challenging. It prevents the boredom that results from doing the same task day after day with no variety.

7.1.6 Personal Growth

Personal growth refers to an employee's need to learn new skills, enhance abilities, and grow as a person. It can be very rewarding to master a new competency on the job, particularly one that you were not trained to do. Similar to change, personal growth prevents the boredom associated with doing the same thing day after day without developing any new skills.

Supervisors should encourage the personal growth of their employees as well as their own. Learning should be a two-way street on the job site; supervisors should teach their crew members and learn from them as well. In addition, crew members should be encouraged to learn from each other.

7.1.7 Rewards

Rewards refer to employees' need to be compensated for their hard work. The rewards can include a crew member's base salary, or they can go beyond that to include bonuses, give-aways, or other incentives. Rewards can be monetary in nature, such as salary raises and holiday bonuses, or they can be non-monetary, such as free merchandise or other prizes (shirts, coffee mugs, and jackets, for example).

To increase motivation in the workplace, supervisors must individualize how they motivate different crew members. It is important that supervisors get to know their crew members and determine what motivates them as individuals. Once again, as diversity increases in the workforce, this becomes even more challenging; therefore, effective communication skills are essential.

Some tips for motivating employees:

- *Keep jobs challenging and interesting* – boredom is a guaranteed de-motivator.
- *Communicate your expectations* – people need clear goals in order to feel a sense of accomplishment when they are achieved.
- *Involve your employees* – feeling that their opinions are valued leads to pride in ownership and active participation.
- *Provide sufficient training* – give employees the skills and abilities they need to be motivated to perform.
- *Mentor your employees* – coaching and supporting your employees boosts their self-esteem, self-confidence, and ultimately their motivation.
- *Lead by example* – become the kind of leader that employees admire and respect, and they will be motivated to work for you.
- *Treat employees well* – be considerate, kind, caring, and respectful. Treat employees the way that you want to be treated.
- *Avoid using scare tactics* – threatening employees with negative consequences can backfire, resulting in employee turnover instead of motivation.

Classroom

Ask trainees to describe how they help or would help crew members develop performance goals and if they know the rate of success of these goals.

Ask trainees to fill out the checklist and share any insights from ways they may have de-motivated employees and how they can change their attitude.

Show Transparency 11 (Figure 4).

It is easy for new supervisors to follow what they have been taught. Whatever management style your supervisor used, you are likely to fall back on the same strategies. It is likely that some of these strategies will not serve to motivate employees. The *Employee De-Motivator Checklist* in *Figure 4* is a reminder of what to avoid if you want to maintain employee morale and motivation.

Look over the following checklist and give an honest assessment of what de-motivators you may fall back on from time to time.

7.2.0 Mentoring

Mentoring is defined as a sustained relationship between two people. Through continued involvement, a mentor offers support, guidance, and assistance as the other person faces new challenges, works to correct earlier problems, or goes through a difficult period. It is easy to see how mentoring can fit into the context of construction. A supervisor can be a mentor to crew members. Alternately, if supervisors are either unavailable or unable to provide responsible guidance to each employee on a regular basis, experienced crew members can fill this gap by mentoring those new to a job, crew, or company.

The two types of mentoring are *natural mentoring* and *planned mentoring*. Natural mentoring occurs through friendship, collegiality, teaching, coaching, and counseling. In contrast, planned mentoring occurs through structured programs in which mentors and participants are selected and matched through formal processes. Planned mentoring is usually initiated and advertised by the employer.

Mentoring has grown in popularity in recent years, in part due to the resounding success of those involved. The benefits claimed by the participants are an increase in positive attitude, congeniality among fellow crew members, and enhanced job performance and productivity. The rebound effects for the employer are better recruiting through employee word-of-mouth, decreased employee turnover, lower absenteeism, and less supervision required by those employees in the program.

Employee De-Motivator Checklist

In the past 30 days, I've sometimes found myself:

_____ Talking down to employees, especially apprentices.

_____ Ignoring suggestions, not listening to ideas offered by employees, or always choosing my own way of doing things over theirs.

_____ Ridiculing employees, especially in front of others.

_____ Overloading good employees with more than their fair share of work instead of training other workers to share the load.

_____ Finding something wrong on every job.

_____ Making employees rush to get a job done when it has been delayed because of an error I made, such as failing to request material deliveries on time.

_____ Ordering employees to work instead of asking them.

_____ Choosing not to explain changes in plans because they only need to know what to do, not why.

_____ Playing on their fear by keeping my distance and not letting them get to know me.

_____ Telling them that I could do their jobs better and faster.

If you placed a check next to more than one of the above items, you have a tendency to use fear as a motivator.

Figure 4 • Employee De-Motivator Checklist

Instructor's Notes:

7.3.0 Confrontation

The key to positive confrontation is for supervisors to believe that they are helping employees take responsibility for their own:

- Behavior
- Production
- Quality
- Safety
- Adherence to company rules

Keep in mind that punishment seldom works. All too often, punishment humiliates people and encourages the punished parties to find a way to repay the people who humiliated them. A payback may take the form of sabotage, tattling to the boss, or the employees attempting to get away with violations of job procedures and rules.

Punishment usually attacks personality—the part of a person that cannot be changed. Positive confrontation remolds poor behavior—the part of a person that *can* be changed. The key, then, is for the supervisor to deal with behavior by focusing confrontations on the job to be done rather than on the person doing it.

7.4.0 The Employee Responsibility Model

The following employee responsibility model shows the application of proper confrontation techniques. The model involves a confrontation between a supervisor and an employee named Mike, who is late for work on a semi-regular basis. The steps to follow are numbered and include examples of the supervisor's actions and the most likely employee reactions.

1. *Identify the problem and select an outcome.*

   Supervisor: "Mike is often late. I want Mike to be on time for work every day."

2. *Determine how often the problem occurs.*

   Supervisor: "Over the past three weeks Mike has been late five times."

3. *Call the employee aside and ask questions.*

   Supervisor: "Mike, you seem to be having a problem getting here in the morning. What's going on?"

   Mike: "Yeah, I was late today. So what? The traffic was crazy."

4. *Paraphrase what the employee says, then state the problem.*

   Supervisor: "The traffic is bad, Mike, but you've been late five times in the past three weeks. I schedule work for all of us to start at the same time. When you're late, it hurts all of us."

   Mike: "I don't do it on purpose. It's the dang traffic."

   Supervisor: "You and I don't control that. Mike, you need to get here on time."

5. *Restate the problem and get agreement from the employee that there is a problem.*

   Supervisor: "Mike, I agree with you that heavy traffic is a problem. Can you agree with me that when you're not here on time, it presents a problem for us on the job?"

   Mike: "Yeah. I guess it's a problem."

6. *Explain why change must occur.*

   Supervisor: "Mike, it's not just a matter of company rules; it's a matter of productivity. You have to get here on time."

7. *Ask the employee what he can do to solve the problem.*

   Supervisor: "Mike, what can you do to get here on time?"

8. *Put the responsibility on the individual. Let the employee tell you how the problem can be solved.*

   Mike: "I guess I could leave home ten minutes earlier."

9. *Rephrase the agreement to make it a commitment.*

   Supervisor: "Okay then, Mike. I'll count on your leaving home earlier to get here on time. Thanks, Mike."

10. *Follow up.*

    Supervisor (the next morning): "I see you made your way through the traffic on time. Thanks, Mike."

When a supervisor uses the employee responsibility model, the ball is put in the employee's court. If, for instance, Mike had been late again, the supervisor can refer to the agreement they made in the above confrontation.

Show Transparency 12 (Confrontation Issues). Have trainees give scenarios for each confrontation issue and how they dealt with it.

Role-play a confrontation situation with a trainee about lack of quality workmanship.

Show Transparency 13 (Figure 5).

Supervisor: "Mike, you and I had an agreement. What happened?"

This makes it clear that the employee has violated an agreement and makes necessary disciplinary action acceptable. Remember, it is the employee's responsibility to follow the rules of the job. It is the supervisor's responsibility to help them do so.

Figure 6 is the Employee Responsibility Model Checklist. Notice that it is based on the Eleven Commandments of Confrontation (*Figure 5*). Following this checklist will help you confront employees to achieve a positive outcome.

SECTION 8

8.0.0 CONFLICT RESOLUTION

Conflict is inevitable. It is a natural part of any human endeavor, arising from the fact that different people have different viewpoints. An estimator wants to be a successful bidder, so the bid is a tight one. Supervisors want to run a successful job, so they want a degree of slack. A natural conflict exists between the two. The trick is to make sure that the conflict remains healthy and constructive.

8.1.0 Managing Conflict

The supervisor should see conflict as a sign of healthy vitality in the organization – a means for:

- Team participation in decisions
- Exploring new methods and directions
- Heading off bitter controversies
- Learning

However, conflict must be handled constructively.

8.1.1 Factors Affecting Conflict Management

Several factors affect how conflict is managed in a construction firm. Among them are:

- *Leadership style* — An organization's leaders, such as its president, supervisors, general superintendent, and office man-

---

The Eleven Commandments of Successful Confrontation

1. Thou shalt not start a confrontation without a goal. Figure out what you want to happen before you confront.
2. Thou shalt not forget to state the problem. Inform the employee of the problem and explain it.
3. Thou shalt not jump to conclusions. Get as much information as you can. Find out what the problem is.
4. Thou shalt not forget to discuss. Find out, for example, if the problem is one of skill, schedule, or simple error.
5. Thou shalt not ignore decoding styles. Identify the employee's decoding style by using the six-pack described earlier in the module.
6. Thou shalt not use close-ended questions. Ask only questions that demand more than a yes-or-no answer.
7. Thou shalt not attack an employee's personality. Focus on the job, not the person.
8. Thou shalt not ignore thy employee's reply. Listen to what the employee has to say.
9. Thou shalt not forget to paraphrase. Rephrase in your own words what the employee says.
10. Thou shalt not forget to follow up. Make sure the employee's behavior changes and you do what you say you will.
11. Thou shalt not dwell on past problems. Once poor work habits or poor performance are corrected, forget about it.

Figure 5 • Eleven Commandments of Successful Confrontation

Instructor's Notes:

Employee Responsibility Model Checklist

In every confrontation, the supervisor should:

_____ Select an outcome.

_____ Determine how often the problem occurs.

_____ Call the employee aside and ask questions.

_____ Paraphrase.

_____ State the problems, and then get agreement.

_____ Explain why change must occur.

_____ Ask the employee to suggest a solution.

_____ Put the responsibility on the employee.

_____ Rephrase the agreement.

_____ Follow up.

Figure 6 • Employee Responsibility Model Checklist

agers, influence how conflict is resolved. They set the tone or climate. Research indicates that companies that welcome conflict and that openly and directly deal with problems are usually the most successful.

The leaders in such companies do not deny or avoid the problems. They do not suppress differing points of view. They do not compromise just to settle a matter or allow themselves the suicidal luxury of win-lose outcomes. Instead, they present their employees with useful models for communication and conflict resolution on the job site.

- *The role of feedback* — Few people enjoy receiving negative feedback. However, an effective supervisor must encourage negative as well as positive peer, customer, and employee feedback. If a supervisor, like the rulers of ancient times, continually kills the messenger who bears bad news, then the supervisor is cutting off the vital flow of information needed to assess and improve individual and team performance.

- *Healthy expression of feelings* — Supervisors must openly encourage the expression of feelings, especially when the feelings are negative. Job sites where employees feel free to express their views and concerns are healthy ones; those where fear rules are not.

- *Follow up* — When it comes to conflict resolution, follow-up procedures and accountability are crucial. Everyone involved in a conflict must perceive that the steps needed are being taken. They must have visible evidence that their concerns are not "falling through the cracks."

Consequently, the supervisor must inform those involved about:

- What was decided
- Who will do what
- When it will be done
- How they will know when the proper steps have been taken
- What is the expected benefit
- How effective was the decision

8.1.2 Strategies for Dealing With Conflict

There are five common strategies for dealing with conflict. Some are effective; some are not. None applies to all situations or management styles, and the supervisor must select the one that is appropriate in a given set of circumstances. Avoid the temptation to use only one method to deal with all the conflicts that arise.

The five strategies for resolving conflict are:

1. *Denial* – In this strategy, supervisors try to deny that a problem exists. They hope it will go away without having to face it. Unfortunately, conflict seldom goes away of its own accord. As a result, denial results in the conflict growing to the point where it becomes unmanageable.

2. *Smoothing over* – Here the supervisor admits that a conflict exists, but instead of resolving it, tries to bury it under a smooth layer of good feelings. This strategy insults those involved in the conflict and plants the seeds of deep resentment. In turn, the resentment leads to a complete stoppage of effective communication.

Ask trainees to give an example of how they used each of the five strategies to resolve conflicts.

Show Transparency 14 (Figure 6).

Ensure that you have all the necessary materials to teach the course. Check the Materials and Equipment list at the front of the module. Prepare for teaching Session VI by reading Sections 9.0.0-9.3.2.

Ask trainees to give an example of organizational decisions and operational decisions they have recently encountered in their workplace.

Assign reading of Module MT202 Sections 9.0.0-9.3.2.

3. *Pulling rank* – In this ploy, the supervisor tries to dominate the situation with personal or positional power. Often the result is a win-lose situation, and in such a situation, the losers are unlikely to go along with the supervisor's solution to the conflict. Instead, they begin looking for an opportunity for payback.

   Although generally ineffective, pulling rank can be effective and even necessary in situations where conflict must be resolved immediately, where important rules are being challenged, or where other methods of conflict resolution have failed.

4. *Cutting a deal* – The supervisor tries to please both parties in the conflict by cutting a deal that gives a little to each. The danger in cutting a deal is that the solution is so watered down that neither side is satisfied. Cutting a deal does have its time and place, particularly when a conflict has reached an impasse.

5. *Negotiation* – In negotiation, the supervisor approaches the conflict with an open mind but has a well-defined outcome as a target. This strategy demands that each party be willing to modify their position to achieve a mutually satisfactory outcome. If not constrained, negotiation usually resolves conflicts in a way that is in the best interest of the organization.

SECTION 9

9.0.0 PROBLEM SOLVING AND DECISION MAKING

As a supervisor, you are frequently called upon to solve problems and make decisions. The ability to effectively do so often determines the difference between an effective leader and an ineffective one. This portion of your training focuses on problem solving and decision making. It distinguishes between the two terms and provides guidelines for developing skills in these important areas, both of which are crucial for supervisors.

The nature of the decisions and the problems that managers face are directly related to their levels within the company. Upper-level managers, such as owners, presidents, and vice presidents, make decisions and solve problems appropriate to their levels of authority and responsibility. You are expected to do the same, keeping your supervisory role in mind.

Although there are various levels of decision making and problem solving, they are interrelated. For example, decisions made by upper management affect those made by mid-level and first-level management. Consequently, if each level of management fails to perform effectively, the entire company can be adversely affected. The end result is that the company does not achieve its full potential.

9.1.0 Organizational vs. Operational Decisions

Decisions can be divided into two major categories: **organizational decisions** and **operational decisions.**

Organizational decisions deal with issues concerning the organization as a whole. Upper management normally makes these decisions. An example of an organizational decision is determining if a new office should be opened in another city. Another example is deciding whether or not to develop an in-house sheet metal fabrication shop or continue to rely on sheet metal subcontractors.

Operational decisions deal with day-to-day operations. Supervisors and other mid-level and first-level managers usually make these decisions. An example of an operational decision is determining who should be assigned a specific job task.

9.2.0 Decision Making

Decision making and **problem solving** are not the same. However, the two terms are often confused and used interchangeably.

Decision making is the process of choosing a course of action from among alternative possibilities; whereas, *problem solving* is the process of deciding how to close the gap between the way things are and the way they should be.

Part of the reason that the two terms are confused is due to the fact that decision making often involves solving problems. It has been said, "... not every decision involves a problem, but every problem involves a decision." For

Instructor's Notes:

This is a two-page transparency. Show Transparency 15 (Participant Activity Case 1) and then show Transparency 15 (Participant Activity Case 2).

Work on the Participant Activity together in class. The correct answers are located at the end of this module.

PARTICIPANT ACTIVITY

Study the two situations below. After each one is a list of suggestions based on the five strategies for dealing with conflict in Section 8.1.2. Label each suggestion with its corresponding number.

Case 1:

Sam is a job-site foreman for a crew of six carpenters, two laborers, and two apprentices. The work is exacting, and inattention would cause a costly mistake or serious injury. Sam suspects that one of his carpenters is taking drugs on site or before coming to work. Sam has some strong concerns but no evidence:

____ "Talk to the guy. Tell him what you think and why, and that you're concerned for him and the rest of the guys."

____ "Tell him what life is about. Tell him drugs are illegal as well as unsafe and that when you catch him, he's history."

____ "Watch him like a hawk."

____ "Tell the guy to control himself and keep his habit off the job."

____ "Don't confront him. It might drive him underground."

Case 2:

Tom is your foreman. Recently, he's noticed that an individual from another sub's crew has been coming over and "shooting the breeze" with the glazier on his crew. The glazier's efficiency is falling off, and there have been some problems with the guy's work in the past. Tom believes there's some resentment among the rest of his crew.

____ "Put the crew at ease. This is an important job. Everyone has to get along."

____ "Don't say anything. Don't make a big deal of it. We're on schedule."

____ "Talk to the other foreman. Tell him we will keep his guy in line."

____ "Confront both men the next time you see them together. Find out what's happening. Then tell your guy what you expect."

____ "Talk to the other guy. Tell him to get to work."

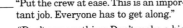

example, when a sheet metal supervisor decides how joints of duct should be assembled, no problem solving is required. This is because the various methods of assembling ductwork are already known. In this case, applying problem solving techniques is a waste of time, since the alternatives are obvious. Problem solving is best suited to identifying alternatives in new or unfamiliar situations.

Decision making involves four basic elements: choice, alternatives, goals, and consequences.

9.2.1 Choice

Choice is the opportunity to select among alternatives. If there is no choice, then there is no decision to be made. Decision making is the process of choosing, and many decisions involve a broad range of choices. For example, you may choose crew members to assign to a task, materials to use in a particular construction phase, and tasks to be completed first.

As a supervisor, policies, procedures, methods, safety, and similar concerns limit your choices. In addition, you must consider the different types of people and personalities that you deal with on a daily basis in identifying your choices.

9.2.2 Alternatives

Alternatives are the possible courses of action from which choices are made. If there are no alternatives, then there is no choice, and, therefore, there is no decision. Most situations offer alternatives. If you cannot see the alternatives,

Ask trainees to give examples of trade-offs they have made or expect to make as a supervisor when making a decision and the results of those decisions.

perhaps you should study the situation more carefully.

People often limit themselves to either/or alternatives. While this simplifies complex problems, it also blinds them from other — perhaps even better — alternatives. It is best not to limit yourself to two alternatives. Rather, you should consider a multitude of possibilities in an effort to find the best solution.

9.2.3 Goals

Goals are the end results that one seeks to achieve. Therefore, they can be used to determine which of several alternatives to follow in a given situation.

In order to achieve project goals, you must sometimes make decisions that are not ideal for you or your crew members but may be best for the company or the project as a whole. Contractors face such trade-offs daily in making decisions when conflicts arise.

Trade-offs are sometimes necessary because companies are trying to attain several goals at the same time. Some objectives are more important than others to the company overall, but importance also varies from person to person and from crew to crew.

9.2.4 Consequences

No decision is an isolated event. Just as problems have multiple causes, decisions have multiple consequences, both intended and unintended. The skillful supervisor considers the consequences of every decision. For example, allowing one subcontractor to install work before another has a direct impact on the work of the second subcontractor. Some of the consequences are obvious, while others are not. Being able to identify and properly weigh obvious and obscure consequences is the mark of an expert decision maker.

9.2.5 Variables

Variables are elements that change from situation to situation, and they profoundly affect decision making. In making decisions, you must first identify the variables involved. Then, you must try to understand their meaning in the context of the situation and use this information to determine the best and most satisfactory alternative to choose.

Depending on the amount and quality of information available, decisions are made under one of three conditions: certainty, risk, and uncertainty.

Conditions of Certainty — Under conditions of certainty, you have enough information to predict reliably the outcome of choosing among alternatives. Often these decisions are simple. For example, if you decide to meet with an employee in the afternoon, chances are good that the employee will show up. Of course, nothing is entirely certain. The employee may become ill or you may be called away from the job site, causing the meeting to be cancelled.

Relatively few decisions are made under conditions of complete certainty. However, under conditions of certainty people behave as though such problems cannot occur. It is important to keep in mind that this is not the case.

Conditions of Risk — Conditions of risk are common. A condition of risk exists when a decision must be made on the basis of incomplete or unreliable information. In such cases, the decision maker assumes, on the basis of experience in similar situations, that a particular decision will yield relatively predictable results. Under conditions of risk, you develop alternatives and estimate the probability of each alternative leading to the desired result.

Probability is the percentage of times a specific outcome occurs if an action is taken a large number of times. For example, there is a very high probability that fresh concrete, heavily rained on, will not result in a structurally sound product. Therefore, it is also likely that it will have to be replaced. If thunderstorms are in the area, you must decide whether or not to pour the concrete based upon the probability of rain in the work-site area.

Conditions of Uncertainty — A great many decisions are made under conditions of uncertainty. Predictions of the future are never certain. However, under conditions of uncertainty, the probabilities attached to available alternatives are even less well known than under conditions of risk. As a result, problem solving comes into play. Even though there is no basis on which to calculate probabilities, they can still be estimated.

Instructor's Notes:

For example, you may order materials to be delivered to the job site three weeks from now. Based on the current state of the job, the probability might be high, perhaps 75 percent, that the crew will be ready for the materials when they arrive. Be aware that a lot can change in three week's time. Work delays, availability of materials, or the supplier's ability to deliver may change the situation drastically by delivery time. Consequently, there is virtually no way to predict these variables in advance.

9.2.6 Limits to Rational Decision Making

You may make a decision when an apparently satisfactory choice is identified, even though you haven't considered all of the alternatives. As a result, the *satisfactory* choice may not be the best one. This approach to decision-making is called **satisficing.** *Satisficing* is the practice of choosing a satisfactory rather than an optimal decision. Since the alternatives do not receive the most rational evaluation, satisficing has also been called *the science of muddling through*.

If you look for the best, or optimal, decision in every situation, few decisions would be made because each one would take so much time. Consequently, the need to make decisions quickly is one reason to resort to satisficing. Another involves value judgments or the fact that people tend not to consider alternatives they do not like.

With complex decisions, alternatives may be so numerous that you cannot evaluate them all. The decision must be reduced to a level where the alternatives can be handled. This technique is known as **bounded rationality.**

It is important to distinguish between satisficing and bounded rationality. Bounded rationality is based on the natural limit of a person's ability to handle complex situations. Satisficing is a deliberate choice to limit the number of alternatives considered. Bounded rationality is naturally imposed, while satisficing is purposefully selected.

9.3.0 Influence of Schedules and Budget

The pressures of schedules and budget often influence decisions. Companies need to remain flexible at all times and be able to change decisions when required to stay on track with budget and time allocations.

9.3.1 Schedules

Dealing with schedules in the decision-making process raises two issues. The first issue is the fact that a schedule is often based on many small decisions that are made without satisfactorily weighing all available alternatives. In effect, they are based on satisficing. Therefore, their validity is questionable. The second issue is the restrictions schedules place on decision making throughout the job. Schedules rarely allow sufficient time for decision making. As a result, important project decisions often suffer.

Supervisors are very concerned with the issue of scheduling as it influences their decisions. The sequence and timing of tasks have a direct effect on decisions. It is for this reason that scheduling aids, such as the critical path method (CPM) of scheduling, exist.

CPM helps you visualize the tasks your crew members are expected to accomplish and when they are to be done. It also helps you select alternatives when making job-related decisions. As a supervisor, your input to a schedule is very important to the success of a job.

9.3.2 Budget

The budget is a financial plan that outlines the amount of money provided to accomplish a specific activity. The basic purpose of a budget is to improve operations by focusing ongoing efforts on the resources available.

The budget is a major variable in every decision you make, and the cost of a choice often determines its suitability. If extra workers are needed to complete a task on time, you must weigh the increased labor cost against the schedule in making a final decision about expanding the workforce. Considering only the schedule would be inappropriate in this case.

9.4.0 Steps in Decision Making and Problem Solving

Not all situations on the job site require you to make decisions. It is up to you to determine, based on your experience, knowledge, and skills, when decisions are required. For example, you may watch your crew pour concrete without seeing a situation that requires that you make

Ask trainees to give examples of decisions they made under conditions of uncertainty and the outcome of those decisions.

Ask trainees to give examples of changes they made in a schedule and a budget and how their decision affected the project.

Ensure that you have all the necessary materials to teach the course. Check the Materials and Equipment list at the front of the module. Prepare for teaching Session VII by reading Sections 9.4.0-9.4.4.

Assign reading of Module MT202 Sections 9.4.0-9.4.4.

a decision regarding their work. On the other hand, you *may* notice a problem, making it necessary to step in and decide on a proper solution based on the alternatives available.

9.4.1 Types of Decisions

Part of decision making involves determining whether the decision to be made is **routine (programmed)** or **non-routine (non-programmed)**. *Figure 7* illustrates the role of both types of decisions in decision making and problem solving.

A *routine (programmed)* decision is one in which the problem is highly structured and repetitive and involves conditions of high certainty. It is based on established rules, regulations, policies, and procedures.

A *non-routine (non-programmed)* decision is one in which the problem is complex and infrequent and involves conditions of high uncertainty. It is often unique and unstructured and falls outside of set procedures, rules, and policies.

Routine Decisions — A routine decision develops through repetition that establishes the most appropriate and legitimate choice for a given set of circumstances. The development is aided by company policies, work procedures, and established industry rules and regulations. Supervisors make a large percentage of routine decisions. Because they are routine, they can be made quickly.

Knowledge, experience, and confidence determine your ability to make appropriate routine decisions. For example, a supervisor who is new to the company may have considerable knowledge of industry standards but little familiarity with company policies or procedures. This limits the supervisor's ability to make the best routine decision. In this case, becoming more knowledgeable about the company's policies will improve the supervisor's ability to make correct routine decisions quickly.

Non-Routine Decisions — Non-routine decisions require you to use judgment in choosing one course of action from among several alternatives. It is your ability to make non-routine decisions that characterizes you as an effective decision maker.

Different degrees of non-routine decisions exist, depending on the complexity and certainty involved. Generally, you will be expected to deal with only less complex forms of non-routine decisions.

9.4.2 Authority for Decision Making

Whether a decision is routine or non-routine, you must make sure that you have the authority to make it. All members of the management team must be aware of their level of decision-making authority. The limits and range of your decision-making authority should be established early in the relationship with other managers and they must be updated as those relationships change.

Your authority must be specific and cover areas such as personnel matters, material use and selection, ordering and purchasing, equipment usage and requisition, methods of work, and related issues. You and your manager must mutually understand your limits of authority in these areas. The limits to your authority may include how much material you can order, your decision-making responsibility for dealing with other trades on the work site, and your ability to hire and fire personnel. The more specific your authority, the more efficiently you can do the job. If you do not know the limits and responsibilities of your authority, ask your manager.

9.4.3 Group Involvement in Decision Making

There are ways to improve the decision-making process. One method is to consider input from others, including subordinates. One of the most important decisions you can make is whether to include others in the decision-making process.

Studies indicate that group decision making is useful when the following conditions exist:

- The costs of being wrong are high
- Uncertain or inadequate information is available
- Numerous alternatives are possible, but only a limited number are easily identified
- No single "best" decision may exist
- Knowledge of the success or failure of the decision made will not be available for an extended period of time

Using the assistance of others in decision making has several distinct advantages:

- The different backgrounds, experiences, and knowledge of others in the group provide more information about the problem at hand.
- A group should be able to generate a greater number of alternatives than an individual.
- Participation in group decision making frequently increases group acceptance of and commitment to the decision.
- Participation in decision making leads to everyone's better understanding of the reasons it was made, thus allowing smoother implementation.

There are also disadvantages in using others in the decision-making process:

- A group frequently takes more time to make a decision than an individual. The larger the group, the greater the difficulty in reaching a decision.
- The cost of having several individuals participate may be prohibitive.
- The decision may be the product of only a few members who hold sway over the others.
- Certain interpersonal problems, such as personality conflicts and status pressure, may inhibit group decision making.

Group decision making is especially valuable when more than one organizational unit or trade is involved. In such cases, others frequently resist a decision made by one unit or trade if they have not had a say in the outcome. This is particularly true with units orga-

Show Transparency 16 (Figure 7).

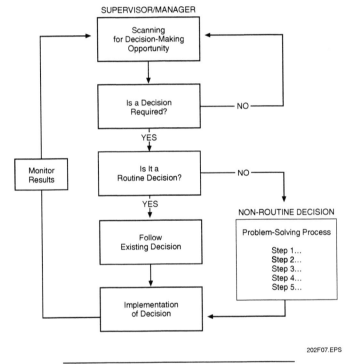

Figure 7 • Routine and Non-Routine Decision Process

Have trainees answer aloud some of the questions from both Tasks I and II. Allow 10-15 minutes.

This is a two-page transparency. Show Transparency 17 (Participant Activity Task I) and then show Transparency 18 (Participant Activity Task II).

PARTICIPANT ACTIVITY

Task I

Discuss the following:

1. Why is it important for a supervisor to make decisions?
2. Which level of management (Top-Mid-First) should be most involved in making organizational decisions? Why? What can occur if they do not meet this responsibility?
3. Of the three management levels, which would be most involved in making operational decisions? Why?
4. From your experience, what is the difference between problem solving and decision making?
5. What decisions in your personal life have you made today that did not require problem solving?
6. What decisions in your job have you made today that did not require problem solving? How were you able to make these decisions without problem solving?
7. How are choices involved in decision-making?
8. What are alternatives?
9. What does it mean when a decision is made under a condition of certainty?
10. In decision making, when does a condition of risk exist?
11. How are limits of rational decision making influenced by satisficing? By bounded rationality?
12. How do schedule, budget, and quality influence a supervisor's decision making?

Task II:

Use your own job in answering the following questions:

WHO MAKES THE DECISION TO . . .

1. Order lumber for the framing crew?
2. Change the routing of a duct?
3. Determine the location of storage for fixtures?
4. Replace a defective power cord on a hand tool?
5. Allow a carpenter to take off early one afternoon to take care of personal business?
6. Call to order rental equipment?
7. Upgrade the quality of paneling being installed?
8. Change the slope of a parking lot surface being formed and prepared for a pour?
9. Reprimand crew members that are horsing around?
10. Sign a work contract with a subcontractor?
11. Report or remove a defective ladder from use?

PARTICIPANT ACTIVITY

Instructor's Notes:

nized under different supervisors. However, resistance to collaboration can be avoided by including representatives of all the units in the decision-making process.

9.4.4 Problem Solving for Non-Routine Decision Making

Most decisions that you face on the job are routine or programmed. However, you also encounter situations where routine decisions are inadequate. This is generally the case when the problem is one that you have not encountered before or which includes unfamiliar variables. Such situations require non-routine or non-programmed decisions.

Figure 8 illustrates the specific steps involved in problem solving and non-routine decision making.

Step 1 Recognize and identifying the problem

One of the biggest difficulties that you face is determining if the problem has been correctly identified. Many supervisors have a tendency to rush into problem solving without properly pinpointing the problem to be solved.

Any three people are likely to define a specific problem three different ways. In management, a problem is the difference between the actual state of affairs and a desired state of affairs. Normally, an organization's goals and objectives determine the desired state of affairs or the results that the supervisor's crew is expected to produce.

For example, an electrical contractor has used 60 percent of the allotted materials on a project but has completed only 40 percent of the work. What is the problem? Many would say that the problem is that 60 percent of the materials have been used but this is only the symptom of the problem. In fact, the problem is the 20 percent overrun. So focusing only on the figure *60 percent* is misleading. The problem solver must focus on the excess 20 percent and discover the reasons for it. Other areas where problems may be unclearly defined include increased scrap and waste, reject levels and rework amounts, and increased labor costs coupled with decreased productivity.

Many supervisors focus on symptoms instead of causes. Therefore, they never get to the root of a problem. To avoid this, you must monitor and evaluate job-related performance accurately so that correct information is available when needed. You must also understand where the work effort is headed, and you must be able to recognize when it is not headed toward the desired results.

In addition, you must avoid the common trap of looking for someone to blame when a problem arises. Instead, you must try to determine what went wrong and why. If you begin the problem-solving effort by comparing actual results with desired results, causes can be more readily identified instead of culprits.

Step 2 Gather and analyze information

Once you have identified the problem, the next step is to collect and analyze information on the causes of the problem. Most of the necessary information can be gathered by simply monitoring the work. The rest can be obtained from others, including employees and other supervisors who might be more familiar with certain aspects of the job. Job-site records, progress reports, and other documentation can also assist you in gathering facts.

You must be as thorough as possible. Sound decisions can only be made if you have a sufficient amount of valid information—information that is based on fact, not assumption. Short-cuts and haste at this point inevitably result in problems later.

In the example of the materials overrun, gathering information could involve reviewing many sources. For example, you could inspect the installed work and check invoices and vouchers for material ordered and received. In addition, you could review waste control, material check-out, storage, and distribution procedures. You could also consult employees and other site supervisors. All these sources of information would contribute to an understanding of the possible cause(s) of the problem.

Step 3 Develop alternative solutions

Defining the problem and gathering and analyzing information lead to the next step: generating alternate solutions. At this point, you should ignore restrictions or restraints that limit choices. Remember that the greater the

Ask trainees to give an example of a non-routine decision they faced and how they solved it.

Ask trainees if they have used an informal chart to solve a problem and whether it was effective in resolving the problem.

Show Transparency 19 (Figure 8).

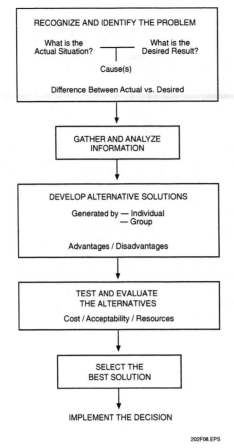

Figure 8 • Problem-Solving Process

number of potential solutions you consider, the more likely that you will find the best solution.

The magnitude of the problem may indicate whether you should include others in this process. Generally, the more complex the problem, the more valuable the participation of others becomes. In the example of the electrical contractor with the 20 percent overrun, the problem is complex. However, the supervisor should consider the advantages and disadvantages of group participation before inviting input from others.

Relative to the example of the electrical contractor, some of the alternate solutions that might be considered are as follows:

- If all the materials haven't been ordered, determine if the materials can be ordered elsewhere for a smaller cost.
- Increase the amount of supervision of the crew.

Step 4 Test and evaluate the alternatives

Each alternative identified in *Step 3* must be considered in view of its potential effectiveness, its cost, and its demand on resources. Establishing a minimum level of acceptability is a good starting point for this step; it provides a yardstick for eliminating or accepting each alternative.

The minimum level of acceptability is usually based on your experience or the information provided by the project supervisor or project manager. It may also be influenced by industry standards, governmental regulations, and local ordinances.

Alternatives must also be compared on the basis of the cost of their implementation, including the cost of labor, materials, supplies, transportation, equipment (rented or purchased), time, and impact on other work activities.

A cost analysis of each alternative should be done before comparing one alternative to another. This information could be organized by using the chart shown in *Figure 9*.

An appropriate solution must also be evaluated considering available resources. If the resources required to implement an alternative are unavailable, the alternative cannot be considered an appropriate choice.

The process of using minimal acceptability, costs, and resources as criteria serves to reduce the initial number of alternatives.

Step 5 Selecting the Best Alternative

Selecting the best alternative should be a logical result of the previous steps. If all the alternatives are identified and properly evaluated, the final choice should be fairly easy. Keep in mind that the final choice can only be as good as the information upon which it was based.

PARTICIPANT ACTIVITY

A decision-making checklist such as the one below will help you to assess your own decision-making style. Take a few moments to answer the questions posed in the checklist. This will provide a review of the essential steps of decision making. When you are finished with your self-evaluation, share the results with a partner and discuss your strengths, weaknesses, and what you will change about your current approach to decision making.

Typically, When You Make a Decision Do You...

1. _____ Identify the exact nature of the problem?
2. _____ Compare current information to decisions made in the past?
3. _____ Identify the results that you want your decision to accomplish?
4. _____ Identify at least three alternatives to choose from?
5. _____ Seek input from the people who will be affected by the decision?
6. _____ Evaluate each alternative by listing its advantages and disadvantages?
7. _____ Make a clear choice among the alternatives?
8. _____ Communicate your choice in person to the people involved?
9. _____ Develop a specific means of implementing your decision?
10. _____ Provide a means for collecting feedback from those affected by the decision?
11. _____ Alter your decision when it becomes clear that it is not working?
12. _____ Inform those affected by the decision about the change?

— PARTICIPANT ACTIVITY —

Therefore, it is important that you make sure that the information collected was correct and accurate.

9.4.5 Implementing a Decision

Choosing an alternative is not the end of the process. Proper implementation is an integral part of successful problem solving and decision making.

The following five-step approach is helpful in implementing a decision:

Step 1 *Decide how to communicate the decision. Should it be verbal, written, or both? Written decisions have the advantage of serving as references in the future.*

Step 2 *Determine to whom the decision needs to be provided. Everyone who is directly affected by the problem is equally affected by the decision. This may include supervisors of other trades, as well as other crews of the same company.*

Step 3 *If the entire group of people affected by it did not make the decision, explain the reasoning behind the decision. Knowing why a decision has been made often contributes to employee and management acceptance.*

Step 4 *Obtain feedback from the affected individuals so that you are certain they understand the decision.*

Step 5 *Express desire for future feedback from the affected individuals regarding the success or failure of the decision.*

This approach is appropriate for implementing both routine and non-routine decisions, but the greatest benefits are realized in connection with non-routine decisions.

Have trainees complete the Participant Activity.

Show Transparency 20 (Participant Activity).

Ensure that you have all the necessary materials to teach the course. Check the Materials and Equipment list at the front of the module. Prepare for teaching Session VI by reading Sections 9.4.5–9.5.0.

Assign reading of Module MT202 Sections 9.4.5–9.5.0.

Ask trainees to recall a situation where they had to compare the total cost of alternatives.

Show Transparency 21 (Figure 9).

Complete the total cost of alternatives chart based on an example given by a trainee.

Comparing Total Cost of Alternatives

Alternative	Labor	Materials	Supplies	Transportation (If Required)	Equipment (Rental or Purchase)	Amount of Time Required	Impact on Other Work Effort	Total Cost
Choice 1 Costs:								
Choice 2 Costs:								
Choice 3 Costs:								
Choice 4 Costs:								
Other								

Figure 9 • Comparing Total Cost of Alternatives

9.5.0 Evaluating Decisions

Supervisors often feel that once a decision has been implemented, their job is completed. However, there remains one final task. Once implemented, the decision must be monitored and evaluated.

Time is the true test of any solution or decision. Until a particular solution or decision has had time to prove its worth, you can only guess its effectiveness. If the solution *does* work as planned, you should make others aware of its value so that it can be used in similar situations in the future.

Through a study or review of the decision, you can also increase your own awareness of effective decision making and problem solving. It is very important that you recognize others who assisted in the decision process for their contributions. Such recognition motivates employees to actively participate in future decision-making activities.

A good supervisor also learns something from decisions that produce undesired or unexpected results. Failures should be studied to prevent future reoccurrences. You should neither be defensive about making a wrong decision, nor should you attempt to pass blame for the failure. Instead, you should admit the error and gather as much information as possible to determine where the decision went wrong. If you are honest and open with your employees about poor decisions, you will be more likely to succeed. In addition, you will be less likely to repeat the errors if you learn from your mistakes.

Keep in mind that successful and unsuccessful solutions help you build a reliable source of alternatives for future use. These alternatives are helpful for new and different circumstances that may arise because they can be adapted to meet the situation.

SUMMARY

The transition from craftworker to supervisor can be very difficult. New supervisors are evaluated and judged not on their own skills, but on the productivity and cost-efficiency of their teams.

Instructor's Notes:

PARTICIPANT ACTIVITY

Discuss the following:

1. Most decisions made by supervisors are routine. Why?
2. Upper management should be concerned with making more non-routine (non-programmed) decisions than any other level of management. Why?
3. Why is it important for supervisors to clearly understand their own decision-making authority?
4. What are the benefits associated with involving others in decision-making activities?
5. What are the disadvantages of involving others in the decision-making process?
6. List the steps involved in problem solving and discuss each step.
7. In evaluating alternatives to a problem, what three elements must be considered before selecting a solution?
8. Develop a set of problem alternatives for a work-related situation. Using the form shown in *Figure 9*, select the alternative solution with the lowest cost.

PARTICIPANT ACTICITY

Show Transparency 22 (Participant Activity).

New supervisors must move from the position of being master craftworkers in their individual specialties to being apprentices in the new realm of management. Many new skills must be mastered, including planning, time management, and all aspects of human relations.

Supervisors have the responsibility of getting the job done by utilizing the skills and talents of their teams. Their abilities to analyze, communicate, motivate, confront, and train their employees become far more important than any technical skills they possess. Learning to be a leader will be the most challenging job a new supervisor will have, and being a good leader will be the most rewarding.

Decision making and problem solving are universal management activities. Decision making is choosing the specific solution to a problem from among various alternatives. Problem solving also involves choice, but it is a specialized form of decision making.

You must monitor all types of decisions after they are implemented and be prepared to make necessary corrections. If the decision is found to be inappropriate, a different decision must be developed and implemented. An organization's success or failure depends greatly upon its managers' ability to solve problems and make necessary decisions.

Instructor's Notes:

MODULE MT202

Review Questions

Have the trainees answer the review questions. The correct answers are located at the end of this module in the Instructor's Guide.

1. Supervisors' performances are evaluated by _____.
   a. their craftsmanship
   b. their self-perception
   c. work accomplished by their crews
   d. their crews' technical skills

2. New supervisors' usual "comfort zone" includes _____.
   a. supervising others
   b. doing most of the work themselves
   c. mastering management techniques
   d. delegating work

3. A common psychological barrier to a supervisor's success is _____.
   a. insecurity about supervising former co-workers
   b. overconfidence about management skills
   c. maintaining a positive attitude
   d. worrying about technical accuracy

4. Tasks must be assigned priority according to _____.
   a. who wants them done
   b. urgency and importance
   c. what we like to do
   d. which tasks can be postponed

5. One advantage of delegating is that it _____.
   a. eliminates the need for training
   b. enhances the team's effectiveness
   c. reduces the supervisor's workload
   d. reduces lawsuits

6. By understanding how your listener decodes your message you can _____.
   a. tailor your words to say what you mean
   b. rely on your message being understood by the listener
   c. confirm what the listener understands
   d. keep your messages shorter

7. Active listening involves _____.
   a. frequently interrupting speakers to ask questions
   b. formulating a response to everything the speakers say while they are talking
   c. letting the speakers know that you agree with what they are saying
   d. paying attention to the speakers by showing interest and providing reflection

8. An example of physical attendance is _____.
   a. nodding your head in agreement
   b. asking a good question
   c. looking down at your notes as you listen
   d. rephrasing what the speaker has just told you

9. The supervisor's major responsibility is _____.
   a. teaching technical skills
   b. developing and employing human resources
   c. conducting performance reviews
   d. conducting new employee orientation

10. Performance evaluations are ____.
    a. tools to teach, motivate, and praise employees
    b. optional
    c. held to calculate raises and bonuses
    d. checklists that you give to the employee

11. Training is ____.
    a. the sole responsibility of the supervisor
    b. a key to improving job quality
    c. best delivered in the classroom
    d. best delivered to every employee in the same fashion

12. An indicator of a supervisor's success is ____.
    a. employees relying on the supervisor for direction in all matters
    b. employees discussing only positive things with the supervisor
    c. employees who are not afraid to confront the supervisor
    d. employees taking responsibility for directing their own development

13. The first step in solving a performance problem is to ____.
    a. confront the worker who is not performing
    b. find out who is to blame for the problem
    c. identify the root cause
    d. ask everyone for help in identifying the problem

14. When confronting employees, supervisors should focus first on ____.
    a. a list of things that they don't like about the employees
    b. retaining the employee
    c. what the employee wants to do about the situation
    d. defining the problem

15. The purpose of confrontation is to ____.
    a. encourage the employee to develop responsibility
    b. punish the employee for wrongdoing
    c. keep the upper hand with employees
    d. discourage protests or workplace violence

16. A successful model that supervisors use to confront workers includes ____.
    a. making an example of them to discourage similar behavior in other employees
    b. comparing them with other workers to set high standards
    c. agreeing on a solution and asking them to paraphrase
    d. focusing on their mistakes and documenting them

17. The employer reaps the following benefits from mentoring *except* ____.
    a. fewer layoffs
    b. better recruiting
    c. decreased employee turnover
    d. lower absenteeism

18. Constructive conflict contributes to ____.
    a. learning in the organization
    b. increased profits
    c. a supervisor's power in the organization
    d. an individual's power in the organization

19. Supervisors who use ____ as a strategy for handling conflict normally ignore what is going on and hope it will go away.
    a. cutting a deal
    b. smoothing over
    c. pulling rank
    d. denial

Instructor's Notes:

Module MT202 ◆ Review Questions

20. Supervisors who use _____ as a strategy for handling conflict often try to cover the conflict with positive feelings and end up causing deep resentment among team members.
    a. smoothing over
    b. pulling rank
    c. negotiation
    d. denial

21. It is important for supervisors to encourage positive and negative feedback from _____.
    a. customers and employees
    b. other work groups
    c. experienced employees
    d. customers they can trust

22. An example of an organizational decision is _____.
    a. merging two departments for greater efficiency
    b. reserving two parking spaces for employees-of-the-month
    c. reassigning work crews
    d. establishing new procedures for archiving construction site photographs

23. An example of an operational decision is _____.
    a. asking workers to work overtime during the month of April
    b. adding a new top management position
    c. deciding to pursue ISO certification
    d. sending the crew to CPR training

24. _____ is the process of changing a course of action from among alternative possibilities.
    a. Risk management
    b. Conflict management
    c. Decision making
    d. Delegating

25. Decisions are often influenced by the pressures of schedules and _____.
    a. management
    b. budgets
    c. delegation
    d. policies

Have trainees read the Glossary for unfamiliar terms. Answer trainee questions.

Have trainees prepare for the Module Examination.

Administer the Module Examination. Be sure to record the results of the Exam on Craft Training Report Form 200 and submit the results to the Training Program Sponsor.

NATIONAL CENTER FOR CONSTRUCTION EDUCATION AND RESEARCH

GLOSSARY

Trade Terms Introduced in This Module

Active listening: The act of retaining what you hear as well as listening for the ideas and feelings that the sender is trying to communicate.

Bounded rationality: Reducing the complexity of a decision to a level where alternatives can be considered.

Comfort zone: The area of activity in which a person feels most secure and confident.

Communication: The transferring of ideas, thoughts, feelings, or facts between people.

Decision making: The process of choosing a course of action from among alternative possibilities.

Decode: Translating the information sent from one person to another.

Directive style: A management style characterized by little if any involvement of employees.

Discrepancy: An outcome that is unexpected or different from what it should be.

Encode: Translating thoughts, perceptions, ideas, feelings or facts into a form that others can understand.

Involved style: The management style that encourages high levels of communication between supervisors and their teams.

Morale: The individuals' attitude toward the tasks that they are expected to perform. High morale means employees will be motivated to work hard and have a positive attitude about doing their jobs.

Motivation: An individual's need to perform an action. It is also the amount of effort that a person is willing to put forth to accomplish something.

Non-routine (non-programmed) decision: Decision in which the problem is complex and infrequent and involves conditions of high uncertainty. It is often unique and unstructured falling outside of set procedures, rules, and policies.

Organizational decisions: Related to issues concerning the organization as a whole.

Operational decisions: Related to day-to-day operations.

Performance evaluations: Regular appraisals of workers' job performance.

Problem solving: Process of deciding on how to close the gap between the way things are and the way they should be.

Routine (programmed) decision: Decision in which the problem is highly structured, repetitive, and involves conditions of high certainty. It is based on established rules, regulations, policies, and procedures.

Task analysis: Identifying the tasks and skills involved in a job.

Think outside the box: Look at situations with a new "eye" that does not depend on how you have always done it; see things from another perspective and use limitless thinking to approach your work.

Satisficing: Practice of choosing a satisfactory rather than an optimal decision.

Uninvolved style: The management style characterized by employees making decisions with little input by the supervisor.

Variables: Elements that change from situation to situation that profoundly affect decision making.

Instructor's Notes:
_____
_____
_____
_____
_____
_____

MOD MT202-02—TEACHING TIPS

The following are suggested activities or instructional methods to help you teach the material in this AIG.

Section 6.4.2 The Employee Response Model

Discuss the consequences of missed maintenance on equipment. For example, point out the importance of maintenance to prevent project schedule delays, safety hazards, and equipment failure.

NATIONAL
CENTER FOR
CONSTRUCTION
EDUCATION AND
RESEARCH

MODULE MT202

Answers to Review Questions

	Answer	Section Reference
1.	c	2.0.0
2.	b	2.1.1
3.	a	2.1.0
4.	b	2.3.1
5.	b	2.3.2
6.	b	4.1.1
7.	d	4.2.1
8.	a	4.2.1
9.	b	5.0.0
10.	a	5.5.0
11.	b	6.1.0
12.	d	6.1.0
13.	c	6.4.1
14.	d	7.4.0
15.	a	7.4.0
16.	c	7.4.0
17.	a	7.2.0
18.	a	8.1.0
19.	d	8.1.2
20.	a	8.1.2
21.	a	8.1.1
22.	a	9.1.0
23.	a	9.1.0
24.	c	9.2.0
25.	b	9.3.0

NATIONAL CENTER FOR CONSTRUCTION EDUCATION AND RESEARCH

MODULE MT202

Participant Activities

Participant Activity Section 4.2.1
Here are some possible "translations" of what is being said. Have the group come up with their own, but use these when they get stuck or as alternate answers.

"This is the most stupid, mixed-up job I ever worked on" could mean "I don't understand what I am supposed to do and how my work contributes to the project."

"It's impossible, but I'll have it done for you by Thursday afternoon" could mean "I'm just telling you I'll have it done to appease you."

"The laborer you assigned to work with the electricians is really sharp" could mean "I like the way you manage" or "I want to let you know what a good job you are doing."

"I guess the other guys must be too busy to do the job you gave me" could mean "I don't think you have confidence in me" or "I'm looking for affirmation of my abilities."

"Maybe you'd better talk more with your people about this" could mean "There is a problem but I don't want to get involved."

"Oh, I thought I was working as fast as expected" could mean "I'm not sure what the expectations are." or "You never gave me clear directions."

"How come I always get the grunt work?" could mean "I'm not feeling like a valued member of the team."

"What's wrong with kids these days? Don't they want to work?" could mean "I don't know how to communicate with younger members of the team."

"That architect doesn't know what he is doing." could mean "I made a mistake and I don't want anyone to know." Or "I'm confused about the plans but don't want to admit it."

"Sorry I'm late. Things have been going real bad at home" could mean "I need help with my home situation" or "I'm unhappy with my job and am using problems at home as an excuse."

Participant Activity Section 9.2.0
Case 1: 2, 3, 1, 4, 1
Case 2: 2, 1, 3, 5, 3

CONTREN™ LEARNING SERIES — USER UPDATES

The NCCER makes every effort to keep these textbooks up-to-date and free of technical errors. We appreciate your help in this process. If you have an idea for improving this textbook, or if you find an error, a typographical mistake, or an inaccuracy in NCCER's Contren™ textbooks, please write us, using this form or a photocopy. Be sure to include the exact module number, page number, a detailed description, and the correction, if applicable. Your input will be brought to the attention of the Technical Review Committee. Thank you for your assistance.

Instructors – If you found that additional materials were necessary in order to teach this module effectively, please let us know so that we may include them in the Equipment/Materials list in the Instructor's Guide.

Write: Curriculum Revision and Development Department
National Center for Construction Education and Research
P.O. Box 141104, Gainesville, FL 32614-1104

Fax: 352-334-0932

E-mail: curriculum@nccer.org

Craft _____ Module Name _____

Copyright Date _____ Module Number _____ Page Number(s) _____

Description _____

_____

_____

_____

_____

(Optional) Correction _____

_____

_____

_____

(Optional) Your Name and Address _____

_____

_____

_____

Project Supervisor

Module MT203-01

Safety

Safety
Instructor's Guide

Module MT201

MODULE OVERVIEW

This module introduces the project supervisor trainee to safety. This module will enable the trainee to determine the real cost of accidents, perform accident investigations and complete required forms, and learn how to conduct effective safety meetings.

PREREQUISITES

There are no prerequisites for the module; however, prior to training with this module, it is recommended that the trainee complete the following modules:
  Project Supervision, Modules MT201 and MT202

LEARNING OBJECTIVES

Upon completion of this module, the trainee will be able to:

1. Describe the safety responsibilities of supervisors.
2. Determine the real cost of accidents, including *direct* and *indirect costs*.
3. Identify the basic components of a safety program.
4. Explain how to conduct a safety inspection and employee observation.
5. Describe how to confront and deal with a worker who was observed performing an unsafe act.
6. Explain how to perform an accident investigation and complete the necessary report forms.
7. Identify the components of effective safety meetings.
8. Define what is meant by a *qualified person* and a *competent person*.

PERFORMANCE OBJECTIVES

This is a knowledge-based module – there is no performance profile examination.

NCCER STANDARDIZED TRAINING PROGRAM

The National Center for Construction Education and Research (NCCER) provides a standardized national program of accredited craft training. Key features of the program include instructor certification, competency-based training, and performance testing. The program provides trainees, instructors, and companies with a standard form of recognition through a National Craft Training Registry. The program is described in full in the Guidelines for Accreditation, published by the NCCER. For more information on standardized craft training, contact the NCCER in writing at P.O. Box 141104, Gainesville, FL 32614-1104; calling 352-334-0911; or e-mailing info@nccer.org. More information may be found at our Web site at www.nccer.org.

HOW TO USE THIS ANNOTATED INSTRUCTOR'S GUIDE

Each page presents two sections of information. The larger section displays each page exactly as it appears in the Trainee Module. The narrow column ties suggested trainee and instructor actions to each page and provides icons to call your attention to material, safety, audiovisual, or testing requirements. The bottom of each page includes space for your notes.

 If you see the Teaching Tip icon, that means there is a teaching tip associated with this section. Also refer to the suggested teaching tips at the end of the module.

PREPARATION

Before teaching this module, you should review the Module Outline, Learning Objectives, and the Materials and Equipment List. Be sure to allow ample time to prepare your own training or lesson plan and gather all required equipment and materials.

MATERIALS AND EQUIPMENT LIST

Materials:

Transparencies

Markers/chalk

Module Examinations*

Sample job safety analyses**

Sample company policy statement regarding safety, OSHA regulations, state construction codes, city and council planning and building codes**

Equipment:

Overhead projector and screen

Whiteboard/chalkboard

*Located in the Test Booklet packaged with this Annotated Instructor's Guide.
**If available on loan from your workplace or other resource.

ADDITIONAL RESOURCES

This module is intended to present thorough resources for task training. The following reference works are suggested for both instructors and motivated trainees interested in further study. These are optional materials for continued education rather than for task training.

National Safety Council

The following references are available from the National Safety Council, 1121 Spring Lake Drive, Itasca, IL 60143-3201 or visit their website at www.nsc.org.

Accident Prevention Manual for Business and Industry – Administration and Programs, 2000.

Accident Prevention Manual for Industrial Operations, Administration, and Programs, 2000.

Accident Prevention Manual for Industrial Operation, Engineering and Technology, 1997.

Fundamentals of Industrial Hygiene, 1996. Barbara A. Plog (ed.), 1996.

Other Resources

Job Hazard Analysis - OSHA Publication No. 3071, U.S. Department of Labor.

Job Safety Analysis - Safety Manual No. 5, U.S. Department of the Interior, Mining Enforcement and Safety Administration.

TEACHING TIME FOR THIS MODULE

An outline for use in developing your lesson plan is presented below. Note that each Roman numeral in the outline equates to one session of instruction. Each session has a suggested time of 2 1/2 hours. This includes 10 minutes at the beginning of each session for administrative tasks and one 10-minute break during the session. Approximately 7 1/2 hours are suggested to cover *Safety*.

Topic	Planned Time

Session I. The Supervisor's Role in Safety
- A. Introduction   _____
- B. The Supervisor's Role in Safety   _____
    1. Supervisors' and Team Leaders' Responsibilities   _____
    2. Employee Responsibilities   _____
- C. Accident Causes, Costs, and Controls   _____
    1. Useful Definitions   _____
    2. Accidents vs. Incidents   _____
    3. Causes of Accidents   _____
    4. Unsafe Acts   _____
    5. Three Levels of Accident Causation   _____
    6. Costs of Accidents   _____
    7. Participant Activities   _____

Session II. Basic Components of a Construction Safety Program
- A. Basic Components of a Construction Safety Program   _____
    1. Management Support and Policy Statement   _____
    2. Policy on Alcohol and Drug Abuse   _____
    3. Assignment of Responsibilities   _____
    4. Employee Screening, Selection, and Placement   _____
    5. Safety Rules   _____
    6. Orientation and Training   _____
    7. Safety Meetings and Employee Involvement   _____
    8. Emergency Reporting and Response Including First Aid   _____
    9. Inspections, Employee Observations, and Audits   _____
    10. Accident Investigation and Analysis   _____
    11. Records   _____
    12. Program Evaluation and Follow-Up   _____
- B. Employee Safety Training   _____
    1. Employee Safety Orientation   _____
    2. Job Safety Training   _____
    3. Job Hazard Recognition   _____
- C. Job Safety Analysis   _____
    1. Performing a Job Safety Analysis   _____
    2. Identifying Hazards and Potential Accidents   _____
- D. Conducting Effective Safety Meetings   _____
    1. Toolbox Safety Talks   _____
    2. Practice the Five "Ps" for Successful Safety Talks   _____
    3. How to Organize a Successful Safety Meeting   _____
    4. First Aid and CPR Training   _____
    5. Training Documentation   _____

Session III Job-Site Safety; Module Examination
    A. Job-Site Safety Inspections, Audits, and Observations   _____
        1. The Supervisor's Role   _____
        2. Elements of Job-Site Inspections   _____
        3. Key Elements of Employee Observation   _____
    B. Accident Reporting and Investigation   _____
        1. The Supervisor's Role   _____
        2. Conducting the Investigation   _____
    C. Government Regulations   _____
        1. City, Country, and State Regulations   _____
        2. The Occupational Safety and Health Act   _____
    D. Summary   _____
        1. Summarize Module
        2. Answer Review Questions
    E. Module Examination   _____
        1. Trainees must score 70% or higher to receive recognition from the NCCER.   _____
        2. Record the testing results on Craft Training Report Form 200 and
           submit the results to the Training Program Sponsor.   _____

Safety

Instructor's Notes:

Instructor's Notes:

_____
_____
_____
_____
_____
_____

ACKNOWLEDGMENTS

The NCCER wishes to acknowledge the dedication and expertise of Steve Pereira, the original author and mentor for this module on safety development.

Steven P. Pereira, CSP

President, Professional Safety Associates, Inc.

Denham Springs, LA

We would also like to thank the following reviewers for contributing their time and expertise to this endeavor:

J.R. Blair

Tri-City Electrical Contractors

An Encompass Company

Mike Cornelius

Tri-City Electrical Contractors

An Encompass Company

Dan Faulkner

Wolverine Building Group

David Goodloe

Clemson University

Kevin Kett

The Haskell Company

Danny Parmenter

The Haskell Company

Course Map

This course map shows all of the modules of the *Project Supervision* curriculum. The suggested training order begins at the bottom and proceeds up. Skill levels increase as you advance on the course map. The local Training Program Sponsor may adjust the training order.

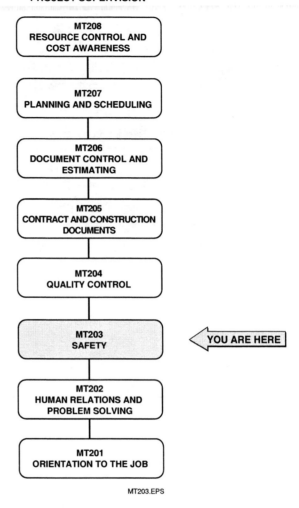

Instructor's Notes:

NATIONAL CENTER FOR CONSTRUCTION EDUCATION AND RESEARCH

MODULE MT203

TABLE OF CONTENTS

1.0.0	**INTRODUCTION**	3.1
2.0.0	**THE SUPERVISOR'S ROLE IN SAFETY**	3.2
2.1.0	Supervisors' and Team Leaders' Responsibilities	3.4
2.2.0	Employee Responsibilities	3.4
3.0.0	**ACCIDENT CAUSES, COSTS, AND CONTROLS**	3.5
3.1.0	Useful Definitions	3.5
3.2.0	Accidents vs. Incidents	3.6
3.3.0	Causes of Accidents	3.6
3.4.0	Unsafe Acts	3.6
3.5.0	Three Levels of Accident Causation	3.8
3.6.0	Costs of Accidents	3.9
3.6.1	*Calculating Accident Costs*	3.10
3.6.2	*Accident Experience vs. Future Insurance Costs*	3.10
3.6.3	*Cost of Administering an Effective Safety Program*	3.14
4.0.0	**BASIC COMPONENTS OF A CONSTRUCTION SAFETY PROGRAM**	3.14
4.1.0	Management Support and Policy Statement	3.14
4.2.0	Policy on Alcohol and Drug Abuse	3.14
4.3.0	Assignment of Responsibilities	3.14
4.4.0	Employee Screening, Selection, and Placement	3.15
4.5.0	Safety Rules	3.15
4.6.0	Orientation and Training	3.15
4.7.0	Safety Meetings and Employee Involvement	3.16
4.8.0	Emergency Reporting and Response Including First Aid	3.16
4.9.0	Inspections, Employee Observations, and Audits	3.16
4.10.0	Accident Investigation and Analysis	3.17
4.11.0	Records	3.17
4.12.0	Program Evaluation and Follow-Up	3.18
5.0.0	**EMPLOYEE SAFETY TRAINING**	3.18
5.1.0	Employee Safety Orientation	3.18
5.2.0	Job Safety Training	3.19
5.3.0	Job Hazard Recognition	3.20
5.3.1	*Unsafe Acts*	3.20
5.3.2	*Unsafe Conditions*	3.20

6.0.0	**JOB SAFETY ANALYSIS**	3.23
6.1.0	Performing a Job Safety Analysis	3.23
6.1.1	*Methods of Conducting JSAs*	3.23
6.1.2	*Selecting Jobs to Be Analyzed*	3.23
6.1.3	*Job Observation*	3.24
6.1.4	*Common Errors in Performing a Job Analysis*	3.24
6.2.0	Identifying Hazards and Potential Accidents	3.24
6.2.1	*Accident Types*	3.24
6.3.0	Writing Instructions for JSAs	3.25
6.4.0	Recommended Safe Job Procedures	3.25
7.0.0	**CONDUCTING EFFECTIVE SAFETY MEETINGS**	3.26
7.1.0	Toolbox Safety Talks	3.26
7.2.0	Practice the Five "Ps" for Successful Safety Talks	3.27
7.3.0	How to Organize a Successful Safety Meeting	3.28
7.4.0	First Aid and CPR Training	3.28
7.5.0	Training Documentation	3.28
8.0.0	**JOB-SITE SAFETY INSPECTIONS, AUDITS, AND OBSERVATIONS**	3.28
8.1.0	The Supervisor's Role	3.28
8.2.0	Elements of Job-Site Inspections	3.39
8.3.0	Key Elements of Employee Observation	3.31
8.3.1	*Reasons for Unsafe Behavior*	3.31
8.3.2	*Changing Unsafe Behavior*	3.32
8.3.3	*Confronting Unsafe Behavior*	3.32
9.0.0	**ACCIDENT REPORTING AND INVESTIGATION**	3.33
9.1.0	The Supervisor's Role	3.33
9.2.0	Conducting the Investigation	3.33
10.0.0	**GOVERNMENT REGULATIONS**	3.36
10.1.0	City, County, and State Regulations	3.36
10.2.0	The Occupational Safety and Health Act	3.37
10.2.1	*Inspections*	3.37
10.2.2	*Violations*	3.38
10.2.3	*Citations and Penalties*	3.38
10.2.4	*Training Requirements*	3.39
	SUMMARY	3.40
	REVIEW QUESTIONS	3.41
	GLOSSARY	3.45
	APPENDIX	3.47
	ADDITONAL RESOURCES	3.56

Instructor's Notes:

LIST OF FIGURES

Figure 1	•	Duties of a Safety Coordinator	3.3
Figure 2	•	The Heinrich Triangle	3.7
Figure 3	•	Levels of Accident Causation	3.8
Figure 4	•	A Delicate Balance	3.9
Figure 5	•	Hidden Accident Costs	3.10
Figure 6	•	Typical Worker's Compensation Premium Rates	3.11
Figure 7	•	Accident Cost Worksheet	3.13
Figure 8	•	Records and Retention Schedule	3.17
Figure 9	•	Horseplay Causes Accidents	3.21
Figure 10	•	Pre-Job Safety Briefing/Task Hazard Analysis	3.22
Figure 11	•	Job Hazard Analysis Form	3.25
Figure 12	•	Safety Meeting Pre-Planning Tool	3.27
Figure 13	•	Safety Sampling	3.30
Figure 14	•	Incident/Accident Investigation Report	3.34
Figure 14	•	Incident/Accident Investigation Report (Continued)	3.35

Instructor's Notes:
_____
_____
_____
_____
_____
_____
_____

NATIONAL CENTER FOR CONSTRUCTION EDUCATION AND RESEARCH

MODULE MT203

Safety

OBJECTIVES

Upon the completion of this module, you will be able to do the following:

1. Describe the safety responsibilities of supervisors.
2. Determine the real cost of accidents, including *direct* and *indirect costs*.
3. Identify the basic components of a safety program.
4. Explain how to conduct a safety inspection and employee observation.
5. Describe how to confront and deal with a worker who was observed performing an unsafe act.
6. Explain how to perform an accident investigation and complete the necessary report forms.
7. Identify the components of effective safety meetings.
8. Define *qualified person* and *competent person*.

SECTION 1

1.0.0 INTRODUCTION

This module of the project supervisor's training deals with safety and loss prevention. Its purpose is:

- To provide a basic understanding of how safety performance affects the profit and loss of a company today and the company's ability to remain competitive in the future.
- To serve as a guide and resource for supervisors in carrying out their safety and loss prevention duties.

Did you know that construction has the highest worker injury rate of any major industry in the United States? Here are a few statistics from the Building and Construction Trades Department of the AFL-CIO that you should know:

In 1999, the construction industry reported the largest number of workplace fatalities, with 1,190 work-related deaths—20 percent of all workplace fatalities in the country. This is an increase from the number of deaths in the construction industry the previous year.

Our industry is dangerous. On a typical working day, four to five construction workers will die from on-the-job injuries and nine hundred will be seriously injured. That means more than one thousand construction workers are killed on the job every year.

Ensure that you have all the necessary materials to teach the course. Check the Materials and Equipment list at the front of the module. Prepare for teaching Session I by reading Sections 1.0.0–3.6.3.

Show Transparency 1 (Course Objectives).

Assign reading of Module MT203, Sections 1.0.0–3.6.3.

Copyright © 2003 National Center for Construction Education and Research, Gainesville, FL 32614-1104. All rights reserved. No part of this work may be reproduced in any form or by any means, including photocopying, without written permission of the publisher.

SAFETY — INSTRUCTOR'S GUIDE MODULE MT203

Ask trainees to describe the type of safety program at their workplace.

Ask trainees to describe the type of tool and equipment inspection program in place at their company and its rate of compliance.

According to recent studies, work-site injuries in all industries cost the nation $150 billion annually. While workers' injuries and deaths in the construction trades have dropped dramatically since the Occupational Safety and Health Administration (OSHA) came into existence in 1970, these numbers are still too high.

A safety and loss prevention program is a management tool. It is used to achieve the overall goals of the organization. There are certain elements that are basic to all safety and loss prevention programs, regardless of the size of the company and scope of the operations. The degree to which these basic elements are incorporated into your program will depend on the size and scope of your operation and your company's management structure.

Your company's safety and loss prevention program should provide the basic framework for operating in a safe, efficient, and profitable manner. It must be a way of doing business, not an afterthought. Safety and loss prevention should be incorporated into all phases of the operation and involve all employees at every level, including management.

As a supervisor, you are responsible for the safety of your crew. In order to work toward creating a safe environment, you need to be *proactive* rather than *reactive*. Setting up systems, proper procedures, and safety checks and balances will go a long way toward establishing a safe and productive work environment.

There are many regulations that companies have to comply with to meet local, state, and national guidelines. Safety regulations are established to achieve an accident-free worksite. Many safety policies and procedures that are in place are established to help you and your company to comply with external regulations. It is important to remember that there is a good reason why each of the regulations exists. Following good safety practices helps you to develop a quality organization.

This module contains suggestions, checklists, and audit forms that may be useful to you as a frontline supervisor. Keep in mind that these are resources and references that you may modify, expand, or condense to suit your specific needs.

2.0.0 THE SUPERVISOR'S ROLE IN SAFETY

One of the supervisor's most important responsibilities is to ensure the safety of people on and around the job site. This includes employees, subcontractors, other trades, and the public. Supervisors are concerned with safety because people are the supervisor's primary resource for getting the job done. As a result, the health, welfare, and morale of the people with whom the supervisor works and has contact must be maintained.

Construction has hundreds of specific work activities, many of which pose a hazard to the personal safety of those performing them. Not surprisingly, statistics show that each year a large number of construction workers are injured on the job. Furthermore, billions of dollars are lost annually as a result of these injuries and the delays associated with them. These statistics only tell part of the story—that of the accidents which are reported. Many workplace accidents are not reported because they do not involve personal injury, so the losses could be even higher.

The supervisor's role is changing from simply being skilled in a specific craft. Now, the supervisory role includes being a trainer, motivator, inspector, planner, and counselor. The supervisor is now considered a key player in the organization as well as in maintaining a safe job site.

The focus of the workplace is also shifting. There has been a movement towards quality improvement coupled with efficient productivity. Therefore, safety must be integrated into all portions of the work.

The supervisor is concerned for the employees' health and welfare as well as the company's interests. With this in mind, it is important that supervisors recognize and eliminate unsafe conditions and practices and by training others—both actively and through example—about the importance of safety.

Supervisors are in the best position to promote safety because they work closely with crew members and know what they are doing, what they are wearing, and, with some limitations, what their concerns are. In addition, supervisors know the hazards associated with specific tasks through experience. In short,

Instructor's Notes:
_____
_____
_____
_____
_____
_____

Discuss Figure 1.

supervisors are *safety conscious,* and it is their duty to make everyone else on the job safety conscious as well.

Often, a safety coordinator is appointed for crews that have five or more members. The assignment of a safety coordinator helps the supervisor to make the job safer and to maintain quality. It is usually the responsibility of the superintendent and project manager or supervisor to determine who will play the role of the safety coordinator, and to determine how much time will be allotted for these duties.

Figure 1 is a sample job description for a safety coordinator.

DUTIES OF A SAFETY COORDINATOR

Daily Duties:
- Visually check all temporary wiring and repair if needed.
- Check ladders for missing or broken parts.
- Look over project for any and all hazards that may exist.
- List all hazards on Weekly Safety Report.
- Notify Safety Director immediately of any hazard that poses a threat of a serious injury.

Weekly Duties:
- Test all GFCI circuit breakers or receptacles and record on GFCI test card.
- Review Weekly Safety Reports before weekly toolbox meeting. Select several and use these as basis for Weekly Safety Meeting. Have at least one (1) employee comment on job-site safety conditions.
- During your review of the Weekly Safety Reports, if you notice that an employee is not listing hazards on his/her report, arrange for a training session with the Safety Director.

Monthly Duties:
Use the following color codes to indicate regular inspection of tools and equipment used on a project:

January	Red	July	Brown
February	Blue	August	Purple
March	Green	September	Red & Blue
April	White	October	Green & White
May	Yellow	November	Yellow & Orange
June	Orange	December	Brown & Purple

- Test all extension cords, putting monthly safety color on approved cords.
- Visually examine tools for missing or broken parts. Place monthly safety color on approved tools.

As Required:
- Coordinate with Project Superintendent the training of inexperienced employees. Arrange for a training session with the Safety Director if this duty becomes too time consuming.
- Communicate any and all job-site hazards to the Project Superintendent. If we can correct the hazard, then do it immediately. If it is not our responsibility to correct the hazard, then the Project Superintendent or Project Manager should document information about the hazard to the General Contractor. If this cannot be done or the General Contractor does not correct the hazard, then notify the Safety Director immediately.

Source: Tri-City Electrical Contractors, Inc., Altamonte Springs, FL, an Encompass Company

Figure 1 • Duties of a Safety Coordinator

Teaching Tip

Distribute sample copies of a job safety analysis and a standard operation procedure and discuss their role in a safety program.

While all duties outlined are important, careful attention must be paid to new employee training. Over 50 percent of all accidents happen to new employees. Having the safety coordinator spend a few minutes with all employees who are new to a project will help to determine the employees' capabilities. Making sure that everyone is briefed and qualified for the jobs they will perform will decrease injuries on a project. Someone with several years experience still requires quite a bit of training in the areas of job-site safety. It does not hurt to ask employees if they have ever performed the task, even if you know they have three or four years of experience.

The principal reasons to have an effective safety program are:

- Employers have a moral and legal obligation to provide their employees with a safe place to work that is free from recognized hazards.
- Employers want to minimize losses due to accidents. Increased insurance costs due to accidents can affect future bids. Likewise, the hidden or uninsured costs of accidents or injuries can rob a company of its profits.
- Employers want to avoid costly legal problems such as civil suits, criminal sanctions, or OSHA citations. OSHA penalties range from $7,000 up to $500,000 per violation.

Failure to have an effective safety and loss prevention program can result in:

- Needless pain and suffering for injured workers and their families
- Economic hardship for workers and the company
- Delays and reassignments of work due to the loss of the injured employee
- Loss or damage to equipment or property
- Lost production time as the supervisor and others prepare required reports and investigate the accident
- Direct medical costs
- Insurance costs, medical compensation, and general liability
- Citations and fines for violations of safety and health regulations

2.1.0 Supervisors' and Team Leaders' Responsibilities

Supervisors and team leaders have the following responsibilities:

- Check employee assignments to make certain employees are properly trained and qualified for the task.
- Establish correct work methods that will not encourage unsafe behavior. This may involve the development, review, or revision of **Job Safety Analyses (JSAs)** or **Standard Operation Procedures (SOPs)**.
- Make sure proper tools, materials, and equipment will be available when required.
- Make all required specifications, plans, instructions, and JSAs available.
- Survey the job site/task. Anticipate and identify safety hazards. Eliminate those safety hazards to the extent possible and/or required.
- Estimate and address the employee's hazard awareness and carefulness in avoiding hazards.
- Estimate and address the employee's attitude toward working safely. (*Attitude* is a predisposition to behave. In other words, it's how you would act if no one were watching.)
- Periodically observe work in progress, noting worker adherence to proper work methods, JSAs, and procedures; especially note worker behavior.
- Use proper coaching techniques to correct unsafe behavior and reward safe behavior.
- Notify other groups with whom work must be coordinated.

2.2.0 Employee Responsibilities

The following list gives the employees' responsibilities for a safe workplace.

- All employees are expected to follow rules, policies and procedures, both written and implied. This includes JSAs and SOPs.
- No employee is expected to undertake a job until he or she has reviewed the appropriate procedure or received job instructions on how to do the task properly and has been authorized to perform that job.

Instructor's Notes:

- No employee should undertake a job that appears unsafe or that will unnecessarily place the employee at risk.
- No employee should undertake a job that he or she is incapable of doing or has insufficient knowledge to accomplish.
- Employees should never place themselves or co-workers at risk by engaging in any unsafe behaviors.
- Before starting a job, each employee should ask and answer the following questions:
  — What are the hazards associated with this job?
  — Where is the energy and how can it be eliminated or controlled?
  — What could possibly go wrong?
  — What must I do to protect myself and my co-workers from personal injury?
- Stay focused on the job at all times. Anticipate the unexpected and protect yourself and your co-workers.
- Immediately report all job-related incidents, near misses, and injuries to the supervisor.
- Immediately report all noted hazards to the supervisor. Correct those conditions that are within your control.
- Caution and counsel fellow employees when unsafe or hazardous acts are observed.

Note: When dealing with experienced, fully qualified workers, it may not be necessary or appropriate to establish the exact work method or procedure to be followed. Many tasks require that procedures be developed as the work progresses. If this is the case, the worker must conduct a job safety analysis (risk assessment) of the work to be performed to determine if it is safe or to establish the safe procedures to follow. Workers are also expected to look after the safety of their co-workers.

SECTION 3

3.0.0 ACCIDENT CAUSES, COSTS, AND CONTROLS

A proactive accident prevention program demands that:
- Hazards be identified before they lead to an accident
- The worst identified hazards be addressed first
- The control of hazards take the form that will ensure proper and lasting protection

This hazard recognition and control effort can be a formal program conducted in an organized manner combined with an informal process of hazard recognition promoted by the safety-conscious worker.

Safety-conscious workers have the ability to recognize a hazard and will take it upon themselves to do something about it. Being a safety-concious worker requires that you stop and talk to a co-worker when at-risk or hazardous behavior is observed. This must be done in a non-threatening way to be effective.

3.1.0 Useful Definitions

Here are some of the basic definitions and concepts necessary to implement an effective hazard recognition program.

Safety is a general term denoting an "acceptable level of risk;" of relative freedom from and low probability of harm; and control of recognized hazards to attain an acceptable level of risk.

Risk is a measure of both the probability and the consequences of all hazards of an activity or condition. It is a subjective evaluation of relative failure potential. It is the *chance* of injury, damage, or loss.

A **hazard** is defined as a condition, a changing set of circumstances, or a behavior that presents a potential for injury, illness, or property damage.

An **unsafe act** is a behavioral departure from an accepted, normal, or correct procedure or practice. It is an unnecessary exposure to a hazard or conduct that reduces the degree of safety. It is sometimes called *unsafe behavior* or *placing yourself at risk* (hazardous acts).

Ask trainees to describe the type of hazard recognition program in place at their company and its effectiveness.

If trainees are not aware of such a program at their workplace, ask them to discuss the matter with their supervisor and report back to the class what they learned.

Ask trainees to describe a recent accident at their workplace, what was the cause, and how it could have been prevented.

Unsafe conditions are any physical states that deviate from those that are acceptable, normal, or correct in terms of their past production or potential future production of personal injury and/or damage to property. This is any physical state that results in a reduction in the degree of safety normally present.

An **accident** is defined as an unplanned and sometimes injurious or damaging event that interrupts the normal progress of an activity. It is invariably preceded by an unsafe or hazardous act, an unsafe condition, or both. Events that do not result in property damage or personal injury are sometimes called *incidents*, *near misses,* or *near hits.*

An *event,* as defined in the context of safety management, is an unplanned release of energy.

3.2.0 Accidents vs. Incidents

The only difference between an accident with injury and a near miss is the outcome, which is usually a matter of timing and/or bad luck. Near misses are like free warnings. It is estimated that for every serious or fatal accident, there are 29 minor accidents and 300 near misses. This exponential relationship is shown in *Figure 2.* We can and must do something about them to prevent future events.

A safety-conscious worker is one that has the ability to recognize a hazardous condition or behavior and a willingness to do something about it. A safety-conscious worker is one who sees a spill and cleans it up, spots a frayed cord on a tool and turns it in for repair, or notices a co-worker lifting a heavy box and offers to help.

Contrast this with the worker who sees the spill and avoids it, but does nothing to clean it up. This worker was safe, but not safety conscious. This worker took care of himself but did not protect co-workers. Compare the safety-conscious worker with the worker that just walks past the spill without noticing it at all. Again, this safe worker did not have an accident by accident. A safety-conscious worker looks—and looks hard—for hazards and does something about them. Safety-conscious workers look after their co-workers.

3.3.0 Causes of Accidents

We know that accidents and injuries result in needless pain and suffering as well as financial hardship for employees, their families, and the company. We also know that we have moral, legal, and financial obligations to prevent accidents and injuries. To do this, we must understand what causes accidents. Accidents are caused by:

- Unsafe acts
- Unsafe conditions
- Combinations of unsafe acts and conditions
- Acts of God

William Heinrich, a noted psychologist, suggested that accidents resulted from unsafe acts 88% of the time, unsafe conditions 10% of the time, and acts of God 2% of the time. Later studies suggest that most accidents involve both unsafe acts and unsafe conditions. Many people tend to focus their attention on unsafe conditions, often overlooking the unsafe acts. Most safety professionals today agree that we should not ignore unsafe conditions, but a greater focus should be on the actions of people. *Figure 2* highlights that unsafe acts and conditions are the foundation upon which accidents are built.

Blending the definitions of hazard, accident, and incident, the art of hazard recognition deals with identifying conditions or changing sets of circumstances that could result in the unplanned event.

3.4.0 Unsafe Acts

The following is a list of the classic, most common or hazardous unsafe acts:

- Failing to use personal protective equipment
- Failing to warn co-workers of potentially hazardous conditions or unsafe behaviors
- Failing to follow instructions, procedures, and/or sound advice
- Using defective equipment
- Lifting improperly
- Taking an improper working position
- Making safety devices inoperable

Instructor's Notes:

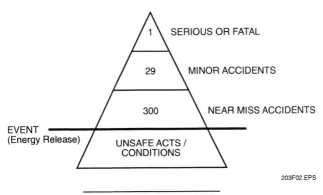

Figure 2 • The Heinrich Triangle

Ask trainees to give examples of an accident where an unsafe act and an unsafe condition would cause an accident.

Show Transparency 2 (Figure 2).

- Operating equipment at improper speeds
- Operating equipment without authority
- Servicing equipment while it is in motion or energized
- Loading or placing equipment or supplies improperly or in a dangerous way
- Using equipment improperly
- Working while impaired by alcohol or drugs, legal or illegal
- Engaging in horseplay

Example:

If Brad uses an angle grinder without the required guard, he places himself at risk unnecessarily and thus commits an unsafe act. His behavior is a departure from the expected procedure. There is a reduction in the degree of safety normally present in the task of grinding.

The vast majority of all accidents are primarily the result of an unsafe act, which includes hazardous acts or behaviors, although many accidents have both unsafe acts and unsafe conditions present.

Example:

Diane is using a ladder known to have a cracked rung. On descent, she puts weight on the cracked rung, the rung breaks, and she falls. This accident combines both an unsafe act with an unsafe condition. Diane elected to use the ladder, knowing that rung was defective.

Unsafe conditions are physical states that deviate from that which is acceptable, normal, or correct. This is any physical state that results in a reduction in the degree of safety normally present.

The following is a list of the most common unsafe conditions:

- Congested workplaces
- Defective tools, equipment, or supplies
- Excessive noise
- Fire and explosive hazards
- Hazardous atmospheric conditions
  — Gases
  — Dusts
  — Fumes
  — Vapors
- Inadequate supports or guards
- Inadequate warning systems
- Poor housekeeping
- Poor illumination
- Poor ventilation

It should be noted that many of the unsafe conditions listed are the result of poor planning or poor execution of a task, which may be thought of, to some degree, as an unsafe act or behavior. Poor housekeeping, for example, is the result of sloppy work, which may be a behavioral issue.

Ask trainees to give an example of each level of accident causation.

Show Transparency 3 (Figure 3).

3.5.0 Three Levels of Accident Causation

To prevent accidents, we need to know how they occur. There are three levels of accident causation, as depicted in *Figure 3*.

Level I – Direct causes of accidents. These are the accident events—the release of energy—that may or may not cause an injury or damage. In accident prevention and investigation, look for the energy sources. These energy sources generally fall into one of three categories: accident types, power sources, and/or hazardous materials.

For example, when the ladder rung broke, Diane fell to the ground. The fall was a "release of energy," the event. The fact that Diane was not injured was a matter of chance. Other examples of released energy include being struck by something, caught in between two objects, out of control electrical or thermal energy, detonated explosives, and being cut or abraded by a jagged edge.

Level II – Indirect causes of accidents. These are the hazardous acts and conditions that must be recognized and eliminated or controlled to prevent an incident. In the past, many people have looked only for these unsafe acts or conditions that preceded an accident. When found, the act or condition was corrected. When we find that these hazardous acts or unsafe conditions are just symptoms of a greater problem, we classify them as indirect causes.

Level III – Root causes of an accident. The root, or basic, causes of a hazardous act, condition, or incident are the system deficiencies that permitted, encouraged, allowed, or failed to identify and correct, follow-up on, anticipate, or predict the unsafe act or condition that preceded the accident. Root causes have permanent results when corrected. They are weaknesses which not only affect the single accident being investigated, but also might affect many other future accidents and operational problems.

Figure 4 shows that without an effective management or leadership plan, safety is compromised. When management controls are weak or non-existent, the consequences of property damage, near misses, lost workdays, and even fatalities can result. By implementing good controls and management plans, safety is enhanced.

The safety and health management plan coordinators are supposed to:

- Plan, direct, and control the work at hand
- Anticipate, predict, and plan for problems before the fact
- Place the right worker, with the right equipment, in the right work environment to accomplish the task
- Identify and correct the unsafe acts and conditions that invariably precede an accident
- Detect and correct any mismatch in the arrangement of worker, equipment, and environment
- Teach workers to recognize hazards
- Hold workers accountable for their actions

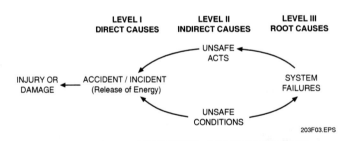

Figure 3 • Levels of Accident Causation

Instructor's Notes:

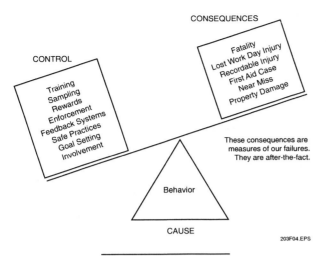

Ask trainees to discuss what works, what doesn't work, and what is missing in their company's safety plan.

Show Transparency 4 (Figure 4).

Figure 4 • A Delicate Balance

If the safety and health management plan coordinators did not do these things, that failure was the cause of the accident. The safety and health management plan typically establishes the policies and procedures, procures the equipment and tools, trains and instructs the workers, and provides the work environment through a variety of activities. Some of these include the safety program, design reviews, quality programs, process safety management program, and employee selection.

For example, through investigation we find there was no procedure established for the storage of ladders. Consequently, Diane's ladder was haphazardly laid flat on the floor of the shop where it could be easily damaged. This failure to provide a proper storage location or procedure was a root cause of the unsafe condition. Further, Diane, with the knowledge and help of her supervisor, had used the ladder in a broken condition on at least three occasions. Both noted that the ladder should be fixed, but neither took action to have the ladder fixed. Did the system not "encourage" a disregard for safe work procedures? It is little wonder that Diane was using a defective piece of equipment. Who else in the shop is using a defective piece of equipment, with a least tacit approval?

What in the "system" failed to prevent Diane's unsafe acts?

Why didn't the supervisor correct Diane's unsafe behavior? Because the supervisor did not take responsibility to observe and correct the workers. Why didn't the supervisor accept this responsibility? Because upper level management and leadership never demanded it. This is a system failure—a root cause. The root action in this case would be to establish an accountability system for supervisors in which they are held accountable for the safety of their workers.

3.6.0 Costs of Accidents

Accidents are very costly. When they occur, everyone involved loses, including the injured worker, the employer, and the insurance company. The only person who usually benefits from an accident is the plaintiff's attorney.

Accident costs are often classified by the costs involved. There are direct (insured) and indirect (uninsured) costs associated with accidents.

Direct or insured costs include medical costs and worker's compensation insurance benefits, as well as liability and property damage insurance payments. Of the three direct costs, worker's compensation insurance benefits are

Ask trainees if they know what their company's indirect/uninsured costs are in relation to the cost of an accident.

Show Transparency 5 (Figure 5).

the most costly. Direct costs of accidents are not generally fixed. Rather, they vary depending upon each company's accident experience. This is reflected in the company's workers' compensation **experience modification rate** (EMR).

Indirect or uninsured costs are the hidden costs involved with an accident. Examples include the costs associated with property damage, equipment damage, production delays, supervisory time, retraining, image, and morale.

The costs associated with accidents can be compared to an iceberg, as illustrated in *Figure 5*. The tip of the iceberg represents the direct costs, which are the costs that can be seen above water. On the other hand, the larger indirect costs are underwater or unseen. Studies indicate that the indirect costs of accidents usually are two to seven times greater than the direct costs.

3.6.1 Calculating Accident Costs

Real dollars are lost when workplace accidents occur. These dollars have a tremendous effect on the company's profit margin. For example, consider a company that operates on a profit margin of 3 percent of its gross income. The company suffers a loss of $50,000 due to accidents. In order to make up for that loss, the company must increase its gross income by $1,716,667!

3.6.2 Accident Experience vs. Future Insurance Costs

What effect, if any, does a contractor's accident experience have on future insurance costs? In a word, significant. Contractors with poor accident experience will generally pay more for worker's compensation (WC) insurance than those with a good record. The difference is in their worker's compensation experience modifier (EMR).

Experience rating is a method of modifying future WC insurance premiums by comparing a particular company's actual losses to the losses normally expected for that company's type of work. The average rate for a particular class of work is called the *book rate* or *manual rate*. Contractors with better than average loss experience have a modifier (multiplier) less than 100 percent, which is a credit factor. Those with worse than average experience have a modifier over 100 percent, which is a penalty factor.

Figure 6 illustrates this. The EMR is multiplied by the total WC cost to determine what the company will actually pay for WC insurance.

The lower half of *Figure 6* illustrates the effect of an EMR of 1.1. In some states, a contractor with poor accident experience can actually be

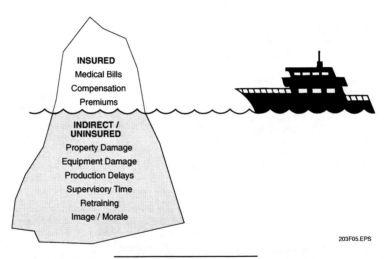

Figure 5 • Hidden Accident Costs

Instructor's Notes:

Module MT203 • Safety

penalized twice. Regular insurance companies may not want to insure the firm. If this happens, the contractor may have to go into the assigned risk pool for coverage. The manual rates for the assigned risk pool are significantly higher than those charged by regular insurance carriers. The right side of *Figure 6* shows what can happen to contractors in the assigned risk pool.

Typical Worker's Comp Premium Rates for a Louisiana Project

		Stock Insurance Companies		Assigned Risk Pool	
Classification	Payroll	WC Rate[1]	Insurance Cost	WC Rate[1]	Insurance Cost
Carpenters	$120,000.00	$37.16/100	$44,592.00	$45.88/100	$55,056.00
Electricians	$62,000.00	$11.80/100	$7,316.00	$17.02/100	$10,552.40
Pipefitters	$61,000.00	$9.24/100	$5,636.40	$13.58/100	$8,283.80
Ironworkers	$72,000.00	$44.39/100	$31,960.80	$58.88/100	$42,393.60
TOTALS	**$315,000.00**		**$89,505.20**		**$116,285.80**
Effects of a 1.1 EMR		EMR ___ x 1.1		EMR ___ x 1.1	
			$98,455.72		$127,914.38
		DIFFERENCE	$8950.52	DIFFERENCE	$11,628.50

[1]Louisiana manual rates as of 9/15/00

Figure 6 • Typical Worker's Compensation Premium Rates

PARTICIPANT ACTIVITY

Using the WC rates listed in *Figure 6*, assume that you and two other contractors are bidding the same job. Your labor and materials costs are essentially the same but you have different workers' compensation experience modifiers.

1. Company A has an experience modifier of 1.0, Company B has an experience modifier of 0.8, and your company has an EMR of 1.6. Use these figures to calculate the insurance costs for:

   Company A _____
   Company B _____
   Your Company _____

As you can see, your EMR can have a major impact on your bid. You should keep in mind that the frequency of job-related injuries generally has a greater impact on your EMR than the severity of the injury.

A company's EMR is based on the first three of the last four years' accident experience. For example, your EMR for the 2000 policy year is based on your losses in 1997, 1998, and 1999. In other words, what a company does today has a significant impact on its operations for years to come.

---- PARTICIPANT ACTIVITY ----

Show Transparencies 6 (Figure 6) and 7 (Participant Activity).

Work out the EMRs in the Participant Activity together. The correct answers can be found at the end of this module.

Ask trainees to research their company's EMR and determine whether their company is in an assigned risk pool.

SAFETY — INSTRUCTOR'S GUIDE MODULE MT203

PARTICIPANT ACTIVITY
(continued)

What follows is a case study of a construction accident, which we will call *Stan's Accident*.

Read the case study carefully, and then calculate the total direct and indirect costs involved using the Accident Cost Worksheet in *Figure 7*.

Most of the accident costs have been stated in this story. Medical costs may have been paid by a worker's compensation fund, to which the employer contributed. In reality, the employer bears all costs directly or indirectly and, therefore, so does the overall cost of construction in the U.S.

STAN'S ACCIDENT

1. On Monday at 8:30 a.m., Stan fell while working. His leg was injured and he struck his head, causing a wound and considerable bleeding.
2. Dave, Sam, Roy (co-workers) and Mack (the construction supervisor) gave Stan assistance and first aid.
3. Stan had a concussion, and it required 18 stitches to close the head wound. His leg was badly bruised with a slight fracture of the temporal base.
4. Mack drove Stan home from the clinic and arrived back at work at 12:50 p.m. He was back to work by 1:00 p.m. Production had stopped due to problems which only Mack could resolve.
5. Pete, a subcontractor, had a crew of three who could not do their scheduled work because Mack's crew had not completed their work. This resulted in an extra four hours charged to the job by the contractor, plus an extra $230.00 for rental of the equipment.
6. Mack spent an additional two hours reviewing, discussing, and preparing the accident report.
7. The insurance clerk took an hour to process the necessary forms.
8. The medical clinic charges were: doctor $180.00; X-rays $90.00; treatment and supplies $156.00; medication $55.00; drug test (required by company for accidents) $55.00; and follow-up visit $100.00.
9. Stan returned to work the following Friday, but was only about 75 percent as productive for the first week back (five workdays). He was given full pay for the day of the accident but no additional pay during the recovery time. He was not off long enough to collect workers' compensation for his lost wages.
10. Mack, Dave, and Roy worked a total of 60 hours overtime to get production back on schedule. Mack was not paid any overtime. Dave and Roy were each paid for 20 hours overtime.
11. Fringe benefit cost for all employees and contractors was 38 percent of base pay.

Rates of pay:

Stan	$17.50/hour
Mack	$3,290/month based on 173 hours/month (no overtime)
Sam	$16.50/hour
Roy	$16.50/hour
Dave	$12.50/hour
Insurance Clerk	$10.00/hour
Subcontract crew	
Two Mechanics	$17.25/hour
Helper	$12.50/hour

PARTICIPANT ACTIVITY

Instructor's Notes:

Module MT203 • Safety 3.13

ACCIDENT COST WORKSHEET

Person Injured: _____

Type of Accident: _____

Nature of Injury: _____

Direct (Insured) Costs		Indirect (Uninsured) Costs	
List Items	Cost in Dollars	List Items	Cost in Dollars
Subtotal A		**Subtotal B**	

Total A + B = _____

203F07.EPS

Figure 7 • Accident Cost Worksheet

Show Transparency 8 (Figure 7).

Read over Stan's Accident and work out the accident cost on the Accident Cost Worksheet with the class. The correct answers are located at the end of this module.

SAFETY — INSTRUCTOR'S GUIDE MODULE MT203

Ensure that you have all the necessary materials to teach the course. Check the Materials and Equipment list at the front of the module. Prepare for teaching Session II by reading Sections 4.0.0-7.5.0.

Distribute copies of a sample policy statement, if available. Ask trainees if they are familiar with their company's policy statement.

Assign reading of Module MT203 Sections 4.0.0-7.5.0.

3.6.3 Cost of Administering an Effective Safety Program

Studies referenced in the Business Roundtable's A-3 report estimate that the cost of maintaining an effective safety program is approximately 2.5 percent of the direct labor costs. This cost includes salaries for safety, medical, and clerical personnel, cost of safety meetings, inspections of tools and equipment, orientation meetings, personal protective equipment, and miscellaneous supplies and equipment. However, most of these items are required by state and federal laws, and contractors are required to pay for them anyway. The likely added cost of a safety program beyond what is legally required is probably closer to 1 percent of the direct labor costs.

Implementing an extensive safety program can actually save your company money. The same studies conclude that for every dollar invested in a safety program, the contractor could save as much as $3.20, which is an excellent return on investment.

SECTION 4

4.0.0 BASIC COMPONENTS OF A CONSTRUCTION SAFETY PROGRAM

There are certain basic elements common to most effective safety programs. The degree and extent to which these elements are included in your program will vary depending on the size and nature of your operations.

4.1.0 Management Support and Policy Statement

A policy statement outlines management's philosophies and goals toward safety and loss prevention. It is signed by the president or chief executive officer (CEO) and communicated to all that are affected.

A good policy statement recognizes management's responsibility for safety and loss prevention without diminishing each employee's role. It expresses a strong, positive commitment to safety that is realistic. Most important to the success of any safety program is management support and commitment.

4.2.0 Policy on Alcohol and Drug Abuse

Most experts agree that at least one out of five, or 20 percent, of construction workers have an alcohol or drug abuse problem. The impact of this problem results in:

- Increased worker's compensation and healthcare costs
- Decreased productivity
- Increased tools and materials costs
- Increased absenteeism

Many owners and contractors have recognized this problem and have implemented plans to combat the situation. If other contractors in your area have drug and alcohol abuse programs and your company doesn't, the company may be getting more than its 20 percent share of the employees who have a problem.

4.3.0 Assignment of Responsibilities

The duties and responsibilities of each employee, from the president of the company to the hourly employees, must be clearly defined and communicated. Along with these duties and responsibilities come accountability and authority. Individuals cannot be held accountable for items over which they have no control.

The following safety responsibilities are typically assigned to frontline supervisors and employees within an organization.

Frontline supervisors or designated site safety supervisors:

- Take responsibility for crew safety
- Train employees in safe work practices
- Observe employees for unsafe acts and take corrective action
- Inspect tools, equipment, and the general work area for unsafe conditions and take corrective action
- Take responsibility for obtaining prompt first aid for injured people on the work site
- Report and investigate all accidents and take corrective action
- Hold crew safety meetings
- Discuss safety with individual employees
- Audit subcontractors' compliance with job safety requirements

Instructor's Notes:

- Report to management items and practices beyond their control
- Serve on safety committees

Employees:

- Work in accordance with accepted safety practices
- Report unsafe conditions and practices
- Observe safety rules and regulations
- Make safety suggestions
- Serve on safety committees
- Do not undertake jobs that they do not understand

4.4.0 Employee Screening, Selection, and Placement

Selecting the right employee for the job is critical. The employee must have the necessary skills and physical capabilities to perform the work safely. A good attitude toward safety is an important trait in any employee.

Although they are in a minority, some people make a living out of getting hurt on the job and suing the employer. Abusers of alcohol and drugs can also cause major problems in the workplace. If you are involved in hiring, you must establish some practical, cost-effective methods of screening and hiring workers. However, you must do so in a non-discriminatory manner.

Some methods open to you include the following:

- Written applications
- Reference checks
- Division of Motor Vehicles (DMV or MVR) record checks
- Pre-employment drug screening
- Pre-job assignment physical exams
- Use of outside firms who review public court records for the names of individuals who have sued their employer. (*Note:* This is not permitted in some states.)
- Physical inspection of driver's license
- Skill testing and/or strength testing

4.5.0 Safety Rules

Safety rules are a necessary part of any program. They should be logical and enforceable, and they should be prepared and presented in terms that are easily understood. Rules are of no value unless they are communicated, understood, and enforced. All parties concerned must also understand penalties for the violation of safety rules. Safety rules must be fair and consistently and uniformly enforced.

Safety rules are generally divided into the following categories:

- General/company safety rules that apply to all employees
- Client-specific, job-specific, or site-specific safety rules
- Craft or special safety rules that apply to a specific type of work operation

4.6.0 Orientation and Training

Safety training can be broken down into two major categories. These include new employee orientation and job-specific safety training. Statistics have shown that employees who have been on the job 30 days or less account for 25 percent of all construction injuries. This clearly illustrates the need for an effective orientation program. Safety training should be conducted when:

- A new employee is hired
- An existing employee is assigned to a new job
- New jobs are initiated

The employee's supervisor usually conducts new employee orientation and job-specific safety training programs.

Safety training should cover the following:

- Correct work procedures
- Care, use, maintenance, and limitations of any required personal protective equipment
- The hazards associated with any harmful materials used. Examples include warning properties, symptoms of over-exposure, first aid, and clean-up procedures for spills
- Where to go for help

Ask trainees to list examples of safety rules in each of the three safety rule categories that they have experienced in the industry.

Ask trainees to describe their company's site emergency reporting and response procedures and what consequences, if any, of failing to act properly.

4.7.0 Safety Meetings and Employee Involvement

Maintenance of employee interest and involvement is a key element in any safety program. Safety meetings are often used for this purpose. Safety meetings can vary from a formal presentation to short five-minute **toolbox talks**.

Safety meetings, when properly conducted and held in a timely fashion, can be used to do the following:

- Exchange information regarding specific safety matters
- Diffuse potential job disruptions by providing an outlet for critical issues
- Provide a written record of the actions taken
- Establish an effective communications link between management and employees

Both federal and state governmental regulations require employers to train their employees on the hazards associated with the work and the necessary safeguards to be taken. Safety meetings can be used as training sessions.

The project manager should hold periodic safety meetings and sit in on some of your meetings and constructively critique your performance.

Some safety meetings will involve problem solving. The following steps should be used in problem-solving meetings:

Step 1 Identify the problem. Make certain that everyone agrees on the problem identified.

Step 2 Analyze why the problem exists. What are the causes? What procedures or work practices need to be changed or corrected?

Step 3 What should be done to correct the causes? Lead the discussion and get practical solutions.

Step 4 Encourage participation, and get buy-in. This comes from having participants agree on the problem and the solutions identified. In addition, they should then agree to take part in corrective actions.

Step 5 Provide feedback. Follow up with the results once the corrective actions have been put in place.

4.8.0 Emergency Reporting and Response Including First Aid

Provisions should be made in advance for reporting and responding to job-site emergencies, including the following:

- Personnel injuries
- Fires or explosions
- Windstorms
- Cave-ins or building collapses
- Civil disturbances

Advance planning can cut down response time to job-site emergencies. In many cases, advance planning can even minimize or reduce the severity of the incident. For instance, the names, addresses, and phone numbers of the nearest medical, fire, police and emergency response agencies should be posted at each job-site phone. In addition, all personnel should be familiar with the site emergency reporting and response procedures. Next, adequate provisions should be made for prompt access to first aid and follow-up medical care. Finally, it is highly desirable to have at least two people on each job site trained in basic first aid and CPR.

One area often overlooked by contractors is dealing with the news media. When serious accidents happen, the press usually responds. How you respond can have a dramatic effect on how the public perceives your company. You should be familiar with your organization's policy on dealing with the news media. If there is no policy, you should request that one be developed.

4.9.0 Inspections, Employee Observations, and Audits

A system of inspections, employee observations, and **audits** is necessary to maintain an effective safety and loss prevention program. Inspections are made to detect unsafe conditions and make plans to correct them. Next, employee observations help detect unsafe work practices and procedures. Finally, safety audits are used to monitor the use and effectiveness of the company's safety polices and procedures.

Frontline supervisors generally conduct safety inspections and record employee observations. Project managers normally review inspection reports and audit compliance with the firm's policies and procedures.

Instructor's Notes:

4.10.0 Accident Investigation and Analysis

All accidents, injuries, illnesses, and near misses should be reported and promptly investigated.

There are several reasons why these investigations are important, including:

- To determine the cause or causes so action can be taken to prevent a re-occurrence (the primary reason)
- To meet insurance company requirements
- To satisfy government regulations
- To document facts and preserve evidence in case lawsuits are filed later
- To detect trends and identify potential problem areas

The frontline supervisor has always performed the primary accident investigation function. It is generally agreed that frontline supervisors are the most familiar with the work area and employees. Therefore, they are best able to determine most of the underlying causes of an accident.

Depending on the nature of the incident and other conditions, accidents may also be investigated by the safety coordinator, the safety committee, or management. In any case, the project manager should be made aware of all reported or suspected incidents as soon as practical. The project manager should review every report for accuracy and completeness, and be certain that the cause(s) of the incident have been properly identified. The project manager should ensure that the corrective action is appropriate and is being implemented accordingly.

4.11.0 Records

Good record keeping is essential to an effective safety program. Without good records, it is virtually impossible to make any kind of analysis or measure the effectiveness of a program. It should also be understood that state and federal safety and health regulations require certain records.

Figure 8 shows a partial listing of the types of records that should be kept. The length of time required to keep each record varies considerably. You should follow your company's internal record keeping system. In addition, you should note the type of safety records being kept, indicate who will keep them, and determine how long they will be kept.

Figure 8 • Records and Retention Schedule

Ask trainees to identify the types of reports their company keeps and how long each report or record is retained.

Refer to the end of the module for a teaching tip.

Ask trainees who are unfamiliar with their company's record-keeping process to find out what is involved.

Remind trainees that many company handbooks contain information about the company's safety policies and that they may have signed a statement acknowledging that they read and understood those policies.

4.12.0 Program Evaluation and Follow-Up

Program evaluation and follow-up are some of the most important, but often neglected, elements of an effective safety program. To be assured that your program is meeting the company's goals and objectives, it must be evaluated on a regular basis and modified as needed. The project manager has the responsibility for evaluating the job-site safety program on a regular basis.

A safety and loss prevention program is a management tool used to help achieve the company's goals and objectives. To be effective, it should be put in writing, communicated to all involved parties, and above all, incorporated into daily activities.

Remember, as a supervisor it is your responsibility to get the job done safely, on time, and within budget. To do this you must identify the hazards, assess and evaluate the risks, and develop a plan of action. Regular employee observations, safety inspections, and audits are needed to evaluate safety performance. Remember, you have to inspect to get what you expect.

SECTION 5

5.0.0 EMPLOYEE SAFETY TRAINING

The supervisor is the critical player in providing employee safety training. That is not to say that others do not play a role, only that the primary responsibility lies with the employees' supervisor.

This responsibility offers opportunity as well as challenge, especially if the supervisor is not a skilled trainer. The challenge involves instilling a greater awareness of safety and a more positive attitude toward safety practices in everyone in the workplace.

The safety training efforts of a supervisor can be divided into several major areas:

- Employee safety orientation
- Job safety training
- Job hazard recognition
- Safety meetings
- First aid and CPR training

5.1.0 Employee Safety Orientation

Safety training begins with the supervisor's first contact with new workers. New workers can be longtime company employees that have been assigned to the supervisor's project or brand new employees that have been hired from outside the company.

As part of their initial conversations with new workers, supervisors should discuss policies such as lunch breaks and parking regulations. They must also discuss safety.

The supervisor should provide the employee with both written and verbal safety policies. Many organizations have employee handbooks or safety manuals that outline these policies. The supervisor should give the employee a personal copy of the handbook or manual to keep and reference in the future.

Often, the company handbook includes a statement that the employee must read and sign. This is an excellent means of making sure that the employee understands and accepts the company's safety policies. Supervisors should not rush their new employees through this process. Rushing communicates that the safety policies of the company are unimportant or a waste of time. By rushing, supervisors are defeating the company's safety awareness efforts, making it hard to get the employee to pay attention to safety concerns in the future.

Before giving a work assignment to a new employee, the supervisor should take time to review and emphasize potential hazards of the job, addressing corresponding safety rules and why they must be followed. Specific safety procedures that every new employee should know are:

- Report all accidents immediately. This includes near misses as well as injuries.
- Report any unsafe conditions immediately.
- Report all emergency situations immediately. Everyone on the job site should know response procedures, evacuation routes, assembly areas, and where to go for help.
- Check all tools and equipment. All tools and equipment, especially ladders and scaffolding, should be in good condition before they are used. Only qualified operators should operate equipment.

Instructor's Notes:

- Obey safe operation rules.
- Wear protective clothing.

These items fall into the category of the four R's:
- Responsibilities
- Rights
- Rules
- Regulations

Supervisors must establish these four R's early in every employee's work experience and require that they be followed. The consequences of not following rules and regulations must be communicated early and often.

5.2.0 Job Safety Training

Job instructions must include the critical elements of job safety. Many accidents can be avoided if the supervisor's initial job instructions include correct safety practices. One well-established and effective approach to training employees is the four-step method of training.

Step 1 *Prepare the Employee*

Nervous employees learn very little, so prepare the employees for what you are going to teach them by putting them at ease. Define the job in detail and show the quality standards that must be met. Determine what they already know about the job task, and build on this foundation. Point out the importance of the specific job and how it fits into the bigger picture of the project.

Step 2 *Present the Information*

Explain and demonstrate the various elements of the job. Tell the employees what elements of the task may be unique or specific to the company. This clarifies how your company wants something done versus how they may have done it for other companies. Where appropriate, don't hesitate to explain why something is to be done in a specified way. Understanding why helps the employees to better understand the job. Provide safety tips that are part of doing the task, and demonstrate necessary safety practices. Integrating safety with skills training underscores the belief that "the safe way is the only correct way."

Step 3 *Try Out Performance*

Have the employees do the job and observe their performance. Prior to each step, have them describe what they are going to do and why. This encourages them to think ahead. Avoid excessive criticism. Demonstrate patience and give encouragement. Compliment their recognition and use of safe work practices. Do not allow them to work unsupervised until you are sure that they have the necessary skill and understand the required safety practices.

Step 4 *Follow Up on the Work*

Allow the employees to work more independently after Step Three has been satisfied. Tell the employees who to contact for further assistance if you are not immediately available. Check back frequently, and give special attention to reinforcing the employees' use of safe work practices. If there are deviations from these practices, correct them immediately. Do not allow any employee to develop unsafe habits, even for a short period of time. After an appropriate interval, based on individual aptitude and performance, allow the employees to work without close supervision.

Supervisors who follow this four-step method of training will find their training efforts to be successful. In addition, the skill levels of the employees will improve rapidly.

Training can be the solution to a safety problem that is caused by a lack of knowledge or skill. The goal of the trainer is to change unsafe behavior by increasing knowledge of the hazards, safe work practices, and skills needed to work safely. The objective is to change behavior and attitude to the extent possible.

Behavior consists of observable actions that can be measured, while *attitude* is the internal predisposition to behave. Attitudes are difficult to change, but behavior is not. For example, a worker can be trained to perform (behave) in a safe manner even though the true safety attitude may be "it won't happen to me." The most efficient method to change attitude is to change behavior.

Ask trainees to give examples of unsafe acts they witnessed in the workplace.

Show Transparency 9 (Job Safety Training).

Demonstrate how to fill out an OSHA 200 log by going through each of the six steps a trainer should know.

Refer to the end of the module for a teaching tip.

People do not learn a great deal by just being told something. Rather, people learn best by a combination of hearing, seeing, and doing. A trainer should therefore do the following:

1. Tell the individuals *what you want them to learn.*
2. Tell them *how to do the task* being taught.
3. *Demonstrate* the skills involved.
4. *Let them try it.*
5. Give *immediate feedback* on their attempts.
6. *Repeat the sequence* until the skill is learned.

5.3.0 Job Hazard Recognition

Hazard recognition is a vital area of safety training. It includes identification of both unsafe acts and unsafe conditions.

Accidents and injuries do not happen by themselves. Instead, they are the result of unsafe acts, unsafe conditions, or combinations of the two. Unsafe acts are actions or behaviors causing unnecessary exposure to a hazard. Unsafe conditions are facilities, tools, and equipment that present exposure to an unnecessary hazard. Unsafe actions can lead to unsafe conditions, so the two types of hazards are closely related.

5.3.1 Unsafe Acts

Extensive research shows that roughly nine out of every ten accidents on the job are caused by unsafe acts. The tenth is caused by unsafe conditions. Among the unsafe acts most likely to cause accidents are:

Operating equipment without authority – Employees should only use equipment for which they are trained and authorized.

Using unsafe equipment – Using tools, equipment, or materials that are defective or otherwise unsafe.

Failure to secure – Not tying down materials on a vehicle or in a work area.

Failure to warn – Failing to signal properly or failing to warn others about potential hazards.

Operating at unsafe speed – Running instead of walking, driving at unsafe speeds, or handling materials too slowly or rapidly.

Bypassing safety devices – Disconnecting, removing, plugging, or blocking safety features.

Unsafe loading – Loading a vehicle, platform, scaffold, ladder, or other apparatus beyond its safe load limit.

Unsafe placing – Placing tools, equipment, or other materials where they obstruct work areas, aisles, or exits; placing hands in, on, or between pieces of equipment.

Taking unsafe positions – Lifting or carrying loads improperly; walking or working on or near unguarded openings or scaffolds; improperly riding vehicles; entering a work area where there are gases, exposed power lines, or excessive temperatures.

Working on dangerous or moving equipment – Oiling, cleaning, or adjusting equipment while it is in motion; working on equipment without turning off the power; getting on or off of moving vehicles.

Failing to wear protective safety devices – Not wearing safety glasses, proper footwear, ear protectors, helmets, or other devices.

Horseplay, teasing, and distracting – Playing within the work environment.

Figure 9 shows how dangerous horseplay can be. Correcting these unsafe actions should be a major focus of every supervisor's safety training effort.

5.3.2 Unsafe Conditions

The best method to use for training employees to recognize unsafe conditions is to point out these conditions as they are discovered in the workplace. This means that you must continually be alert for unsafe conditions, and you must take the time and effort to point them out to your employees. Training, coupled with actual illustration of hazards, will enable employees to become efficient in identifying unsafe conditions.

Part of your safety training and job planning is to focusing on *Task Hazard Analysis*. This is a pre-job safety briefing and analysis that helps workers to identify the work to be done, the tools to be used, and any potential hazards that they anticipate. Using a form such as the one in *Figure 10* is an efficient way to document task hazards and to plan for permit and safety requirements.

Instructor's Notes:

Module MT203 ♦ Safety

Figure 9 • Horseplay Causes Accidents

Among the most common unsafe conditions that you should teach your employees to recognize are the following:

Defects – Equipment, tools, or materials that are worn, torn, cracked, broken, rusty, bent, rotten, or splintered.

Lack of or inadequate guards – Rails missing from platforms, catwalks, scaffolds, site-constructed ladders, open shaft areas, or partially complete stairwells; power tool safety guards that have been modified or removed; power lines or explosive materials that are not fenced off or enclosed.

Poor housekeeping – Cluttered floors and work areas; temporary electrical service; inadequate aisle space; blocked exits; unsafe storage of materials or tools.

Poor lighting – Insufficient or excessive light; lights of the wrong color; glare; light arrangements that produce shadows or too much contrast.

Poor ventilation – Concentrations of vapors, dust, gases, or fumes; wrong capacity, location, or arrangement of ventilation system; impure air source; high temperatures and humidity.

Improper dress or use of personal protective equipment (PPE) – Inadequate personal protective equipment, loose clothing, or long hair.

The best approach to dealing with unsafe conditions is to prevent the conditions from producing an accident or injury. There are several ways to render unsafe conditions less hazardous. The most common methods, in order of effectiveness, are:

1. *Eliminate the hazard* – If possible, remove the hazard. Use a different material, process, or procedure.

2. *Reduce the hazard level* – If possible use a material, process, or procedure that is less hazardous.

3. *Provide safety devices* – If a hazard cannot be removed, shield it with enclosures, barriers, screens, interlocks, or other guarding devices.

Ask trainees to relate accidents they know of that were caused by horseplay and therefore completely preventable.

Show Transparency 10 (Figure 9).

Ask trainees if their workplace conducts pre-job safety briefings. If not, would they be willing to introduce a form like the example in Figure 10 to their supervisor for input?

Show Transparency 11 (Figure 10).

PRE-JOB SAFETY BRIEFING / TASK HAZARD ANALYSIS

Task / work to be performed: _____

Tool / equipment involved: _____

Employee Signature: _____

Potential Hazards — Check all that apply:

Physical Hazards:
- Falls on same elevation ____
- Falls from elevations ____
- Pinch points ____
- Rotating / Moving equipment ____
- Electrical hazards ____
- Hot / cold substances / surfaces ____
- Strains / sprains / repetitive motion ____
- Struck by falling / flying objects ____
- Sharp objects ____

Hazardous Chemicals / Substances:
- Flammable Materials ____
- Reactive Materials ____
- Corrosive Chemicals ____
- Toxic Chemicals ____
- Oxidizers ____
- Hazardous Wastes ____
- Biohazards ____
- Radiation Hazards ____
- Other (specify) _____

Energy Sources and Controls

Where is the energy? (Steam, electricity, pneumatic) _____

What is the magnitude of the energy? (Pressure, voltage) _____

What could happen or go wrong to release the energy? _____

How can it be eliminated or controlled? _____

What am I going to do to avoid contact? _____

Permits Required — Check all that apply:

____ Welding & Burning   ____ Line Entry   ____ Critical Lift
____ Lockout / Tagout   ____ Electrical Hot Work   ____ Vehicle Entry
____ Excavation   ____ Confined Space Entry ____ Other

PPE Requirements — Check all that apply:

Safety Glasses ____   Goggles ____   Face Shield ____   SCBA ____
Respirator (type) _____   Gloves (type) _____
Chemical Suit (type) _____   Head Protection (type) _____
Safety Shoes / Boots ____   Full Body Harness ____   Lanyard ____   Lifeline ____
Other (specify) _____
Special Precautions: _____

Completed By: Date & Time: _____

203F10.EPS

Figure 10 ♦ Pre-Job Safety Briefing/Task Hazard Analysis

Instructor's Notes:

4. *Provide warning* – If guarding is impossible or impractical, warn others about the unsafe condition. Post danger signs, signal lights, horns, whistles, painted striped patterns, red flags, or any other recognized devices for communicating warning messages.
5. *Provide safety procedures / protective equipment* – When controls that are higher in the priority list cannot be implemented to fully reduce the hazard, procedures and practices should be introduced to allow employees to work safely around the hazard. Wearing protective equipment is an element of a procedure. Procedures are the lowest control on the priority list because they depend totally on human behavior to recognize the hazard and take appropriate corrective action.

As part of their safety training, employees should be taught to bring all unsafe conditions to the your attention. You must be prepared to act on such employee reports. Failure to act on an employee's initiative demonstrates a lack of commitment that soon spreads among other employees.

SECTION 6

6.0.0 JOB SAFETY ANALYSIS

Job safety analysis (JSA) is a procedure used to review job methods and uncover hazards that may have been overlooked in the layout or design of the equipment, tools, processes, or work area; developed after production started; or resulted from changes in work procedures or personnel.

6.1.0 Performing a Job Safety Analysis

The four basic steps of performing a job safety analysis are:

Step 1 Select the job to be analyzed.

Step 2 Break the job down into successive steps or activities and observe how these actions are performed.

Step 3 Identify the hazards and potential accidents. This is the critical step because only an identified problem can be corrected or eliminated.

Step 4 Develop safe job procedures to eliminate the hazards and prevent potential accidents.

6.1.1 Methods of Conducting JSAs

The three basic methods for conducting a job safety analysis are direct observation, group discussion, or group discussion using a videotape of a job.

A fast and efficient method of conducting a JSA is through direct observations of job performance. In many instances, however, this method may not be practical or desirable. For instance, new jobs and those that are done infrequently do not lend themselves to direct observation. When this is the case, the JSA can be made through discussions with persons familiar with the job. Individuals often involved in the process include, but are not limited to, frontline supervisors, safety specialists, engineers, experienced employees, and outside contractors.

6.1.2 Selecting Jobs to be Analyzed

When selecting jobs to be analyzed, most people start with the worst first. You should be guided by the following factors:

- *Frequency of Accidents* (including near misses) – A job that repeatedly produces accidents is a candidate for a JSA. The greater the number of incidents associated with a job, the greater its priority claim for a JSA.

- *Production of Disabling Injuries* – Every job that resulted in a serious or disabling injury should be given a JSA.

- *Severity Potential* – Some jobs may not have a history of accidents, but may have the potential for severe injury.

- *New or Revised Jobs* – Jobs created by changes in equipment or in processes obviously have no history of accidents, but their accident potential may not be fully appreciated. Analysis should not be delayed until accidents or near misses occur.

- *Multiple Employee Exposure* – Jobs that expose more than one individual to potential hazards should also be analyzed.

Ask trainees to describe job safety analyses they have performed, what method they chose, and which jobs were selected for analysis and why.

Ask trainees to give an example of each of the eight types of accidents.

6.1.3 Job Observation

It is important to select an experienced, capable, and cooperative person who is willing to share ideas. If the employees have never participated in a job safety analysis, explain to them the purpose, which is to make a job safe by identifying and eliminating or controlling hazards. Show the workers a completed job safety analysis.

Follow these steps to conduct the job observation:

Step 1 Select the right person to observe.
Step 2 Brief the employee on the purpose of the job safety analysis.
Step 3 Observe the person as the job is performed and try to break it down into basic steps.
Step 4 Record each step.
Step 5 Check the breakdown with the person involved.

6.1.4 Common Errors in Performing a Job Analysis

Five common errors that are often made when performing a job analysis are:

- Making the breakdown so detailed that an unnecessarily large number of steps are listed
- Making the job so general that basic steps are not recorded
- Failing to identify the education and experience level of the target audience
- Failing to identify end uses (training, actual procedure, basis for procedure)
- Conducting JSAs for all jobs instead of identifying jobs that require JSAs

6.2.0 Identifying Hazards and Potential Accidents

The purpose is to identify all hazards, both physical and environmental. To do this, ask yourself these questions about each step:

- Is there a danger of striking against, being struck by, or otherwise making harmful contact with an object?
- Can the employees be caught in, on, by, or between objects?
- Is there a potential for a slip, trip, or fall? If so, will it be on the same elevation or to a different elevation?
- Can the workers strain themselves by pushing, pulling, lifting, bending, or twisting?
- Is the environment hazardous to anyone's safety or health?

6.2.1 Accident Types

There are eight main types of accidents. Being aware of these will increase your awareness of potential hazards.

1. *Struck by*
   — moving or flying object
   — falling material

2. *Struck against*
   — stationary or moving objects
   — protruding object
   — sharp or jagged edge

3. *Contact with*
   — acid
   — electricity
   — heat
   — caustic material
   — cold
   — radiation
   — toxic and nocuous substances

4. *Caught*
   — in
   — on
   — between

5. *Fall to*
   — same level
   — lower level

6. *Overexertion*
   — lifting
   — pulling
   — pushing

7. *Rubbed or abraded by*
   — friction
   — pressure
   — vibration

8. *Bodily reaction from*
   — voluntary motion
   — involuntary motion

Instructor's Notes:

Module MT203 ♦ Safety

6.3.0 Writing Instructions for JSAs

Follow these recommendations for writing JSAs:

- Put any qualifying statements first, not last.
- Start each instruction with an action word.
- Each instruction should be observable.
- Each instruction should be measurable.

When evaluating a given procedure, ask, "What should the employee do, or not do, to eliminate this particular hazard or to prevent this potential accident?" The answer must be specific and concrete to be beneficial. General precautions such as "be careful," "use caution," or "be alert" are useless. Answers should state what to do and how to do it.

For example, the recommendation, "make certain that wrench does not slip or cause loss of balance" is incomplete. It does not tell how to prevent the wrench from slipping. Here is a more complete recommendation: "Set the wrench properly and securely. Test its grip by exerting a slight pressure on it. Brace yourself against something immovable, or take a stance with feet wide apart before exerting full pressure. This prevents loss of balance if the wrench slips."

Job safety analyses can be very beneficial if they are performed correctly. They not only result in a safer job, but also increase productivity and eliminate waste. Take the time to do them correctly, and most important, use them in your daily work.

6.4.0 Recommended Safe Job Procedures

The final step in conducting a JSA is to develop a recommended safe job procedure to prevent the occurrence of potential accidents. The principle solutions are:

- Find a new way to do the job.
- Change the physical conditions that create the hazard.
- Try to eliminate remaining hazards by changing work methods or procedures.
- Try to reduce the necessity of doing a risky job, or at least the frequency with which it must be performed.

The use of a job hazard analysis form is helpful to organize these steps and document results of JSAs. *Figure 11* is convenient to use in your daily work to streamline your job hazard identification process.

Figure 11 • Job Hazard Analysis Form

Ask trainees to develop a recommended safe job procedure to prevent the occurrence of potential accidents by type using the job hazard analysis form.

Show Transparency 12 (Figure 11).

Ask trainees to give examples of toolbox talks they have conducted or would like to conduct.

SECTION 7

7.0.0 CONDUCTING EFFECTIVE SAFETY MEETINGS

Employee safety meetings range from short, informal gatherings of the supervisor and crew members to larger, more formal sessions featuring guest speakers. Normally, they are held on a regular basis and welcome employee participation and involvement. Their purpose is to bring employees together in support of the company safety program. Their primary benefit is better group behavior and attitudes toward safety.

As with any other type of meeting, a safety meeting is most successful when the person conducting it focuses on providing useful information to the group. Therefore, it is important that sufficient time and effort be devoted to planning and preparing for the meeting.

Some important points that should be considered are:

Time limit – Plan a definite time limit. Formal sessions should last no more than an hour; informal sessions should be considerably shorter.

Time of day – Make sure that training is conducted during working hours. It should not be conducted during the employees' breaks or lunches.

Comfort of the group – People who are uncomfortable quickly tire and become restless and disinterested. Make certain that the meeting area is neither too hot nor too cold. Ensure that the lighting is proper and there is no background noise. Confirm that everyone can see what is being presented.

Appropriate topic – Most people can absorb only a few ideas at a time; therefore, a safety session should present no more than one or two main ideas. Make sure that the topic is related to work in progress, future activities, or recently completed tasks.

Knowledgeable presenter – Be sure that the individual who conducts the meeting is well-versed in the subject and sincere about the presentation.

Group composition – Employees, supervisors, and upper-level managers should attend safety meetings. The presence of management demonstrates company interest in good safety practices.

Good planning – Handouts, visual aids such as films, slides, or videotapes, and other teaching aids improve presentations. Be sure to gather and prepare all necessary equipment and materials before the session. Check audiovisual equipment to be sure it works.

Time for discussion – Every safety meeting should include time for questions and answers. This time allows employee participation and the opportunity for further clarifications. Keep control of question-and-answer sessions, and keep them focused on safety topics. Be tactful, and keep the discussion on track.

7.1.0 Toolbox Safety Talks

Toolbox talks are informal safety training sessions designed to continuously reinforce the company's safety program. They are often more effective than formal sessions in developing strong safety practices on the job. These meetings usually include supervisors and their work crews. They may, however, include others who are involved in the same work effort.

Some helpful hints for organizing and conducting informal safety sessions are as follows:

- Start with a review of recent safe work and practices.
- Present short talks on topics related to current or upcoming work activities.
- Limit prepared talks to 10 minutes.
- Encourage employee participation.
- Use open-ended questions, ask for opinions, invite suggestions, and provide appropriate follow-up.
- Review recent near misses and workplace injuries, and discuss how they happened and how they could have been prevented.
- Review and discuss hazards encountered while working with other crafts.

Instructor's Notes:

- Look ahead to potential safety hazards involved in upcoming work and remind employees of the proper use of safety equipment and the procedures to be followed in dealing with those hazards.

7.2.0 Practice the Five "Ps" for Successful Safety Talks

This is a tried and tested technique that will help you give better safety talks. The technique is simple, it is effective, and it works. You can easily apply this technique not only to safety talks, but also to vital subjects such as quality, productivity, service, job instruction, and cost improvement.

The Five "Ps" are:

Prepare – Think safety. Write things down for your idea-bank. Read safety materials thoroughly. Listen to others' ideas and attitudes. Organize and outline your talks. Practice.

Pinpoint – Don't try to cover too much ground. Concentrate on one safety rule, one first aid hint, one unsafe practice, or one main idea.

Personalize – Establish common ground with your listeners. Bring it close to home. Make it important in their minds. Make it personal and meaningful to them.

Picture – Create clear mental pictures for your listeners. Appeal to both their ears and their eyes. Help them really "see what you mean." Use visual aids.

Prescribe – In closing your safety talk, answer the question the listeners always have—"So what?" Tell them what to do. Ask for special action. Give a prescription.

In addition to the above technique, make a list of at least five questions that the participants should be able to answer at the end of the meeting. Make certain this information is covered in the meeting.

Figure 12 is a safety meeting pre-planning tool to help you organize your meetings for results.

Safety Meeting Topic:

(Select a topic pertinent to the job at hand)

Feedback Questions

Make a list of at least five questions you want participants to be able to answer at the end of the meeting.

Question 1:

Answer 1:

Question 2:

Answer 2:

Question 3:

Answer 3:

Question 4:

Answer 4:

Question 5:

Answer 5:

Visual aids to be used:

Participant handouts:

Figure 12 • Safety Meeting Pre-Planning Tool

Show Transparency 13 (Figure 12).

Ensure that you have all the necessary materials to teach the course. Check the Materials and Equipment list at the front of the module. Prepare for teaching Session III by reading Sections 8.0.0–10.2.4.

Ask trainees to relate any insights they have gained from witnessing scheduled safety inspections, audits, and observations.

Assign reading of Module MT203 Sections 8.0.0–10.2.4.

7.3.0 How to Organize a Successful Safety Meeting

Follow these guidelines to construct an effective meeting.

Introduction

Introduce the subject or topic. Explain the importance of paying attention. You might say,

"Today's topic is the need for and use of Ground Fault Circuit Interrupters (GFCIs). Failure to use a GFCI when using portable electric tools could be fatal. At the end of the meeting, you should be able to answer these questions:

- What is a GFCI?
- When do I need a GFCI?
- How do I use it?
- How do I test it?
- Where can I get one?
- If I need assistance, who should I see?"

Body

Using the questions as a topical outline, cover the material in detail. When doing so, get feedback from the audience by asking questions that require an explanation such as, "what is a GFCI?" Do not ask questions that can be answered with a simple "yes" or "no."

Conclusion

Summarize the key points of the meeting. Ask the five questions and get answers from the audience. Clarify any misunderstandings or incorrect comments.

Questions and Answers

Give employees time to ask questions about the materials covered. Encourage participants to bring up safety-related issues. If you don't know the answer to a question, say so, but tell the person you will find out and get back to them. Make sure you do so.

Documentation

Be sure each attendee signs the safety meeting report form. Review the names and make sure no fictitious names or inappropriate comments have been listed on the meeting report form.

7.4.0 First Aid and CPR Training

Many construction companies provide both first aid and cardio-pulmonary resuscitation (CPR) training to their employees as a part of their safety-training program. This training may be provided to all employees or only to key individuals. Apart from the obvious value associated with having employees competent in first aid, OSHA regulations require that at least one person be trained in first aid. That way, someone will be available at the work site when area medical facilities are not readily accessible. First aid and CPR training are available through local Red Cross chapters, local chapters of the National Safety Council, and independent training agencies.

7.5.0 Training Documentation

All employee safety training should be appropriately documented. There are several reasons for this. First, you need to know what training your employees have received. Second, training records allow you to identify areas where employee training is required. Finally, safety-training records are especially valuable whenever an OSHA inspector visits the work site. Many OSHA inspectors request to see the employees' training files even before they tour the site. Having complete records available can help with the inspection process.

SECTION 8

8.0.0 JOB-SITE SAFETY INSPECTIONS, AUDITS, AND OBSERVATIONS

The primary reason for conducting safety inspections, audits, and observations is to detect undesirable conditions and work practices before an accident or injury occurs. Prevention enables the appropriate action to be taken so that an accident or injury can be avoided.

8.1.0 The Supervisor's Role

As the supervisor, the responsibility for maintaining a safe and healthy work site rests with you. Inspections, audits, and observations are a part of your job. These activities should be considered as important as productivity and quality control.

Instructor's Notes:

_____
_____
_____
_____
_____

Module MT203 ♦ Safety

The benefits of regular job-site inspections, audits, and observations include the following:

- Being able to monitor changing job-site conditions to identify and eliminate unsafe conditions or practices as they occur
- Instilling and reinforcing a sense of the importance of self-inspection and evaluation in employees
- Contributing to quality and overall effectiveness of the workforce by enforcing good housekeeping
- Monitoring compliance with company and OSHA safety and health regulations

The terms *safety inspections, audits,* and *observations* are sometimes used interchangeably. However, there are some significant differences in their meanings, as outlined below:

- *Safety inspections* refer to checking an area to identify and report hazards to which workers, materials, and equipment are exposed.
- *Safety audits* involve a methodical review of certain safety policies and procedures to see if they are adequate and are actually being implemented.
- *Safety observations* involve observing an employee and making an immediate determination as to whether the employee is working safely or not.

The role of management, including your role as a supervisor, is to complete the job safely, on time, and within budget. To do this, each member of the management team must participate in the safety inspection, audit, and observation process. In each case, you must do five things:

1. Recognize the hazard or unsafe condition.
2. Determine what caused the unsafe condition or why the employee was performing an unsafe act.
3. Take the appropriate corrective action for immediate, temporary control. This includes informing involved or exposed personnel of the nature of the unsafe act or condition. It also involves communicating the corrective measures required.
4. Follow up to see that the corrective action was taken and that the action was proper and adequate. Establish a long-range system fix to address the root cause of the problem.
5. Provide adequate documentation

The frequency and detail of these inspections and audits will vary depending upon the nature of the work, the degree of hazard and risk, and the number of people exposed or involved. In some cases, inspections and audits are done on a regularly scheduled basis. In other cases, they are done randomly and are unannounced. For example, cranes and rigging equipment should be functionally checked and inspected before each use. In addition, more detailed inspections and tests are required at monthly and annual intervals. However, if a series of incidents or near misses occur involving cranes or rigging equipment, an unannounced inspection or audit should be conducted.

Observations, on the other hand, should be done continuously. As you walk through your work area checking on your personnel and their work, you should observe their work practices. If they are working in an unsafe manner, you should stop them and take the appropriate corrective action. Records should be kept noting the number of safe and unsafe observations. This information can be used to identify problem areas where additional training may be required.

Figure 13 is a form for recording safety observations. By doing a safety sampling, you can begin to track and post the results of actual safety practices by crew or by individual. Use of this form helps you to manage the work site and improve safety and the overall quality of the job.

8.2.0 Elements of Job-Site Inspections

The identification of unsafe acts and conditions is included in a good inspection program. However, unless identification is combined with analysis and correction, it is useless. *Analysis* simply means determining why unsafe acts and conditions exist and deciding how to eliminate them. *Correction* means establishing feedback and control mechanisms to eliminate unsafe acts and unsafe conditions. You must remember that safety inspections are only as good as the corrective measures taken. We need to take immediate, temporary action (replace the guardrails) and we need to fix the system (why the guardrails were removed).

Ask trainees to give examples of results from unannounced inspections or audits.

Ask trainees to describe the responses by their crew members, other employees, and other supervisors of a job-site inspection they initiated.

SAFETY — INSTRUCTOR'S GUIDE MODULE MT203

Ask trainees if the safety sampling form they use has additional items that would be useful.

Show Transparency 14 (Figure 13).

SAFETY SAMPLING

01	Using Defective Equipment						
02	Using Defective Material						
03	Using Defective Tools						
04	Using Defective Vehicles						
05	Operating Vehicles Unsafely						
06	Failure to Use Safety Glasses						
07	Failure to Use Goggles/Faceshield						
08	Failure to Use Respiratory Equipment						
09	Failure to Use Hearing Protectors						
10	Failure to Use Safety Belt / Lifeline						
11	Failure to Use Safety Equipment Provided						
12	Failure to Use Personal Protective Equipment						
13	Failure to Recognize Unsafe Situation						
14	Unsafe Lifting						
15	Unsafe Carrying						
16	Working on Unsafe Work Platform						
17	Working in Unsafe Position						
18	Working on Machinery in Motion						
19	Using Tools Unsafely						
20	Using Equipment / Materials Unsafely						
21	Making Safety Devices Inoperative						
22	Failure to Secure						
23	Horseplay						
24	Poor Housekeeping						
25	Working Under Suspended Load						
99	Safe Observation						
	TOTALS						

Supervisor in Charge: _____ Date: _____

Location: _____

Comments: _____
_____

203F13.EPS

Figure 13 • Safety Sampling

Instructor's Notes:

_____
_____
_____
_____
_____
_____
_____

A work-site inspection involves several elements, including the following:

1. *Preparation* – Prepare by reviewing previous safety records, inspection reports, and audits. Make a list of outstanding items. Familiarize yourself with the types of accidents that have occurred before. Make notes of specific items to check. Set a good example as you conduct the inspection by wearing appropriate safety equipment. Carry a pad of paper or other items that may be useful for taking notes during the inspection.

2. *Developing a Safety Checklist* – Develop and use a checklist that covers all areas of the job site. This will prevent you from overlooking critical areas. The *Appendix* illustrates a construction job checklist based on OSHA guidelines. You should add items to it based on your company policy and job conditions.

When using the checklist, do the following:

- *Plan your route* – Know what areas you want to inspect. Stick to your plan so that you don't overlook areas that you have identified in your preview of the safety records. Take notes. Don't trust your memory. Be specific in your note taking regarding location, environmental factors, people, equipment, and possible corrective measures.

- *Correct on the spot* – If you find a problem in your area, immediately correct it. Do not put off correcting items that can be corrected immediately. Waiting increases risk. If you discover a hazard in an area outside your authority, call attention to these conditions. Contact the people who are responsible and advise them that your employees will not be allowed to work in the area until the problem has been eliminated. If you discover a defective piece of equipment that cannot be repaired or removed from the area, attach a "Danger–Do Not Operate" tag on the equipment.

- *Follow up immediately* – Immediately following the inspection, put your notes and inspection sheets in useful order. Initiate requests for corrections of hazards in areas you do not control. Assign responsibility for items in areas that you do control. Establish appropriate completion and follow-up times.

- *Conduct subsequent follow-up* – Check off those items that have been corrected and initiate follow-up requests on outstanding items. This is often the most difficult, most neglected, and most important step.

- *Communicate* – Let your employees know that you are looking for solutions to safety problems. Ask them for their suggestions and comments. Let them know you are concerned about unsafe work practices and conditions, not just checking up on them. Use this opportunity to talk about other phases of the safety program. Make your safety inspection an effective communications channel between you and your employees. Stress the positive and avoid being critical.

A properly executed job-site safety inspection is among the most practical measures you can take to improve safety conditions on the job. It puts into action many of the principles supported by your company and, therefore, sets an excellent example for your employees.

8.3.0 Key Elements of Employee Observation

The objective of employee observation is to identify unsafe behaviors and then analyze and correct them. Unsafe behaviors are associated with the majority of accidents. The organization that replaces unsafe behaviors with safe behaviors will have fewer accidents.

The major error in an employee observation activity involves blaming the employee for the unsafe behavior. Safety performance is a long-term undertaking. Blaming employees is counterproductive. The proper way to change unsafe behavior is to determine the reason for the employee's unsafe behavior and correct the situation.

8.3.1 Reasons for Unsafe Behavior

Some reasons for unsafe behavior include working when the following factors are a problem:

Physical Factors – Noise, hearing, visibility, eyesight, stamina, and strength

Mental Factors – Depression, alcohol or drugs, emotional instability, and stress associated with job pressure, heat, etc.

Training, Education, and Experience Factors – Insufficient knowledge, inadequate training, lack of education, inadequate skills, and lack of hazard awareness

Ask trainees to describe what method(s) they used to confront or change unsafe behavior among their crew members.

Motivational/Attitude Factors – Risk-taking personality, needs motivational encouragement, personal conflicts with co-workers and management

Environmental Factors – Facilities or surroundings, job procedures, weather, and equipment or tools

Do not blame an employee for an unsafe behavior when one or a combination of the above factors creates obstacles to doing the job in a safe manner. Instead, work with the employee to correct or eliminate the factors contributing to the unsafe behavior so the work can be done safely.

8.3.2 Changing Unsafe Behavior

Ways to change unsafe behavior to safe behavior include:

Training – Train employees to recognize unsafe behavior and factors that contribute to it. Training in safe behavior should be ongoing.

Involvement – Employees should be involved in identifying unsafe behavior, contributing factors, and corrective actions. Realize that it is natural for people to resist change. Employee participation, whenever possible, will assist in "buy-in" to changes.

Sampling (observation) – Once an unsafe behavior is changed to a safe behavior, periodic observation of that work practice should be done to assure the safe behavior continues.

Goal setting – Goals should be set for employee observations. They should be periodically reviewed, evaluated, and modified as necessary.

Rewards – A reward system, such as gifts or prizes, may be useful in an employee observation program at the beginning and to foster continued interest. The best reward is positive recognition and feedback for employees correcting unsafe behavior.

Safe practices – In an employee observation program, employees working safely should be given positive recognition for doing so. Behavior that brings reward will be repeated. Behavior that is not rewarded or is punished will not be repeated.

Enforcement – Enforcement of safe behavior must be fair, ongoing, and consistent. An on-again, off-again effort will reduce the effectiveness of the overall safety efforts.

Feedback systems – Immediate praise for safe behavior must be consistently practiced. Employees must also be given periodic feedback on the results of efforts to minimize unsafe behavior and to promote safe behavior.

In summary, an employee observation program involves these key points:

- Ensure knowledge and skill
- Observe behavior
- Encourage and recognize safe behavior
- Discourage and correct unsafe behavior

8.3.3 Confronting Unsafe Behavior

When you talk to an employee about unsafe behavior or work practices you should:

- Do it in a private place, one-on-one (remember to praise in public and criticize in private)
- Express concern for the individual's safety and well-being
- State the nature of the unsafe act or practice
- Find out why they are doing what they are doing. Are there any barriers to safe performance?
- Inform the employee of the appropriate safety rule, practice, or procedure
- Close on a positive note

If you continuously observe an employee performing an unsafe act and do nothing about it, the employee assumes that it is okay or it is not a priority to you. The employee then continues to be exposed to a potential accident or injury.

In addition to informing them when they are doing something wrong or incorrect, you should give recognition for doing a good job. Obviously they do not need a pat on the back all the time, but positive feedback is very important. Remember that behavior that brings reward is repeated and behavior that does not bring reward or is punished is not repeated. This is known as the "law of effect."

Sometimes, employees will ask you to overlook a simple safety violation. In response, you should indicate that asking you to overlook the safety violation is like asking you to compromise your attitude towards the value of that employee's life.

Instructor's Notes:

SECTION 9

9.0.0 ACCIDENT REPORTING AND INVESTIGATION

You must inform the company of the nature and scope of any accidents involving injuries or significant property damage. This largely involves following company procedures developed in response to requirements of insurance carriers, state industrial accident boards, state worker compensation laws, OSHA guidelines, and other governing bodies.

The first report of an accident involving an injury or significant property damage usually must be completed within eight hours of the incident. Some companies set this requirement for accidents that require first-aid treatment, while others require an immediate report only if the worker is taken off-site for medical treatment or if property damage exceeds a specific amount.

9.1.0 The Supervisor's Role

Generally, you should call the company personnel director, safety director, or designated individual to file an accident report as soon as possible. This should be done at once whenever you have basic information about the accident. This informal report alerts the company that an accident with injury has occurred. It also allows the support staff to assist in dealing with the medical needs of the injured employee. In addition, support staff is able to provide further instructions and clerical assistance.

The informal report to the company must be followed by a written report within the time frame specified in company policies. If the accident involves a subcontractor's employee, you must also provide the general contractor with a copy of the report.

Figure 14 includes an example of a typical accident report form. Most construction companies have a standardized form for reporting employee injuries or illnesses.

If your company does not have such a form, you can accurately report an accident simply by listing the following information on a sheet of paper:

- Date of accident
- Time of accident
- Name, address, phone number, and social security number of injured personnel
- Job title or position of injured personnel
- Location of accident
- Nature and extent of injury or property damage
- Description of medical aid provided
- Apparent cause of accident
- Name, address, and phone numbers of any witnesses and/or owners of the damaged or involved equipment
- Recommended corrective action
- Name and position of person completing the report

The more information you can give under each of these headings, the more valuable your report will be. The completed report will be a basis for preparing the formal (follow-up) accident investigation report.

9.2.0 Conducting the Investigation

It is said that the only thing of value that can be salvaged from an accident is knowledge. Knowledge is gained from experiences, both positive and negative. Knowledge alerts others to the possibilities of accidents and helps prevent similar mishaps. Such knowledge can be gathered only through careful investigation. Consequently, all accidents and near misses should be investigated. Remember that a near miss is an incident that did not cause an injury or damage, but had a high potential for doing so.

The reasons for conducting an investigation are as follows:

- Determine direct causes
- Determine indirect (underlying) causes
- Prevent similar accidents/incidents
- Document facts
- Meet insurance company requirements
- Satisfy government regulations
- Detect trends and problem areas
- Provide information on costs
- Promote and improve safety

Discuss what a supervisor might write in an informal accident report.

Show Transparency 15 (Figure 14).

CASE No.	INCIDENT/ACCIDENT INVESTIGATION REPORT	☐ First Aid ☐ Recordable ☐ Lost Time
INVESTIGATED BY: ☐ Individual ☐ Team Names & Department: _____		ACCIDENT TYPE (Check all that apply) ☐ Near Miss ☐ Injury ☐ Property Damage
		DATE & TIME INVESTIGATION STARTED

1. NAME of Injured	2. SOCIAL SECURITY NUMBER	3. SEX ☐ M ☐ F	4. D.O.B.	5. DATE & 6. TIME of ACCIDENT
7. NAME(s) of witness(es)				
8. Employee's USUAL JOB CLASSIFICATION	9. JOB CLASSIFICATION at TIME of ACCIDENT		10. Employment Category ☐ Regular, Full Time ☐ Temporary ☐ Non-employee	
11. Length of employment ☐ Less than 1 Year ☐ 6-10 Years ☐ 1-5 Years ☐ 11 Years or More		12. Time in Occupation at Time of Accident ☐ Less than 1 Month ☐ 6 Months - 5 Years ☐ 1-5 Months ☐ 5 Years or More		
13. Name & Employer of Others Injured or Involved in Same Accident		14. Employee's work schedule at time of Accident Shift _____ ☐ Overtime ☐ Turnaround ☐ Call Out ☐ Other _____		
15. Nature & Extent of Injury or Damage		16. Severity ☐ FATALITY ☐ Lost Workdays - Days Away from Work ☐ Lost Workdays - Days of Restricted Duty ☐ Medical Treatment ☐ Other		
17. Supervised By ☐ Company ☐ Contractor ☐ Other		18. Type of Supervision ☐ Supervision Present ☐ Supv. Not Present		
19. Specific Location of the Accident: On Company Premises ☐ Yes ☐ No				
20. Describe How the Accident Occured (Facts Only) _____				

Figure 14 • Incident/Accident Investigation Report (1 of 2)

Instructor's Notes:

21. ACCIDENT Sequence: Describe in reverse order of occurrence events preceding the injury and accident. Starting with the injury or "near miss" and moving backward in time, reconstruct the sequence of events that led to the injury or "near miss".
  A. Injury Event
  B. Accident Event
  C. Preceding Event #1
  D. Preceding event #2, #3, etc.

22. TASK and ACTIVITY at TIME OF ACCIDENT
  A. General type of task _____
  B. Specify activity _____

  C. Employee was working ☐ Alone ☐ With crew or fellow worker
                           ☐ Other: specify _____

23. CAUSAL FACTORS. Events and conditions that contributed to the accident. Include those identified by the *Guide for Identifying Causal Factors*.
_____
_____
_____
_____
_____
_____

24. CORRECTIVE ACTIONS. Those that have been, or will be, taken to prevent recurrence. Include those identified by the *Guide for Identifying Causal Factors and Corrective Actions*. Include the name(s) of the person(s) responsible for implementing or taking the corrective action and the date action is to be complete. To close the loop, record the actual date(s) the corrective action was taken.
_____
_____
_____
_____
_____

PREPARED By: _____   REVIEWED By:
TITLE _____   _____
Department _____ Date _____   Supervisor   Date

Team Member Review   _____
Name   Title   Dept.   Date   Supervisor   Date
_____
_____   _____
_____   Supervisor   Date

Figure 14 • Incident/Accident Investigation Report (2 of 2)

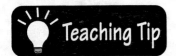

Show examples of OSHA regulations, state construction codes, and city and council planning and building codes

The objective of your investigation is to determine causes and recommend corrective actions. Your objective should not be to find fault or blame someone. Looking for someone to blame only damages your credibility and can influence the nature of information others will give you in the future. If you alienate your employees by blaming them, they will not come to you in the future to report safety issues.

In an accident investigation, you must be objective but thorough. You need to uncover the *when, where, why,* and *how* behind the accident. There are facts that you should establish that will help you to prepare your final report. They are:

- *Physical Description of Accident* – Accident site, actions of injured party, and the nature of the injury or extent of medical needs
- *Contributing Factors* – Work-site conditions such as weather or noise, machinery and/or equipment involved, and the employees or people involved
- *Recommendations for Avoidance* – Nature of recommendation and schedule for implementing recommendation

Obtain information as soon after the accident as possible so that the accident is fresh in the employees' minds. Be sure to interview any witnesses one at a time and practice good listening skills. Remember to use open-ended questions that encourage each witness to talk freely. Ask witnesses for written statements in their own words. Ask them to be detailed in their responses.

Recommendations should be formulated by putting gathered information in good order and basing all statements on known facts. Recommendations should be detailed and should establish definite time limits for making work-site improvements. Be sure to follow up on recommendations to make sure they are implemented. If properly done, an accident investigation provides a wealth of knowledge concerning a given accident. More importantly, it can show employees, supervisors, and managers how to improve their safety performance.

SECTION 10

10.0.0 GOVERNMENT REGULATIONS

As a supervisor, you must keep current on all of the regulatory agencies that govern the work that you do. You are often responsible for making sure that your company is adhering to all of the different regulatory bodies for compliance with regulations and reporting. There are city, county, state, and federal regulations that affect your own company policies and procedures.

10.1.0 City, County, and State Regulations

Cities have regulations that deal with all planning, zoning, building, and code enforcement functions. A typical city might have divisions such as planning, building regulation, and community services. The planning division is responsible for land use decisions, which include the future land use plan of the city, the zoning regulations, review of subdivisions, and site plans. The building regulation division is often responsible for the interpretation and enforcement of the various construction codes adopted by the city. They typically issue building permits and review construction plans that have been submitted with applications for building permits.

Counties also monitor all types of building construction to ensure that they meet minimum code requirements and the expectations of occupants for structural integrity, workmanship, and safety. Counties often receive and review residential and commercial building plans for structural, plumbing, mechanical, and electrical requirements and provide comments and code requirements to builders, developers, and citizens of the county. Counties coordinate and consult with public safety officials, health department, and state building code officials.

States normally have a Construction Codes Board or other regulatory body to oversee state building codes and licensing of contractors and inspectors. The purpose of such regulation is to assure building owners of sound, safe, and sanitary structures, and building users that their health, safety, and welfare are protected. It serves to assure the public that building materials are durable, serviceable, safe, and sanitary.

Instructor's Notes:

Regulation assures contractors and builders of consistent and uniform building standards throughout the state.

Since the structure of these regulatory bodies is different in each area, you should familiarize yourself with your own city, county, and state regulations to make sure that your company policies reflect reporting and regulatory requirements.

10.2.0 The Occupational Safety and Health Act

The field of safety and health was deeply changed when the U.S. Congress passed and the President signed the Williams-Steiger Occupational Safety and Health Act (OSHA) in 1970. Never before had virtually all general industries, including construction, been subjected to federally prescribed mandatory rules for workers' safety and health.

Today, OSHA refers to two things: the law that applies to more than 5 million businesses and over 65 million workers in this country and to the primary government safety agency that establishes and enforces these federal standards for regulating work environments. The OSHA set of rules form the basis for inspections, citations, penalties, and nearly every activity in which OSHA engages.

Included in OSHA are specific requirements directed toward the employer as well as the employee, including the following:

- Each employer shall furnish to his or her employees a place of employment that is free from recognized hazards that are causing or are likely to cause death or serious physical harm to his or her employees.
- Each employer shall comply with the occupational safety and health standards promulgated under this act.
- Each employee shall comply with occupational safety and health standards and all rules, regulations, and orders issued pursuant to this act, which are applicable to his or her own actions and conduct.
- Each employer and employee shall comply with prescribed training requirements.

The construction industry is concerned with Part 1926 of OSHA regulations. However, the construction industry must also comply with certain standards found in Part 1910.

OSHA is not the only federal agency affecting worker safety and industry standards. Other administrations involved in safety efforts include the Mine Safety and Health Administration (MSHA), the Environmental Protection Agency (EPA), and the Consumer Product Safety Commission (CPCS). The authority these agencies exert on your job site depends upon the location of work, type of work, and particular hazards.

10.2.1 Inspections

Originally, OSHA officials could enter a company or other work place at reasonable times by simply presenting basic identification. No court-issued search warrants were required. This situation changed when an Idaho businessman brought suit in a search case and the U.S. Supreme Court ruled in his favor in the Barlow decision of 1978. The Barlow decision allows employers to invoke the fourth amendment of the U.S. Constitution, and it requires OSHA to obtain search warrants to conduct inspections. However, few employees actually feel the need to exercise this right.

To ensure compliance with OSHA standards, inspections are made where and when considered advisable by the agency. Inspections are made without prior notice to the company being inspected.

In general, the following priorities dictate inspections:

- Accidental death or mishap in which five or more individuals are injured
- Verbal or written complaints received by OSHA concerning an imminent hazard
- Routine inspections of industries, including construction, considered especially hazardous
- Routine inspections of other industries
- Follow-up inspections to determine if corrections identified as necessary during an initial inspection have been implemented

It is important to know your own company policy and to adhere to it when responding to or dealing with state, city, local, and federal regulations.

Ask trainees to relate their experiences with OSHA inspections.

Ask trainees to reference any citations witnessed in their workplace and what the company did to prevent a violation in the future.

Inspections conducted by an OSHA compliance officer may be divided into five major parts. They are as follows:

Part 1 *Inspector's Credentials* – An inspection begins when the OSHA compliance officer arrives at the establishment. The compliance officer displays official credentials and asks to meet an appropriate employer representative. Employers should always insist upon seeing the compliance officer's credentials.

Part 2 *Opening Conference* – At the opening conference, the compliance officer discusses with the employer the nature of the inspection and outlines briefly his or her approach to making the inspection. The compliance officer may request the attendance of all appropriate subcontractors as well as representatives of the general contractor.

Part 3 *Selection of Representatives* – An employee representative and an employer representative are selected to accompany the compliance officer on the inspection.

Part 4 *Walk-Around Inspection* – During the inspection, the compliance officer may talk to any employee or listen to any employee regarding work-site safety. The employee representative should make of list of the names of all of the people to whom the compliance officer speaks. In addition, the employee representative should take photos of anything that the compliance officer photographs.

Part 5 *Closing Conference* – The compliance officer conducts a closing conference with management to discuss the conditions and practices observed. The compliance officer discusses all unsafe or unhealthful conditions observed on the inspection and indicates all apparent violations for which a citation may be issued or recommended. The employer is told of appeal rights. The compliance officer does not indicate any proposed penalties. Only the OSHA area director has that authority, and this is done only after a full report is received.

10.2.2 Violations

The compliance officer may specify violations stemming from infractions of the general duty clause of the OSHA act or standards contained in general industry standards, Part 1910, or construction standards, Part 1926. Violations may be in any of the following categories:

- *Other Than Serious Violations* – Violations where safety and health are affected, but they probably would not cause death or serious physical injury.

- *Serious Violations* – Violations where there is substantial probability that death or serious physical injury could result, and the employer knows or should know of the hazard.

- *Imminent Danger* – A situation where there is reasonable certainty that the danger can be expected to cause death or serious physical harm immediately, or before the danger can be eliminated through normal enforcement procedures.

- *Willful Violations* – Violations where the employer intentionally and knowingly commits a violation of the standards or is aware that a hazardous condition exists and makes no reasonable effort to eliminate it.

- *Repeat Violations* – Violations for which an employer has previously been issued a citation.

10.2.3 Citations and Penalties

After the compliance officer reports findings, the area director determines what citations, if any, will be issued. In addition, the area director determines what penalties, if any, will be proposed.

Citations inform the employer and employees of the regulations and standards that have allegedly been violated and of the proposed length of time set for their abatement. The employer will receive citations and notices of proposed penalties by certified mail. The employer must post a copy of each citation at or near the place a violation occurred for three days or until the violations is abated, whichever is longer.

The types of violations that may be cited and the penalties that may be proposed are as follows:

Instructor's Notes:
_____
_____
_____
_____
_____
_____

- *Other Than Serious Violation* – A proposed penalty of up to $7,000 for each violation is discretionary. A penalty for an other-than-serious violation may be adjusted downward by as much as 95 percent, depending on the employer's good faith (demonstrated efforts to comply with the act), history of previous violations, and size of business. When the adjusted penalty amounts to less than $50, no penalty is proposed.
- *Serious Violation* – A mandatory penalty of up to $7,000 for each violation is proposed. A penalty for a serious violation may be adjusted downward, based on the employer's good faith, history of previous violations, the gravity of the alleged violation, and size of business.
- *Willful Violation* – Penalties of up to $70,000 may be proposed for each willful violation, with a minimum penalty of $5,000 for each violation. A proposed penalty for a willful violation may be adjusted downward, depending on the size of the business and its history of previous violations. Usually, no credit is given for good faith.

If an employer is convicted of a willful violation of a standard that has resulted in the death of an employee, the offense is punishable by a court-imposed fine or by imprisonment for up to six months, or both. A fine of up to $250,000 for an individual, or $500,000 for a corporation, may be imposed for a criminal conviction.

Willful violations include:

— *Repeat Violation* – A violation of any standard, regulation, rule or order where, upon re-inspection, a substantially similar violation is found. Repeat violations can bring a fine of up to $70,000 for each such violation. To be the basis of a repeat citation, the original citation must be final; a citation under contest may not serve as the basis for a subsequent repeat citation.

— *Failure to Correct Prior Violation* – Failure to correct a prior violation may bring a civil penalty of up to $7,000 for each day the violation continues beyond the prescribed abatement date.

— *Egregious* – When the hazard is so great it seems intentional. When the egregious policy is applied, the fines will be multiplied by the number of exposed workers. This is the most serious category.

Additional violations for which citations and proposed penalties may be issued upon conviction:

- Falsifying reports, records, or applications can bring a fine of $10,000 or up to six months in jail, or both.
- Violations of posting requirements can bring a civil penalty of up to $7,000.
- Assaulting a compliance officer, or otherwise resisting, opposing, intimidating, or interfering with a compliance officer in the performance of his or her duties is a criminal offense, subject to a fine of not more than $5,000 and imprisonment for not more than three years.

Citations and penalty procedures may differ somewhat in states with their own occupational safety and health programs.

10.2.4 Training Requirements

OSHA requires employers to train employees in the safety and health aspects of their jobs, making it the employer's responsibility to limit certain job assignments to employees who are designated certified, *competent*, or *qualified*. Assigning jobs to workers who have been trained and have demonstrated the ability to complete jobs safely helps to create an accident-free workplace. For this reason, OSHA and other regulatory bodies require that both competent and qualified personnel be designated to perform certain tasks on the job site that may be dangerous.

OSHA Standard 29, Part 1926, *Safety and Health Regulations for Construction* outlines specific training topics that are mandatory for construction specialties such as rigging equipment, noise measurement, lead exposure, powered platforms, electrical installation, fall protection, and excavations. While OSHA sets requirements for training workers in these areas and standards for the levels of training needed to perform specific tasks, the employer is ultimately responsible for interpreting and implementing the regulations and designating personnel as competent or qualified.

A **competent person** is defined as a person who is capable of identifying existing and predictable hazards in the surroundings or working conditions which are unsanitary, hazardous, or dangerous to employees, and who has authorization to take prompt corrective measures to eliminate them.

A **qualified person** is defined as one who, by possession of a recognized degree, certificate, or professional standing or by extensive knowledge, training, and experience, has successfully demonstrated the ability to solve or resolve problems relating to the subject matter, the work, or the project.

Companies must demonstrate that they provide the training described by OSHA or send workers to training offered by third parties. In any case, the employer is responsible for making sure that workers are properly trained, the training is documented, and the workers that they designate as competent or qualified fulfill the requirements for each standard. Workers can earn competent or qualified designations within their company once they have completed training and have demonstrated the ability to carry out the roles as defined by OSHA.

Project supervisors need to recognize that these safety requirements exist and to be able to understand the difference between competent and qualified designations. The application of these regulations varies with each job and within each specific craft.

SUMMARY

Supervisors have the greatest influence on employee safety practices because they most directly influence the employees.

Construction industry safety figures indicate that, even with improved equipment, techniques, and safety awareness, extensive injuries and property damage continue to occur, resulting in tremendous economic losses. Specifically, these losses contribute to the decline of company profits and place many companies in real jeopardy. The increase in insurance rates of all types is producing a cost burden that many organizations cannot continue to bear.

Many companies are implementing safety programs to combat the continued rise in accident rates. The involvement of all members of a company in a coordinated effort to reduce the accident and injury rate has placed a new emphasis on safety programs. In addition, safety in the work environment has become a standard part of most construction contracts, and the role of the supervisor in conducting employee safety training, site inspections, and accident investigations has expanded.

Requirements of federal and state laws have become more intense and industry specific as governmental agencies attempt to reduce accident rates through legal means. City and county governments have more recently added their emphasis on improved safety and health practices in the form of local ordinances.

All of these efforts to improve the health and safety of individuals have their greatest chance for success when competent, caring, and attentive supervisors perform their most important responsibility — providing leadership in job safety. Through leadership, a supervisor can develop qualified and competent personnel that understand the importance of safety.

Instructor's Notes:

_____
_____
_____
_____
_____
_____

Review Questions

1. An unplanned event that may or may not result in personal injury or property damage is ____.
   a. an accident
   b. a near miss
   c. a milestone
   d. a fatality

2. An example of an indirect or uninsured accident cost is ____.
   a. production delays
   b. workers' compensation
   c. medical costs
   d. equipment damage

3. An example of a direct or insured accident cost is ____.
   a. the EMR
   b. damaged equipment
   c. production delays
   d. negative publicity

4. Studies show that the greatest number of accidents is in the ____ category.
   a. property damage
   b. minor injury
   c. serious or disabling injury
   d. near miss

5. The cost of implementing an effective safety program ____.
   a. provides a return on investment
   b. can never be made up
   c. promotes business failures
   d. is an indirect expense to the company

6. ____ cause(s) the greatest adverse impact on a company's EMR.
   a. One high-cost injury
   b. Ten lower cost but lost-time injuries
   c. A near miss
   d. Ten near misses

7. What percentage of the nation's construction workforce is estimated to have an alcohol or drug abuse problem?
   a. 5%
   b. 10%
   c. 20%
   d. 30%

8. Workers who have been on the job less than 30 days account for what percentage of the total construction injuries?
   a. 10%
   b. 25%
   c. 35%
   d. 60%

9. What is the main reason for investigating accidents and injuries?
   a. To prevent a recurrence
   b. To meet insurance requirements
   c. To document facts in case of lawsuits
   d. To satisfy government regulations

10. Each of these is an element that makes up a safety program *except* _____.
    a. hiring procedures
    b. standard operating procedures and policies
    c. employee screening and placement
    d. responsibility assignments

11. _____ are informal safety training sessions designated to continuously reinforce the company safety program.
    a. Drug and alcohol dialogues
    b. Quick tips
    c. Safety suggestions
    d. Toolbox talks

12. Because of its complexity and importance, safety training _____.
    a. should never be hurried
    b. should always be given in groups
    c. should not be handled by the supervisor
    d. should only be given to experienced employees

13. _____, as a part of safety training, includes identification of both unsafe acts and unsafe conditions.
    a. CPR first check reference
    b. Hazard recognition
    c. Human Resources orientation
    d. An OSHA log

14. An example of a typical unsafe act is _____.
    a. trying something new
    b. wearing ear plugs
    c. operating at an unsafe speed
    d. reading safety procedures during work hours

15. An example of a typical unsafe condition is _____.
    a. a high ladder
    b. cluttered floor and work areas
    c. a work area cooled with a fan
    d. employees taking a break

16. The Five "Ps" help you to organize your safety meetings. One P reminds you not to try to cover too much ground, concentrating on one safety rule, one first aid hint, one unsafe practice, or one main idea. The P described here is _____.
    a. purchase
    b. prepare
    c. peruse
    d. pinpoint

17. The Five "Ps" help you to organize your safety meetings. One P reminds you, when closing your safety talk, to answer the question the listeners always have — "So what?" Tell them what to do. Ask for special action. Give a prescription. The P described here is _____.
    a. prescribe
    b. prepare
    c. picture
    d. personalize

18. Each of the following is an OSHA requirement for employers to provide their employees *except* _____.
    a. a hazard-free workplace
    b. prescribed training
    c. an "equal opportunity" workplace
    d. health standards

Instructor's Notes:

19. Each of the following is a reason why OSHA might inspect a work site *except* _____.
    a. written or verbal complaints concerning an imminent hazard
    b. routine inspection of an industry
    c. follow-up inspection after an initial inspection
    d. to determine environmental impact

20. When your work site is having an OSHA inspection, what happens after the opening conference?
    a. Selecting of representatives
    b. Closing conference
    c. Inspection of inspector's credentials
    d. Walk-around inspection

Instructor's Notes:
_____
_____
_____
_____
_____
_____
_____

Trade Terms Introduced in This Module

Accident: An unplanned and sometimes injurious or damaging event (release of energy) which interrupts the normal progress of an activity. Events that do not result in property damage or personal injury are sometimes called *incidents, near misses* or *near hits*.

Assigned risk pool: The higher-priced insurance category for companies who have poor accident experience and who must pay higher insurance rates.

Audit: A methodical examination or review. Safety audits involve a review of certain safety policies and procedures to see if they are adequate and are actually being implemented.

Competent person: A person who is capable of identifying existing and predictable hazards in the surroundings or working conditions which are unsanitary, hazardous, or dangerous to employees, and who has authorization to take prompt corrective measures to eliminate them.

Direct or insured costs: Costs that can be directly related to an accident such as medical costs, worker's compensation insurance benefits, and liability and property damage insurance payments.

Experience modifier (EMR): A numeric factor used in determining workmen's compensation costs. It rises for contractors with poor accident experience and falls for those with good experience or fewer accidents.

Experience rating: A method of modifying future workmen's compensation insurance premiums by comparing a particular company's actual losses to the losses normally expected for that company's type of work

Hazard: A condition, changing set of circumstances, or behavior that presents a potential for injury, illness, or property damage.

Job hazard analysis: A methodology used to identify, track, and remediate job hazards.

Job safety analyses (JSAs): Procedures used to review job methods and uncover hazards.

Qualified person: A person who, by possession of a recognized degree, certificate, or professional standing or by extensive knowledge, training and experience, has successfully demonstrated the ability to solve or resolve problems relating to the subject matter, the work, or the project.

Risk: A measure of both the probability and the consequences of all hazards of an activity or condition.

Safety coordinator: A person designated on a crew of more that five people to be responsible for maintaining safety quality standards on the job. May be a part-time role.

Standard operation procedures (SOPs): A specific plan or list of processes and methods designed to accomplish a task.

Task hazard analysis: A pre-job safety briefing and analysis that helps workers to identify the work to be done, the tools to be used, and any potential hazards that they anticipate.

Toolbox talks: Informal safety training sessions designed to continuously reinforce the company's safety program.

Unsafe act: A behavioral departure from an accepted, normal, or correct procedure or practice which involves an unnecessary exposure to a hazard or conduct that reduces the degree of safety.

Unsafe conditions: Physical states that deviate from that which is acceptable, normal, or correct and which result in a reduction in the degree of safety normally present.

Have trainees read the Glossary for unfamiliar terms. Answer trainee questions.

Have trainees prepare for the Module Examination.

Administer the Module Examination. Be sure to record the results of the Exam on Craft Training Report Form 200 and submit the results to the Training Program Sponsor.

Instructor's Notes:
_____
_____
_____
_____
_____
_____

NATIONAL CENTER FOR CONSTRUCTION EDUCATION AND RESEARCH

APPENDIX

Construction Job Checklist

A check in the "No" column indicates non-conformance to a general policy or safety standard.

User is cautioned that this checklist does not apply to every type of operation and therefore does not list those items covering every hazard relating to a physical or occupational disease exposure. It contains those items where OSHA citations have been issued most frequently.

General	Yes	No	Violation Corrected
Pre-job safety meeting held with subcontractor(s).			
Monthly safety meetings held with supervisor and subcontractors.			
Competent contractor personnel assigned responsibility to inspect job site for safety.			
U.S. Department of Labor "Safety and Health Protection on the Job" poster posted.			
Management safety policy directives posted.			
afety bulletins, accident prevention material posted.			
Copies of log of employee injuries at job site.			
Copies of supplemental Record of Job Injuries posted when required.			
Copies of Worker Injury Investigation Report kept at job site.			
Minutes of job site safety meetings recorded and kept at job site.			
Safety inspection reports by contractor personnel prepared and kept at job site.			

Sanitation, Miscellaneous	Yes	No	Violation Corrected
Portable containers used to dispense drinking water (required types).			
Where single service cups are used for drinking water, containers are provided for new and used cups.			
Adequate number of toilet facilities provided.			
Special washing facilities provided for workers handling materials that may be harmful to their health.			
Illumination in work areas (including offices, shops) adequate.			

Sanitation, Miscellaneous (Con't.)	Yes	No	Violation Corrected
Required portable fire-fighting equipment available, properly located and maintained.			
Approved metal safety cans used for handling and use of flammables.			
In areas where flammables are stored or where operations present a fire hazard, "No Smoking or Open Flame" sign posted.			
Indoor and outdoor storage of flammables in approved containers or cabinets with warning signs posted.			
Form and scrap lumber, and all other debris kept clear from work areas.			
Combustible scrap and debris removed from work areas at regular intervals.			
Containers provided for collection, separation of waste, trash, oily and used rags.			
Solvent waste, oily rags, and flammable liquids kept in fire-resistant covered containers until removed from worksite.			

Personal Protection	Yes	No	Violation Corrected
Hearing protective devices provided for and worn by workers where noise levels are excessive.			
Hard hats provided for and worn by workers.			
Eye and face protection provided for and worn by workers where exposed to potential eye or face injury.			
Workers required to wear footwear adequate for their assigned work.			
Respiratory protective equipment, provided and worn when workers exposed to harmful dust, fumes, gases.			

Hand and Power Tools	Yes	No	Violation Corrected
Hand-held powered tools (saws, air impact) equipped only with constant pressure switch.			
Hand-held power tools (drills, tappers, fastener drivers, disc and belt sanders, angle grinders) provided with momentary contact "on-off" switch with lock-on control only if turnoff is by single motion of same finger/fingers that turn it on.			
Devices provided on air power tools to prevent tools from becoming accidentally disconnected from hose.			
Air-driven nails operating more than 100 psi provided with safety device on the muzzle to prevent ejection unless muzzle in contact with work surface.			
Air hose connections secured across each such connection between air receiver and tool.			

Instructor's Notes:

Hand and Power Tools (Con't.)	Yes	No	Violation Corrected
Airless spray guns (operating at 1000 psi) equipped with safety device preventing pulling of trigger until safety device is manually released.			
Powder-actuated tools provided with safety shield/guard and operator has evidence of special training in their use.			
Portable power circular saws provided with proper functioning automatic return lower guard and fixed upper guard.			
All fixed power woodworking tools provided with a disconnect switch that can be locked or tagged in the OFF position.			
Defective tools, equipment tagged as unsafe, controls locked in OFF position, or physically removed from job site.			

Scaffolding	Yes	No	Violation Corrected
All open sides and ends of platforms more than 10' above ground on floor level, provided with top rails, 42" high midrails, toeboards (4" high).			
Where workers pass or work under scaffolds, screens, or other overhead protection provided.			
Platform planks laid together tight preventing tools, etc. from falling through.			
Scaffold guyed or tied to structure.			
Scaffolding set plumb with adequate foundation bearing plates.			
Planks secured to prevent dislodgment.			
Overhead protection on scaffolds where workers exposed to overhead hazards.			
Ladders used to gain access to scaffold work platforms.			
Horse scaffolds not more than two tiers or 10' in height.			
Horse scaffolds spaced not more than 4' for medium duty; not more than 8' for light duty.			
Top rails, midrails, toeboards provided on horse scaffolds 10' above ground or floor level.			
Rolling Scaffolds (Manually Propelled)			
Wheel brakes set while in use.			
No riders on work platform while moving.			
Work levels 10' or more above ground or floor level have guardrails and toeboards.			
All cross and diagonal bracing in place and properly connected.			
Height does not exceed 4 times least base dimension unless outriggers are used.			
Ladder access to work platform.			

Scaffolding (Con't.)	Yes	No	Violation Corrected
Swinging Scaffolds – 2 Point Suspension			
Roof hooks or irons securely installed and anchored to a sound structural part of the building. (Beware of ornamental parapets.)			
Sheaves fit size of rope used.			
Equipped with standard guard rails and toe boards.			
Load limit marked on each scaffold.			
Pump Jack Scaffolds			
Poles not greater than 30' high			
Poles rigidly supported at top and bottom by triangle bracing at midpoints as necessary.			
Poles bear on firm, level foundation.			
No more than two persons on the scaffold between supports.			
Equipped with standard guardrail or workers have safety belts attached to lifelines.			

Electrical	Yes	No	Violation Corrected
For power circuits, exposed or concealed, where accidental contact by tool/equipment may be hazardous, warning signs posted and all workers advised of hazard. Regular inspections made to assure effective grounding of non current-carrying metal parts of portable and/or plug connected equipment or GFCIs installed on all 110-120V temporary circuits.			
Temporary lights equipped with guards to prevent accidental contact with bulb.			
Receptacles, attachment plugs not interchangeable on circuits of different voltages, current ratings or types of current (AC or DC).			

Ladders	Yes	No	Violation Corrected
Ladders regularly inspected and destroyed when found defective.			
Side rails extend 36" above landing or provision of grab rails.			
Top of ladders tied-in to prevent displacement.			
Double cleat ladders provided for working area having 25 or more workers and two-way traffic is expected.			
Double cleat ladders not exceeding 24' in length.			
Single cleat ladders not exceeding 30' in length.			
Cleats inset into edges of side rails or filler blocks used.			

Instructor's Notes:

_____
_____
_____
_____
_____
_____

Module MT203 ♦ Appendix

Ladders (Con't.)	Yes	No	Violation Corrected
All job-built ladders constructed to conform to standards.			
No metal ladders used within 10' of electrical lines.			
Stepladders used only in full open position.			
Stepladders of sufficient height so that the top two steps do not have to be used to perform work.			
All manufactured single and extension ladders equipped with ladder shoes.			

Cranes	Yes	No	Violation Corrected
Annual inspection made of crane hoisting machinery and record kept of dates and results.			
On tower cranes, daily inspection made of rigging.			
On tower cranes, weekly inspections made of supporting parts, safety appliances and record kept of repairs required.			
No part of a crane or its load closer to energized electrical lines than the following distances: for lines rated 50 KV or below–10'; for lines rated over 50 KV–10' plus.			
0.4 inch for each 1 KV over 5 KV or twice the length of the line insulator not less than 10'.			
Cranes in transit with no load and boom lowered, equipment clearance not less than 6' for voltages less than 50 KV; 10' for voltages over 50 KV to 345 KV; and 16' for voltages up to 750 KV.			

Excavations, Trenching & Shoring	Yes	No	Violation Corrected
Excavated material effectively stored and retained at least 2' or more from edge of excavations.			
Utility company contacted and advised of proposed excavation work to determine underground utility exposures or when overhead power lines are involved.			
Substantial stop logs or barricades when mobile equipment working adjacent to excavation.			
Trenches over 5' in depth shored to standard, laid back to stable slopes or provided with other equivalent protection where hazard of moving ground exists.			
Trenches less than 5' protected where hazardous ground movement exists.			
Trenches in hard or compact soil, trench above the 5' level sloped (slope not steeper than a 1' rise to each 1/2' horizontal).			
Trenches 4' deep or more provided with ladder located no more than 25' of lateral travel.			

Welding, Cutting, Burning	Yes	No	Violation Corrected
Proper personnel protective equipment provided, in good condition and being used.			
All oxygen and acetylene gauges in working condition.			
Suspension lines inspected before first use and frequently during use.			
Each employee equipped with safety belt attached to lifelines.			
Lifeline attached at top to independent support and extending to ground level.			
Suspension lines capable of supporting six times rated load.			
Power cables for electric hoist machines secured to platform to prevent snagging or excess tension and located so as to prevent snagging between power supply point and platform.			

Floor and Wall Openings	Yes	No	Violation Corrected
Wall openings (30" high, 18" wide or greater) from which there is a drop of 4 feet or more and bottom of opening less than 3' above working surface provided with guardrails.			
Bottom of wall openings less than 4" above work surface provided with standard toeboard (4" high).			
Open-sided floors 6' or more above floor or ground level provided with standard railing and toeboard or other equivalent perimeter protection.			
Stairways when used during construction have handrails on all open sides, guardrails at landings, and filler blocks in all recessed treads.			
During construction, stairways provided with standard railings and guardrails at floor opening.			

Material-Handling Equipment	Yes	No	Violation Corrected
Approved canopy guards on forklift trucks and rollover protection on all earth-moving equipment.			
Rated capacity posted on all lifting, hoisting equipment, clearly visible to the operator.			
Where mobile cranes are regularly used, daily inspections made of critical items and records kept.			
All oxygen and acetylene hoses in good condition and free of grease and oil.			
Mechanical lighters used for lighting torches - no cigarette lighters or matches.			
Oxygen and acetylene cylinders stored in upright position in designated areas with caps in place.			

Instructor's Notes:

___
___
___
___
___
___

Material-Handling Equipment (Con't.)	Yes	No	Violation Corrected
Oxygen and acetylene hoses properly located to avoid damage by moving equipment or creating a tripping hazard.			
Electric arc welding cables in good condition and properly attached by lugs to welding machine.			
Rod holders in good condition.			
Shields provided to protect other workers from flash burns.			
Welders' shields and helpers' goggles in good condition and equipped with proper lenses.			
Fire extinguishers provided within 25' of welding, cutting, or burning operations.			
All flammable material removed from welding, cutting, and burning operations area.			
Protection provided to prevent slag, etc. from falling on workers below.			
All welding cable positioned to eliminate tripping hazards.			

Roofing	Yes	No	Violation Corrected
Pitched roofs over 4:1 pitch.			
Safety belts and lifelines or catch platform.			
Roofing bracket scaffolds secured.			
Crawl or chicken ladders secured and with evenly spaced cleats.			
Ladders extend 36" above eaves and secured top and bottom.			
Flat and low pitch roofs less than 4:1 pitch			
Warning line 6' from roof edge in the work area.			
Protective equipment when handling hot pitch.			
Ladder access same as for pitched roofs.			
Materials stored 6' from edge.			
All openings covered or protected by guardrails.			

Concrete Forming and Pouring	Yes	No	Violation Corrected
Vertical re-steel protected by covering when employees working above.			
Employees wearing safety belts or equivalent when 6' or more above adjacent work surfaces.			
Walk and standing boards when pouring horizontal surfaces on re-steel.			

Concrete Forming and Pouring (Con't.)	Yes	No	Violation Corrected
Employees provided with and using personal protective equipment while pouring concrete.			
No riding concrete buckets.			
All pump concrete lines secured at all joints.			
Shoring and re-shoring adequate for support.			
Properly guarded work platform for walls, columns, beams, etc.			
Power troweling machines equipped with positive on/off switch.			

Steel Erection	Yes	No	Violation Corrected
Decking or safety nets 2 stories or no more than 30 ft. where erection is being performed.			
No riding of steel or the hook.			
Safety railing of no less than 1/2' wire rope 42" high around perimeter of all temporary planked or metal decked floors.			
Employees working on float scaffolds provided with safety belts.			

Demolition	Yes	No	Violation Corrected
Dust controlled by wetting.			
Employees provided with dust respirators and goggles.			
Chutes properly erected and drop area barricaded off.			
Floor openings protected.			
Employee access to building provided with overhead protection.			
Stairways in building used for access properly lighted and maintained.			

Blasting	Yes	No	Violation Corrected
Certified blaster in charge.			
Storage and handling of explosives conforms to Federal requirements and all local ordinances.			
Excess explosives, caps, det-cord, etc., removed to approved storage area prior to blast.			
Standard warning system used.			
Two-way radios turned off.			

Instructor's Notes:
_____
_____
_____
_____
_____
_____

Blasting (Con't.)	Yes	No	Violation Corrected
Signs posted.			
No other work within 110' of loading area.			
Wood poles used for tamping.			
Blasting machines only used in electric blasting.			
Waste explosives and containers disposed of as required by regulations.			
Record of all explosives kept.			
Post blast inspection completed.			
Misfires handled as specified by regulation.			
Access to work area controlled.			

Additonal Resources

This module is intended to present thorough resources for management training. The following reference works are suggested for further study. These are optional materials for continued education rather than for management training.

National Safety Council

The following references are available from the National Safety Council, 1121 Spring Lake Dr., Itasca, IL 60143-3201. http://www.nsc.org.

Accident Prevention Manual for Business and Industry—Administration and Programs, Eleventh Edition

Accident Prevention Manual for Industrial Operations, Administration, and Programs

Accident Prevention Manual for Industrial Operations, Engineering, and Technology

Supervisor's Safety Manual

Fundamentals of Industrial Hygiene.

Other Resources

- *Job Hazard Analysis,* U.S. Department of Labor—OSHA Publication No. 3071
- *Job Safety Analysis,* Safety Manual No. 5, U.S. Department of the Interior, Mining Enforcement and Safety Administration.
- http://www.AFL-CIO.org
- http://www.buildingtrades.org

The booklet *Record Keeping Guidelines for Occupational Injuries and Illnesses* provides detailed information on record keeping. It is available from local OSHA offices or from the U.S. Department of Labor. The address is:

U.S. Department of Labor—BLS
Inquires and Correspondence
Room 2831 A
GAO Building
Washington, D.C. 20212
Telephone (202) 523-1221

Other information relating to OSHA standards and regulations is listed in OSHA Booklet 2019. It may be obtained from local OSHA offices or from the U.S. Department of Labor. The address is:

U.S. Department of Labor—OSHA
Publications Office, Room N3101
200 Constitution Avenue, NW
Washington, D.C. 20210

The National Center for Construction Education and Research offers an extensive range of safety training programs and books. For more information contact us at:

NCCER
P.O. Box 141104
Gainesville, FL 32614-1104
Telephone (352) 334-0911
http://www.nccer.org

Instructor's Notes:

MOD MT203-01—TEACHING TIPS

The following are suggested activities or instructional methods to help you teach the material in this AIG.

Section 5.1.0 **Employee Safety Orientation**

Remind trainees that when they are assigned a new task, their supervisor should indicate the potential hazards and safety procedures. If they fail to do so, the trainee should share the responsibility and ask.

Section 5.2.0 **Job Safety Training**

Substitute an OSHA 200 log with an accident investigation report or a job-site safety inspection form. Make sure the form you use is a transparency so all trainees can view the exercise.

NATIONAL CENTER FOR CONSTRUCTION EDUCATION AND RESEARCH

MODULE MT203

Answers to Review Questions

	Answer	Section Reference
1.	a	3.1.0
2.	a	3.6.0
3.	b	3.6.0
4.	d	3.2.0
5.	a	3.6.3
6.	b	3.6.2
7.	c	4.2.0
8.	b	4.6.0
9.	a	4.10.0
10.	a	4.6.0
11.	d	4.7.0
12.	a	5.2.0
13.	b	5.3.0
14.	c	5.3.1
15.	b	5.3.2
16.	d	7.2.0
17.	a	7.2.0
18.	c	10.2.0
19.	d	10.2.1
20.	a	10.2.1

National Center for Construction Education and Research

MODULE MT203

Participant Activities

Participant Activity Section 3.6.2
WC Rates
    Company A $ 89,505.20
    Company B $ 71,604.16
    Your Company $143,208.32

Stan's Accident
1.   Total direct costs
        Medical clinic charges $636.00

2. Total indirect costs
    List expenses related to Lost Time on the Job and equipment rental:

Mack & crew 4.5 hours	294.84
Mack 2 hours (report)	38.04
Crew of 3 waiting 4 hours	188.00
Insurance clerk (report)	10.00
Stan's pay (day of)	140.00
Stan's lost productivity (5 days; 25%)	$ 175.00
Dave & Roy overtime	1,740.00
Total Wages	$2,585.88
× 38% fringe benefits =	982.63
Total	$3,568.51
Add Equipment Rental	230.00
Total Cost	**$3,798.51**

CONTREN™ LEARNING SERIES — USER UPDATES

The NCCER makes every effort to keep these textbooks up-to-date and free of technical errors. We appreciate your help in this process. If you have an idea for improving this textbook, or if you find an error, a typographical mistake, or an inaccuracy in NCCER's Contren™ textbooks, please write us, using this form or a photocopy. Be sure to include the exact module number, page number, a detailed description, and the correction, if applicable. Your input will be brought to the attention of the Technical Review Committee. Thank you for your assistance.

Instructors – If you found that additional materials were necessary in order to teach this module effectively, please let us know so that we may include them in the Equipment/Materials list in the Instructor's Guide.

Write: Curriculum Revision and Development Department
National Center for Construction Education and Research
P.O. Box 141104, Gainesville, FL 32614-1104

Fax: 352-334-0932

E-mail: curriculum@nccer.org

Craft _____ Module Name _____

Copyright Date _____ Module Number _____ Page Number(s) _____

Description _____

_____

_____

_____

_____

(Optional) Correction _____

_____

_____

_____

(Optional) Your Name and Address _____

_____

_____

_____

Project Supervisor

Module MT204-01

Quality Control

**Quality Control
Instructor's Guide**

Module MT204

MODULE OVERVIEW

This module introduces the project supervisor trainee to quality control. This module will enable the trainee to understand and implement a system of quality control on the job site.

PREREQUISITES

There are no prerequisites for the module; however, prior to training with this module, it is recommended that the trainee complete the following modules:
  Project Supervision, Modules MT201 through MT203

LEARNING OBJECTIVES

Upon completion of this module, the trainee will be able to:

1. Define quality control.
2. Explain the difference between traditional and total quality control systems.
3. Discuss how quality control and safety go hand-in-hand during construction.
4. Explain the supervisor's responsibility for quality control.
5. Explain the benefits of implementing a quality control system.

PERFORMANCE OBJECTIVES

This is a knowledge-based module – there is no performance profile examination.

NCCER STANDARDIZED TRAINING PROGRAM

The National Center for Construction Education and Research (NCCER) provides a standardized national program of accredited craft training. Key features of the program include instructor certification, competency-based training, and performance testing. The program provides trainees, instructors, and companies with a standard form of recognition through a National Craft Training Registry. The program is described in full in the Guidelines for Accreditation, published by the NCCER. For more information on standardized craft training, contact the NCCER by writing us at P.O. Box 141104, Gainesville, FL 32614-1104; calling 352-334-0911, or e-mailing mail info@nccer.org. More information may be found at our Web site at www.nccer.org.

HOW TO USE THIS ANNOTATED INSTRUCTOR'S GUIDE

Each page presents two sections of information. The larger section displays each page exactly as it appears in the Trainee Module. The narrow column ties suggested trainee and instructor actions to each page and provides icons to call your attention to material, safety, audiovisual, or testing requirements. The bottom of each page includes space for your notes.

 If you see the Teaching Tip icon, that means there is a teaching tip associated with this section. Also refer to the suggested teaching tips at the end of the module.

PREPARATION

Before teaching this module, you should review the Module Outline, Learning Objectives, and the Materials and Equipment List. Be sure to allow ample time to prepare your own training or lesson plan and gather all required equipment and materials.

MATERIALS AND EQUIPMENT LIST

Materials:

Transparencies

Markers/chalk

Module Examinations*

Samples of formal and informal quality control, safety, and document control programs**

Equipment:

Overhead projector and screen

Whiteboard/chalkboard

*Located in the Test Booklet packaged with this Annotated Instructor's Guide.
**If available on loan from your workplace or other resource.

ADDITIONAL RESOURCES

This module is intended to present thorough resources for task training. The following reference works are suggested for both instructors and motivated trainees interested in further study. These are optional materials for continued education rather than for task training.

Construction Materials and Building Publication No. 91, International Organization for Standardization (ISO), ISO Online Catalog at www.iso.ch.

Professional Construction Management: Including Contracting C M, Design-Construct, and General Contracting, 1991. Donald S. Barrie and Boyd C. Paulson (Contributor). Upper Saddle River, NJ: McGraw-Hill Higher Education.

Construction Management, 1997. Daniel W. Halpin and Ronald W. Woodhead. New York: John Wiley & Sons.

Construction Operations Manual of Policies and Procedures, 2000. Andrew Civitello, Jr. Upper Saddle River, NJ: McGraw-Hill Professional Book Group.

TEACHING TIME FOR THIS MODULE

An outline for use in developing your lesson plan is presented below. Note that each Roman numeral in the outline equates to one session of instruction. Each session has a suggested time of 2 1/2 hours. This includes 10 minutes at the beginning of each session for administrative tasks and one 10-minute break during the session. Approximately 5 hours are suggested to cover *Quality Control*.

Topic	Planned Time
Session I. Introduction to Quality Control	
A. Introduction	_____
B. Defining Quality Control	_____
C. Managing Quality Control	_____
D. Types of Quality Control	_____
1. Traditional Quality Control	_____
2. Total Quality Control	_____
3. Quality Circles	_____
Session II. Organizing for Quality and Safety	
A. Quality and Safety Concerns in Construction	_____
1. Direct and Indirect Costs	_____
2. Controls in the Design and Planning Stage	_____
3. Safety in the Design and Planning Phase	_____
B. Organizing for Quality and Safety	_____
1. Project Meetings	_____
2. Document Control	_____
3. Quality Assurance Checklists	_____
4. Inspection/Compliance	_____
C. Summary	_____
1. Summarize module	_____
2. Answer review questions	_____
D. Module Examination	_____
1. Trainees must score 70% or higher to receive recognition from the NCCER.	_____
2. Record the testing results on Craft Training Report Form 200 and submit the results to the Training Program Sponsor.	_____

Quality Control

Instructor's Notes:

ACKNOWLEDGMENTS

The NCCER wishes to acknowledge the dedication and expertise of Roger Liska, the original author and mentor for this module on quality control.

Roger W. Liska, Ed.D., FAIC, CPC, FCIOB, PE

Clemson University

Chair & Professor

Department of Construction Science & Management

We would also like to thank the following reviewers for contributing their time and expertise to this endeavor:

J.R. Blair

Tri-City Electrical Contractors
An Encompass Company

Mike Cornelius

Tri-City Electrical Contractors
An Encompass Company

Dan Faulkner

Wolverine Building Group

David Goodloe

Clemson University

Kevin Kett

The Haskell Company

Danny Parmenter

The Haskell Company

Course Map

This course map shows all of the modules of the *Project Supervision* curriculum. The suggested training order begins at the bottom and proceeds up. Skill levels increase as you advance on the course map. The local Training Program Sponsor may adjust the training order.

PROJECT SUPERVISION

- MT208 — RESOURCE CONTROL AND COST AWARENESS
- MT207 — PLANNING AND SCHEDULING
- MT206 — DOCUMENT CONTROL AND ESTIMATING
- MT205 — CONTRACT AND CONSTRUCTION DOCUMENTS
- MT204 — QUALITY CONTROL ⇐ YOU ARE HERE
- MT203 — SAFETY
- MT202 — HUMAN RELATIONS AND PROBLEM SOLVING
- MT201 — ORIENTATION TO THE JOB

Instructor's Notes:

NATIONAL CENTER FOR CONSTRUCTION EDUCATION AND RESEARCH

MODULE MT204

TABLE OF CONTENTS

1.0.0	**INTRODUCTION**	4.1
2.0.0	**DEFINING QUALITY CONTROL**	4.2
3.0.0	**MANAGING QUALITY CONTROL**	4.3
4.0.0	**TYPES OF QUALITY CONTROL**	4.4
4.1.0	Traditional Quality Control	4.4
4.2.0	Total Quality Control	4.4
4.3.0	Quality Circles	4.4
5.0.0	**QUALITY AND SAFETY CONCERNS IN CONSTRUCTION**	4.5
5.1.0	Direct and Indirect Costs	4.5
5.2.0	Controls in the Design and Planning Stage	4.5
5.3.0	Safety in the Design and Planning Phase	4.6
6.0.0	**ORGANIZING FOR QUALITY AND SAFETY**	4.6
6.1.0	Project Meetings and Project Communication	4.6
6.2.0	Document Control	4.6
6.3.0	Quality Assurance Checklists	4.6
6.4.0	Inspection/Compliance	4.8
	SUMMARY	4.8
	REVIEW QUESTIONS	4.9
	GLOSSARY	4.11

LIST OF FIGURES

Figure 1 • Simple QC Checklist	4.3
Figure 2 • Sample Checklist	4.7

Instructor's Notes:

NATIONAL CENTER FOR CONSTRUCTION EDUCATION AND RESEARCH

MODULE MT204

Quality Control

Ensure that you have all the necessary materials to teach the course. Check the Materials and Equipment list at the front of the module.

Ask trainees for examples of the standards they have noticed on specifications and drawings.

Show Transparency 1 (Course Objectives).

Assign reading of Module MT204, Sections 1.0.0 – 4.3.0.

OBJECTIVES

Upon completion of this module, you will be able to do the following:

1. Define quality control.
2. Explain the difference between traditional and total quality control systems.
3. Discuss how quality control and safety go hand-in-hand during construction.
4. Explain the supervisor's responsibility for quality control.
5. Explain the benefits of implementing a quality control system.

SECTION 1

1.0.0 INTRODUCTION

Implementing a system of **quality control** (QC) for construction creates pride in work, gains favorable recognition for the company, and saves the contractor money. To successfully maintain quality control on a project, the supervisor must know the standards required by the company/contractor, the industry, and the customer.

Quality is the one standard that applies to every job. It is specified in the drawings and building specifications in terms varying from the make and model of air handlers to the strength of concrete used in footings and the number of coats of paint on walls and ceilings. Typical contract documents contain hundreds of such quality standards.

Quality is also important in how work is performed. The supervisor is responsible for making sure all work is performed effectively and efficiently. Prior to starting a job activity and while the activity is being performed, the supervisor must make certain that tasks are completed in accordance with the job plan and schedule.

The supervisor and the crew must perform all work in strict accordance with the drawings and specifications. This requires that the supervisor be familiar with the specifications and drawings prior to starting the job and forbids changing these documents without prior approval by the project's designated authority.

Copyright © 2003 National Center for Construction Education and Research, Gainesville, FL 32614-1104. All rights reserved. No part of this work may be reproduced in any form or by any means, including photocopying, without written permission of the publisher.

Ask trainees to give examples of how they incorporated quality control in a decision they made to control resources, complete work on time, and remain within the estimate.

See the teaching tip at the end of this module.

To support the supervisor's efforts, the company should develop and maintain an effective quality management program. Many companies have developed written quality control/assurance policies and programs. Supervisors should determine if such programs exist in their company and become familiar with their role in carrying out the programs in the field. If the company has quality-related teams or committees, the supervisor, if a member, should participate positively to help solve quality-related problems.

Whether or not a company has a formal quality management program, the supervisor is still responsible for seeing that a quality job is completed. This demands not only adherence to drawings and specifications but adequate training of those workers who do the construction. For example, a carpenter must know how to measure, cut, and fit a quality joint that will hold up under specified loads. This knowledge is not addressed directly by the contract documents, but it is certainly implied. It is up to the supervisor to make sure that all work is done by trained and experienced craftworkers using up-to-date and correct methods, tools, and equipment.

Supervisors should never attempt to build the job their own way, especially if construction methods are described in the drawings or specifications. On jobs in which supervisors have little or no experience, they should check with their supervisors prior to starting work regarding the methods they plan to use. If they find the drawings and/or specifications to be unclear, they should not start any work until their supervisor answers questions and clarifies instructions.

A supervisor should never fall into the trap of assuming that the next job can be done exactly like the last one. Too often, this assumption leads to unnecessary rework that drastically delays job progress and dramatically increases project costs. Quality control must be part of every decision the supervisor makes to control resources, to complete the work on time and within budget.

SECTION 2

2.0.0 DEFINING QUALITY CONTROL

Many people think that proof of quality control lies only in the work being accepted by the designated contracting party, such as the owner, architect, or construction manager. They may believe that acceptance makes the completed work a quality product, but this is erroneous thinking. Quality involves a lot more than acceptance of final products. It is an ongoing process of performing in the best and most cost-effective way possible. The acceptance of the final product is merely a benefit of effective quality control. Quality control is doing the job right the first time, thus saving re-work and warranty costs.

Each company sets its own quality standards that cover all aspects of quality control from preplanning the job to the final inspection. By implementing quality standards, the company aims to produce the highest quality installation possible, given the drawings, specifications, and job circumstances that meet or exceed industry standards.

Quality control includes all phases of the work, such as:

- Approving submittals
- Purchasing
- Storing materials and equipment
- Organizing subcontractors' activities
- Inspecting and testing to be sure the required materials are used
- Installing materials as described in the contract

The contractor and the customer both have a role in achieving quality construction as required by the contract. The responsibilities of both parties must balance each other and work in harmony.

The customer is often responsible for:

- Establishing construction standards
- Setting quality control requirements
- Reporting corrected defects

Instructor's Notes:

The contractor is responsible for:
- Completing work on time
- Following the terms of the contract
- Establishing and achieving quality control
- Recording quality control activities

A contractor needs to be able to work with the customer using industry-accepted processes to deliver a quality product.

Customers set the expectations for quality. If you don't deliver, your competitor will. Customers expect:
- Performance
- Reliability
- Competitive prices
- On-time delivery
- Service
- Clear and correct communication

Look at the business from the customer's point of view. A company can improve by looking at its processes from a different angle — the customer's.

People create results. Managing each project successfully with aggressive quality control is a team effort and key to satisfying customers. Involved employees are motivated and focus their talents on being the best. For supervisors, these basics are clear when following a quality control program.

SECTION 3

3.0.0 MANAGING QUALITY CONTROL

Quality control programs range from basic to complex. A good quality control system defines standards, measures efforts to meet these standards, and measures success. Many tools, from computer programs to other less automated systems, help the supervisor record data during the project and then analyze the results against goals or standards. Often, a checklist helps to manage quality control efforts and results. *Figure 1* is an example of a simple quality control checklist.

QUALITY CONTROL CHECKLIST

___ Record maintenance procedures
___ Collect and report test results, certification, and submittals
___ Perform and report inspections
___ Track and resolve poor performance
___ Record quality control activities
___ Maintain correct storage procedures
___ Monitor job progress and reduce delays to meet schedule
___ Ensure on-time delivery materials and proper installation
___ Identify and resolve defects

Figure 1 • Simple QC Checklist

In a small company, one or two people may handle quality control. In a large company, a quality control team may include project managers, supervisors, quality control specialists, architects, and engineers.

It is necessary to discuss and record how a QC team will manage and control all construction. The QC team develops the plan to fit the contractor's and customer's needs.

The components of an effective QC process include the following steps:

Step 1 Develop a plan of action.
Step 2 Establish lines of authority and responsibilities.
Step 3 Record all activities.
Step 4 Make changes if needed.

Show Transparency 2 (Figure 1).

Ask trainees to give examples of other items they would add to the checklist and write them on a blank transparency.

Classroom

Ask trainees to describe what they perceive to be the advantages and disadvantages of the two types of quality control.

Teaching Tip

See the teaching tip at the end of this module.

Homework

Assign reading of sections 5.0.0 – 6.4.0 for the next class.

SECTION 4

4.0.0 TYPES OF QUALITY CONTROL

There are two types of quality control: **traditional quality control** and **total quality control.**

4.1.0 Traditional Quality Control

Traditional quality control in construction sets the minimum standards for materials and workmanship according to original design specifications.

To ensure that these standards are met, random samples are inspected to form the basis for accepting or rejecting work completed or batches of materials. A batch is rejected if it fails the minimum standards or violates the terms of the contract.

Traditional quality control:

- *Assumes an acceptable quality level.* An acceptable quality level allows a small percentage of defective or imperfect items. Materials purchased from suppliers or work performed by a contractor are inspected and passed as acceptable if the estimated defective percentage is within the acceptable quality level.
- *Delays resolving a problem.* For example, problems with materials are corrected after delivery of the product.

Trying to achieve greater quality than what is specified by the acceptable levels of defect would greatly increase inspection costs and reduce worker productivity.

4.2.0 Total Quality Control

Total quality control is a strong commitment to quality involving all parts of a company and typically involves many elements, such as:

- Holding frequent design reviews to ensure safe and effective construction procedures
- Promoting extensive training for supervisors
- Shifting responsibility for detecting defects from QC inspectors to workers
- Maintaining equipment regularly

Total quality control sets a goal of *no* defective items anywhere in the construction process. It is difficult to reach or maintain zero defects permanently. However, by pursuing zero defective items, total QC companies realize many economic benefits, even if the zero-defect level is never achieved.

Material suppliers are also required to ensure zero defects when delivering goods. Initially, all materials from a supplier are inspected, and batches of goods with any defective items are returned. Suppliers with good records can be certified and will not be subject to full inspections at each delivery.

Total quality control is difficult in construction because of the:

- Different nature of each project
- Variety of workers
- Variety of subcontractors
- Education and procedure costs

However, the economic benefits of total quality control include:

- Reduced expenses associated with inventory, rework, scrap, and warranties
- Improved worker enthusiasm and commitment
- Ability to increase prices and profits. Customers often appreciate higher quality work and are willing to pay a premium for good quality.

For these reasons, improved quality control is a competitive advantage, well worth any effort expended.

4.3.0 Quality Circles

The key to successful quality control is to involve all employees at all levels in achieving the highest quality possible. This worker involvement is often formalized in **quality circles,** which are groups of workers who meet regularly to make suggestions for quality improvement. Quality circles originated from the total quality management (TQM) method of quality control. Supervisors are sometimes responsible for forming a quality circle. They pass along QC suggestions for improvement from the quality circles to management.

Instructor's Notes:

Materials

Ensure that you have all the necessary materials to teach the course. Check the Materials and Equipment list at the front of the module.

SECTION 5

5.0.0 QUALITY AND SAFETY CONCERNS IN CONSTRUCTION

Quality control and safety are increasingly important to supervisors. Costs for quality and safety are discussed and set during the design and planning stages and during construction.

Good supervisors try to ensure that the job is done right the first time and that no accidents occur on the project. Each project site that has or will have more than five employees should have a safety coordinator. The purpose of this assignment is two-fold:

- To help make the job safer
- To make the project supervisor's job easier

Project superintendents may elect to assume this responsibility themselves or may select someone who is competent in the particular area of construction and has a significant number of years of experience and training.

During the project kick-off meeting it is usually the responsibility of the superintendent and project manager or supervisor to determine who this safety coordinator will be and how much time will be allotted for these duties. While this position is not a full-time job, it is extremely important. The safety coordinator and the supervisor must carry out all safety standards and procedures in order to have a safe job site and prevent OSHA fines.

Remember that while all duties outlined are important, careful attention must be paid to new employee training. Over 50 percent of all accidents happen to new employees. Having the safety coordinator spend a few minutes with all employees who are new to a project will help to determine the employees' capabilities. Making sure that everyone is briefed and qualified for the jobs they will perform will decrease injuries on a project. Even someone with several years' experience still requires quite a bit of training in the areas of job-site safety. It does not hurt to ask new employees if they have ever performed a task even if you know they have three or four years of experience.

5.1.0 Direct and Indirect Costs

Poor quality control and poor safety procedures lead to rising direct and indirect costs. Reducing direct and indirect costs are often part of a supervisor's job.

Three reasons for rising direct costs are:

- Defects or failures that result in rebuilding or redesigning
- Failures that cause personal injuries or fatalities
- Accidents that result in personal injuries and large costs

If direct costs rise, indirect costs such as insurance, inspection, regulation, and delays in the project schedule follow.

5.2.0 Controls in the Design and Planning Stage

Two important decisions are made in the design and planning stages rather than during construction. They are **cost control** and **quality control.**

Cost control is the practice of keeping costs to a minimum. Cost control is important from the pre-planning stage through the fulfillment of the project. Cost control is a consideration when creating specifications, choosing materials, and authorizing overtime labor. Excellent safety standards and quality job performances are both forms of cost control as well.

Supervisors and inspectors record and maintain quality control procedures through the use of checklists and performance standards. When the project is completed, the QC team reviews the effects of its decisions. How closely did the team follow the original design and planning decisions? Sometimes, this measure gets a lot of attention. If so, then specifying this quality requirement in the design and in the contract becomes very important.

Supervisors, if part of the QC team, should take notes for their teams' next projects. Quality requirements should be clear and detailed so that all team members know how to meet the target for quality. By reviewing lessons learned and evaluating the outcomes, supervisors can learn from past mistakes and apply what they know to achieve success.

Ask trainees for examples of safety in the design and planning phases.

Distribute samples from a non-proprietary document control system and discuss its merits.

5.3.0 Safety in the Design and Planning Phase

Safety during the construction project is also influenced in large part by the decisions made in the planning and design process. When comparing designs, some construction plans are naturally difficult and dangerous to carry out. Other construction plans are designed to considerably reduce possible accidents. For example, supervisors who separate traffic from construction zones during roadway repair can greatly reduce the chance of motor vehicle accidents.

Beyond these design decisions, safety during construction largely depends upon education, alertness, and cooperation. Workers who carefully watch their surroundings are aware of danger spots and avoid unnecessary risks.

SECTION 6

6.0.0 ORGANIZING FOR QUALITY AND SAFETY

There are many ways to organize a construction project to promote quality and safety. A common model is to have one group responsible for quality control and another group responsible for safety within the company. In large companies, there is a separate department for quality control and safety. For smaller projects, the project manager or a supervisor might assume these and other responsibilities. In either case, ensuring safe, quality construction is a concern of the person who has overall responsibility for the project. Organizing for quality and safety requires excellent pre-planning, continuous communication, and an established quality assurance system.

6.1.0 Project Meetings and Project Communication

Project meetings are an important component of any quality control program. Project supervisors keep good communication channels open for their teams by setting up a schedule of meetings. These may include:

- *Bid review meetings* – to firm up the estimate prior to the actual bid date
- *Turnover meetings* – scheduled after the award of a project
- *Job management meetings* – help to keep the details of the project on track
- *Full project team meetings* – progress meetings to check actual dates and cost against project budget
- *Post-project meetings* – to evaluate the project and compile a list of lessons learned

By following this schedule of meetings, the project supervisor provides a framework for communication about the project with the different groups that are involved. These meetings help to keep the project on track and provide a quality assurance framework that can be repeated on every project for consistency and high quality.

6.2.0 Document Control

A major component of quality assurance is a **document control system.** Through all stages of a construction project, documents must be maintained, updated, controlled, and archived. Whether it is a manual or a computerized system, the project record is an invaluable piece of any quality assurance system. Each company has a system in place for document control. Generally, records must be kept for:

- Original job schedule and job progression
- Correspondence sent and received about the project
- Projected schedule and labor plan
- Change proposal requests and/or change directives
- Requests for Information
- Copies of material packing slips and Bills of Materials for fixtures and gear, including ship dates
- Estimate folder
- Meeting minutes
- As-built drawings or photos updated daily to reflect actual installation

6.3.0 Quality Assurance Checklists

The management of each job requires tremendous coordination and timing. One way of organizing the tasks associated with the mobilization of a project is with a **Job Management Checklist.** It identifies who, what, when, where, and how each task is to be completed. It also indicates to what standard or level of inspection each task is to be completed.

Instructor's Notes:

A quality assurance program may consist of a collection of checklists that specify quality and safety standards for each part of the job. *Figure 2* is an example of a checklist for constructing decks at an electrical contracting company.

Checklists serve as quality standards for ensuring that all work is completed in a standardized way and that the appropriate managers have signed off on the work.

Classroom

Ask trainees to write a Job Management Checklist for a job currently handled by one of their crew members and then swap that list with other trainees to get feedback and added suggestions.

SAMPLE JOB MANAGEMENT CHECKLIST

UNDERGROUND AND ABOVEGROUND DECKS

To ensure that all work has been installed as planned, the job superintendent or the supervisor in charge of the work being installed will personally inspect the following on the day before the pour.

PIPE WORK

_____ All underground pipe will be blown out with an air bottle to make sure that it goes from Point A to Point B as planned. Also, make sure that both ends are marked as planned.

_____ All aboveground pipe work done on the decks will be walked from Point A to Point B to make sure it is installed as planned, and ensure that all pipe work is tied properly.

_____ *Stub-ups* – Count all stub-ups going to the next floor. Make sure they are marked as planned and they are duct taped.

_____ *Sleeves* – Count all sleeves. Make sure the sleeves are in as planned and they are properly nailed down.

_____ *Boxes* – Count all boxes. Make sure they are nailed down properly and they are duct taped.

_____ *Lightning Protection* – Count all down leads. Make sure all down leads are duct taped.

This checklist will be completed and signed for all underground and aboveground decks by the superintendent or the lead person in charge of the work.

Bldg.: _____ Floor: _____

Signed: _____

Source: Tri-City Electrical Contractors, Inc., Altamonte Springs, FL, an Encompass Company

Figure 2 • Sample Checklist

Review the objectives of the module and answer any questions the trainees might have.

6.4.0 Inspection/Compliance

On a project, inspectors and quality control personnel represent different organizations. Some are internal to the company; others are external. There are several types of inspectors:

- Contractors from specialized quality assurance companies hired by the owner, the engineer/architect, or various subcontractors
- Local government's building department
- Environmental agencies
- Occupational health and safety agencies

Even though many construction companies have inspectors on site, they are only a formal check on quality control. The inspectors who are external to the company seek to ensure that the company follows or complies with the various local, state, and federal regulations.

Quality control should be a top goal for all the members of a project team. Supervisors should take responsibility for maintaining and improving quality control. Owners should promote good quality control by seeking out contractors who maintain high standards and rewarding employees who introduce new ideas.

Most important of all, improved quality leads to improved productivity. Good quality control pays for itself by:

- Uncovering new, more efficient work methods
- Avoiding fines or unnecessary and expensive regulation
- Avoiding long-term problems

Although many organizations may be involved, issues of quality control arise in virtually all the functional areas of construction activities. Therefore, quality performance is maintained by:

- Accurate and useful information
- Document control including changes during the construction process
- Procurement
- Field inspection and testing
- Final checkout of the facility

SUMMARY

Quality control is a very important part of the construction process. A supervisor may be part of a specialized QC team or wholly responsible for quality control.

Successful quality control meets the standards of the company/contractor, the industry, and the customer. A QC program must strongly emphasize that both quality and safety come from planning and prevention. QC and safety during construction are influenced in large part by the decisions made during the initial planning and design stages.

Safety largely depends upon education, awareness, and cooperation. Inspectors and QC specialists are on site to monitor quality control and safety; however, supervisors and the project manager carry the responsibility for safety rather than outside parties.

Instructor's Notes:

Review Questions

1. Quality of construction _____.
   a. gains favorable recognition for the company
   b. saves the government money
   c. saves the customer money
   d. creates delays in the schedule

2. To successfully maintain Quality Control (QC) on a project, the supervisor must know the standards required by all of the following *except* _____.
   a. building codes
   b. the company
   c. the customer
   d. the competition

3. A simple QC program would include all of the following *except* _____.
   a. reducing expenses
   b. collecting test results
   c. recording maintenance procedures
   d. monitoring installations

4. In a large company, a QC team might include all of the following *except* _____.
   a. architects
   b. QC specialists
   c. supervisors
   d. customers

5. A QC plan will not work if _____.
   a. the supervisor is the only one responsible
   b. a large team is appointed
   c. there is a great deal of competition
   d. the company has recently been sued

6. In a traditional quality control system, work completed or a batch of materials is accepted or rejected based on a _____.
   a. supplier invoice
   b. company memo
   c. random sample
   d. supplier catalog

7. Usually, a batch is rejected because _____.
   a. it does not meet the specifications
   b. too much material was delivered
   c. it is completed too late
   d. of cost overruns

8. Traditional quality control is _____.
   a. the same as total quality control
   b. an international standard
   c. set by the construction industry
   d. a measure of acceptable quality levels

Have trainees complete the Review Questions. Discuss the correct answers found at the end of this module.

9. The following are all elements of total quality control *except* ____.
   a. design reviews and inspections
   b. shifting the responsibility for finding defects from QC inspectors to workers
   c. use of high quality materials
   d. shifting the blame for defects from the supplier to the workers.

10. All of the following are reasons for rising indirect costs *except* ____.
    a. government regulation
    b. equipment replacement
    c. international standards
    d. insurance premiums

11. All of the following costs may increase because of an accident *except* ____.
    a. materials
    b. training
    c. inspection
    d. government regulation

12. The most important decisions about quality and safety during construction are made when ____.
    a. hiring good workers
    b. designing and planning
    c. installing materials
    d. recording routine maintenance

13. Two decisions made during the design and planning stages are ____.
    a. pay rates and vacation time
    b. cost and quality control
    c. materials and inspection schedules
    d. purchasing and storing equipment

14. The ____ meeting is held immediately after the award of the contract.
    a. turnover meeting
    b. bid review meeting
    c. status meeting
    d. job management meeting

15. Quality control may add cost to a project, but pays for itself by ____.
    a. shortening the length of the project
    b. keeping wages under control
    c. preventing long-term problems
    d. increasing the importance of document control

Instructor's Notes:

GLOSSARY

Trade Terms Introduced in This Module

Acceptable quality level: A quality level that allows a small percentage of defective or imperfect items.

Cost control: The practice of keeping financial costs to a minimum. Safety standards and quality job performance contribute to cost control.

Document control system: A system to maintain, update, control, and archive documents.

Job Management Checklist: Identifies who, what, when, where, and how each task is performed and to what standard a task is to be completed.

Quality: Indicates the degree of excellence. In construction, quality can refer to the ongoing process of performing in the best and most cost-effective way possible.

Quality circles: Groups of workers who meet regularly to make suggestions for quality improvement.

Quality control: Defines standards and measures efforts to meet these standards, in an effort to assure high quality products.

Total quality control: Sets a goal of no defective items anywhere in the process.

Traditional quality control: Sets the minimum standards for materials and workmanship according to original design specifications.

Have trainees read the Glossary for unfamiliar terms.

Have trainees prepare for the Module Examination.

Administer the Module Examination. Be sure to record the results of the Exam on Craft Training Report Form 200 and submit the results to the Training Program Sponsor.

Instructor's Notes:

MOD MT204-01—TEACHING TIPS

The following are suggested activities or instructional methods to help you teach the material in this AIG.

Section 1.0.0 **Introduction**

Discuss examples of formal quality control programs trainees have witnessed in the workplace. If you have access to samples of formal quality control programs from your workplace, the library, or government resources, distribute a brief sampling to the trainees of those programs. An example of a government quality control plan for energy can be found at http://www.hsrd.ornl.gov/~lfz/rpchklst/qa.htm#upa

Section 4.3.0 **Quality Circles**

Divide trainees into small groups, present them with a quality control problem that might arise in the workplace, and have a representative of each group report their groups' suggestions for improvement.

Examples of quality control problems:
- A supervisor who consistently submits incorrectly completed purchase orders
- A crew leader who takes shortcuts by installing materials differently than what is described in the contract
- A worker who fails to record maintenance checks regularly

NATIONAL CENTER FOR CONSTRUCTION EDUCATION AND RESEARCH

MODULE MT204

Answers to Review Questions

	Answer	Section Reference
1.	a	1.0.0
2.	d	3.0.0
3.	a	1.0.0
4.	d	3.0.0
5.	a	3.0.0
6.	c	4.1.0
7.	a	4.1.0
8.	d	4.1.0
9.	d	4.2.0
10.	c	5.1.0
11.	a	5.1.0
12.	b	5.3.0
13.	b	5.2.0
14.	a	6.1.0
15.	c	6.4.0

CONTREN™ LEARNING SERIES — USER UPDATES

The NCCER makes every effort to keep these textbooks up-to-date and free of technical errors. We appreciate your help in this process. If you have an idea for improving this textbook, or if you find an error, a typographical mistake, or an inaccuracy in NCCER's Contren™ textbooks, please write us, using this form or a photocopy. Be sure to include the exact module number, page number, a detailed description, and the correction, if applicable. Your input will be brought to the attention of the Technical Review Committee. Thank you for your assistance.

Instructors – If you found that additional materials were necessary in order to teach this module effectively, please let us know so that we may include them in the Equipment/Materials list in the Instructor's Guide.

Write: Curriculum Revision and Development Department
National Center for Construction Education and Research
P.O. Box 141104, Gainesville, FL 32614-1104

Fax: 352-334-0932

E-mail: curriculum@nccer.org

Craft _____ Module Name _____

Copyright Date _____ Module Number _____ Page Number(s) _____

Description _____

_____

_____

_____

(Optional) Correction _____

_____

_____

(Optional) Your Name and Address _____

_____

_____

Project Supervisor

Module MT205-01

Contract and Construction Documents

Contract and Construction Documents
Instructor's Guide

Module MT205

MODULE OVERVIEW

This module introduces the project supervisor trainee to contract and construction documents. This module teaches the trainee how to interpret construction drawings and specifications, how and why construction documents are used, and how work is obtained in the construction industry.

PREREQUISITES

There are no prerequisites for the module; however, prior to training with this module, it is recommended that the trainee complete the following modules:
  Project Supervision, Modules MT201 through MT204

LEARNING OBJECTIVES

Upon completion of this module, the trainee will learn and be able to:

1. Define construction documents and project manuals.
2. Read and interpret construction drawings.
3. Recognize the types (components) of working drawings and specifications.
4. Explain the methods of obtaining work in the construction industry.
5. Identify the types of contracts used in the industry.
6. Explain the need for documentation and the types of documents used on a project.
7. Identify the documents necessary to close out a project.

PERFORMANCE OBJECTIVES

This is a knowledge-based module – there is no performance profile examination.

NCCER STANDARDIZED TRAINING PROGRAM

The National Center for Construction Education and Research (NCCER) provides a standardized national program of accredited craft training. Key features of the program include instructor certification, competency-based training, and performance testing. The program provides trainees, instructors, and companies with a standard form of recognition through a National Craft Training Registry. The program is described in full in the Guidelines for Accreditation, published by the NCCER. For more information on standardized craft training, contact the NCCER by writing us at P.O. Box 141104, Gainesville, FL 32614-1104; calling 352-334-0911; or e-mailing info@nccer.org. More information may be found at our Web site, www.nccer.org.

HOW TO USE THIS ANNOTATED INSTRUCTOR'S GUIDE

Each page presents two sections of information. The larger section displays each page exactly as it appears in the Trainee Module. The narrow column ties suggested trainee and instructor actions to each page and provides icons to call your attention to material, safety, audiovisual, or testing requirements. The bottom of each page includes space for your notes.

 If you see the Teaching Tip icon, that means there is a teaching tip associated with this section. Also refer to the suggested teaching tips at the end of the module.

PREPARATION

Before teaching this module, you should review the Module Outline, Learning Objectives, and the Materials and Equipment List. Be sure to allow ample time to prepare your own training or lesson plan and gather all required equipment and materials.

MATERIALS AND EQUIPMENT LIST

Materials:

Transparencies

Markers/chalk

Module Examinations*

Sample project manual and its contents, including basic contract and construction documents, blueprints, and closeout documents**

Sample log books**

Sample insurance contracts**

Equipment:

Overhead projector and screen

Whiteboard/chalkboard

*Located in the Test Booklet packaged with this Annotated Instructor's Guide.
**If available on loan from your workplace or other resource.

ADDITIONAL RESOURCES

This module is intended to present thorough resources for task training. The following reference works are suggested for both instructors and motivated trainees interested in further study. These are optional materials for continued education rather than for task training.

Construction Materials and Building Publication No. 91, International Organization for Standardization (ISO), ISO Online Catalog at www.iso.ch.

Professional Construction Management: Including Contracting C M, Design-Construct, and General Contracting, 1991. Donald S. Barrie and Boyd C. Paulson (Contributor). Upper Saddle River, NJ: McGraw-Hill Higher Education.

Construction Management, 1997. Daniel W. Halpin and Ronald W. Woodhead. New York: John Wiley & Sons.

Construction Operations Manual of Policies and Procedures, 2000. Andrew Civitello, Jr. Upper Saddle River, NJ: McGraw-Hill Professional Book Group.

TEACHING TIME FOR THIS MODULE

An outline for use in developing your lesson plan is presented below. Note that each Roman numeral in the outline equates to one session of instruction. Each session has a suggested time of 2 1/2 hours. This includes 10 minutes at the beginning of each session for administrative tasks and one 10-minute break during the session. Approximately 5 hours are suggested to cover *Contract and Construction Documents*.

Topic   **Planned Time**

Session I. Project Manuals and Drawings
- A. Introduction   _____
- B. Project Manual   _____
- C. Drawings   _____
  1. Types of Drawings   _____
  2. Understanding Drawings   _____
  3. Reading Drawings   _____
  4. Participant Activity   _____

Session II. Technical Specifications and Contract Documentation
- A. Technical Specifications   _____
  1. Participant Activity   _____
- B. Submittals   _____
  1. Managing Submittals   _____
  2. Processing a Submittal   _____
  3. Reviewing Submittals   _____
- C. Methods of Obtaining Work   _____
- D. Contracts   _____
  1. Types of Contracts   _____
  2. Relationships   _____
- E. Documentation   _____
  1. Criteria for Good Documentation   _____
  2. Communication   _____
  3. Daily Log   _____
  4. Meetings   _____
  5. Project Photographs   _____
  6. Schedule   _____
  7. Daily Time Reports   _____
  8. As-Built Drawings   _____
  9. Closeout Documents   _____
  10. Change Orders   _____
  11. Insurance   _____
  12. General and Special Conditions   _____
  13. Back Charges   _____
- F. Summary   _____
  1. Summarize module   _____
  2. Answer review questions   _____
- G. Module Examination   _____
  1. Trainees must score 70% or higher to receive recognition from the NCCER.   _____
  2. Record the testing results on Craft Training Report Form 200 and submit the results to the Training Program Sponsor.   _____

 **Project Supervision – Module MT205**

Contract and Construction Documents

Instructor's Notes:

Instructor's Notes:
_____
_____
_____
_____
_____
_____

ACKNOWLEDGMENTS

The NCCER wishes to acknowledge the dedication and expertise of Phil Copare, the original author and mentor for this module on contract and construction documents.

Philip B. Copare, MBA
President, Construction Services Enterprise
Education and Safety Consultant
Zellwood, FL

We would also like to thank the following reviewers for contributing their time and expertise to this endeavor:

J.R. Blair
Tri-City Electrical Contractors
An Encompass Company

Mike Cornelius
Tri-City Electrical Contractors
An Encompass Company

Dan Faulkner
Wolverine Building Group

David Goodloe
Clemson University

Kevin Kett
The Haskell Company

Danny Parmenter
The Haskell Company

Course Map

This course map shows all of the modules of the *Project Supervision* curriculum. The suggested training order begins at the bottom and proceeds up. Skill levels increase as you advance on the course map. The local Training Program Sponsor may adjust the training order.

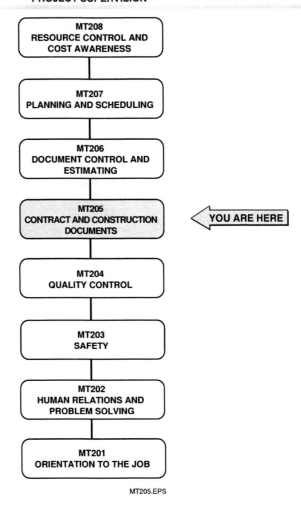

Instructor's Notes:

MODULE MT205

TABLE OF CONTENTS

1.0.0	**INTRODUCTION**	5.1
2.0.0	**PROJECT MANUAL**	5.2
3.0.0	**DRAWINGS**	5.2
3.1.0	Types of Drawings	5.2
3.2.0	Understanding Drawings	5.3
3.2.1	*Interpreting Drawings*	5.8
3.2.2	*Dimensioning Conventions*	5.11
3.2.3	*Parts of Drawings*	5.11
3.3.0	Reading Drawings	5.16
4.0.0	**TECHNICAL SPECIFICATIONS**	5.19
5.0.0	**SUBMITTALS**	5.20
5.1.0	Managing Submittals	5.20
5.1.1	*Submittal Data Schedule*	5.20
5.1.2	*Submittal Log*	5.20
5.2.0	Processing a Submittal	5.20
5.3.0	Reviewing Submittals	5.20
6.0.0	**METHODS OF OBTAINING WORK**	5.22
7.0.0	**CONTRACTS**	5.23
7.1.0	Types of Contracts	5.23
7.1.1	*Lump Sum Contracts*	5.23
7.1.2	*Cost of Work Plus a Fee Contracts*	5.23
7.1.3	*Unit Cost Contracts*	5.23
7.1.4	*Construction Management Contracts*	5.23
7.1.5	*Guaranteed Maximum Price*	5.23
7.2.0	Relationships	5.23
8.0.0	**DOCUMENTATION**	5.24
8.1.0	Criteria for Good Documentation	5.25
8.2.0	Communication	5.25
8.3.0	Daily Log	5.25
8.4.0	Meetings	5.28
8.5.0	Project Photographs	5.28
8.6.0	Schedule	5.31
8.7.0	Daily Time Reports	5.31
8.8.0	As-Built Drawings	5.31
8.9.0	Closeout Documents	5.31

8.9.1	Certificate of Occupancy	5.31
8.9.2	Certificate of Substantial Completion	5.31
8.9.3	Punch List	5.33
8.9.4	Operation and Maintenance Manuals	5.33
8.9.5	Warranties and Guarantees	5.33
8.10.0	Change Orders	5.33
8.10.1	Budget Revisions	5.33
8.10.2	Design Defects and Deficiencies	5.33
8.10.3	Revised Project Needs	5.33
8.10.4	Minor Change Orders	5.33
8.10.5	Time and Material Change Orders	5.34
8.11.0	Insurance	5.34
8.11.1	Workers' Compensation Insurance	5.34
8.11.2	General Liability Insurance	5.34
8.11.3	Vehicle Insurance	5.34
8.11.4	Builder's Risk/Installation Floater Insurance	5.34
8.11.5	Equipment Floater Insurance	5.36
8.11.6	Property Insurance	5.36
8.11.7	Medical/Dental/Life/Disability Insurance	5.36
8.12.0	General and Special Conditions	5.36
8.13.0	Back Charges	5.37
	SUMMARY	5.37
	REVIEW QUESTIONS	5.39
	GLOSSARY	5.41
	APPENDIX	5.43

Instructor's Notes:

LIST OF FIGURES

Figure 1	•	Common Symbols Found on Drawings	5.3
Figure 2	•	Lines of Construction	5.4
Figure 3	•	Architectural Symbols	5.4
Figure 4	•	Civil and Structural Engineering Symbols	5.5
Figure 5	•	Mechanical Symbols	5.5
Figure 6	•	Plumbing Symbols	5.6
Figure 7	•	Electrical Symbols	5.6
Figure 8	•	Abbreviations	5.7
Figure 9	•	Contour Lines	5.8
Figure 10	•	Drawing Center Lines	5.9
Figure 11	•	Dimension Lines	5.10
Figure 12	•	Dimensioning Conventions	5.12
Figure 13	•	Typical Floor or Framing Plan One	5.13
Figure 14	•	Typical Floor or Framing Plan Two	5.13
Figure 15	•	Typical Elevation Drawings	5.14
Figure 16	•	Typical Section Drawings	5.15
Figure 17	•	Detail of Cornice	5.15
Figure 18	•	Typical Schedule	5.16
Figure 19	•	Submittal Data Schedule	5.21
Figure 20	•	Study of Industry Delays	5.24
Figure 21	•	Request for Information	5.26
Figure 22	•	Typical Daily Log Form	5.27
Figure 23	•	Typical Job-Site Meeting Agenda	5.28
Figure 24	•	Typical Job-Site Meeting Minutes	5.29
Figure 25	•	Job Photos	5.30
Figure 26	•	Typical Supervisor's Daily Time Report	5.32
Figure 27	•	Time and Material Change Order	5.35

Instructor's Notes:

NATIONAL CENTER FOR CONSTRUCTION EDUCATION AND RESEARCH

MODULE MT205

Contract and Construction Documents

OBJECTIVES

Upon the completion of this module, you will be able to do the following:

1. Define construction documents and project manuals.
2. Read and interpret construction drawings.
3. Recognize the types (components) of working drawings and specifications.
4. Explain the methods of obtaining work in the construction industry.
5. Identify the types of contracts used in the industry.
6. Explain the need for documentation and the types of documents used on a project.
7. Identify the documents necessary to close out a project.

SECTION 1

1.0.0 INTRODUCTION

Supervisors must be familiar with all contract documents, including those that describe the project as a whole and those that pertain to their work or specific area of responsibility. These documents go beyond the plans and specifications necessary to build a project. They also include drawings and a wide variety of instructions and documents.

Supervisors are responsible for all of the company documents that reflect the work being done on the job site under their control. This includes basic and complex documents alike. For example, daily logs of the work performed and time reports are basic documents. Complex documents, such as technical specifications, insurance contracts, change orders, and drawings, are just as important to the supervisor.

Ensuring that documents are well maintained is a large part of a supervisor's daily work. Neglecting to update documents on time can be costly to the company and put crews at risk for job security. Regular meetings with a set agenda and proper recording of the minutes help the supervisor stay on track and provide backup for legal representation, if required.

Ensure that you have all the necessary materials to teach the course. Check the Materials and Equipment list at the front of the module.

Ask trainees to list some basic documents that they have seen or used in their workplace and to rate their usefulness.

Show Transparency 1 (Course Objectives).

Assign reading of Module MT205, Sections 1.0.0 – 3.3.0.

Copyright © 2003 National Center for Construction Education and Research, Gainesville, FL 32614-1104. All rights reserved. No part of this work may be reproduced in any form or by any means, including photocopying, without written permission of the publisher.

Distribute and discuss some samples of documents found in a project manual including blueprints. Try to have at least one set for every 4-5 trainees.

SECTION 2

2.0.0 PROJECT MANUAL

Supervisors must be familiar with many contract documents, both basic and complex.

When construction was not as complex as it is today, the graphic representation of the project was referred to as the project's **plans**. Today, plans are more often referred to as **drawings**. In addition, the term *spec book,* which is an abbreviated form of *book of* **specifications**, is changing to **project manual.**

The project manual, or spec book, is an organized listing of the basic requirements for construction materials, dimensions, and other details relating to the project. The supervisor must thoroughly understand the contents of the project manual in order to complete the project according to the terms of the contract. A typical project manual contains the following:

- *Invitation to Bid* – Advises prospective bidders about the project.
- *Instruction to Bidders* – Advises each bidder how to prepare the bid, thereby ensuring that each one submitted is in the same format.
- *Bid Form* – Provides a uniform arrangement that makes it easier for the owner to compare the bids received.
- *Bid Bond* – A legal document posted by the bidder as an assurance that the bidder will sign the contract if awarded. The owner usually requires bidders to furnish bid bonds at the time of bidding.
- *Payment & Performance Bond* – A bond posted by the contractor that guarantees the owner that the contractor will perform the work in accordance with the terms of the contract and pay all bills and obligations incurred under the contract.
- *Form of Agreement* – In today's terminology, the term *contract* is being replaced by *agreement*. Various standard forms of agreement between owners and contractors are available from the American Institute of Architects (AIA) and other professional associations.
- *General Conditions* – Define the contractual relationships and procedures relative to the project, particularly the interrelationship of the owner, architect/engineer, contractor, and subcontractor, and the details of how the contract will be administered.
- *Supplementary Conditions* – Lists with necessary modifications to the general conditions as they apply to the specific conditions of the particular project.
- *Technical Specifications* – Written technical information that supports and describes the details of construction drawings
- *Drawings* – Show the physical dimensions and details of the work.
- *Addenda* – Used to modify the contract documents before the bidding process. When the contract documents are modified after the contract has been awarded, the addenda are called *change orders*.

SECTION 3

3.0.0 DRAWINGS

The average person outside the construction industry is bewildered by the maze of lines, dimensions, and notations associated with construction drawings. The drawings give the supervisor a picture of how the project is to be assembled or erected. There is an old saying that a picture is worth a thousand words, and this is especially true in the construction industry.

It is important that supervisors be able to read and understand construction drawings so they can communicate the information to others.

Today, most prints are created by computer-aided drafting (CAD), and they have blue or black lines on a white background. These prints are referred to as *blueprints*.

3.1.0 Types of Drawings

The various types of construction drawings are noted and defined below:

- **Preliminary Drawings** – Beginning drawings prepared by the architect/engineer during the promotion stage of the building's development

Instructor's Notes:
_____
_____
_____
_____
_____
_____

Module MT205 ◆ Contract and Construction Documents

- **Presentation Drawings** – Architectural perspective views of the building used to make formal presentations
- **Working Drawings** – All drawings needed by the various tradesmen to complete a project

3.2.0 Understanding Drawings

The ability to read drawings comes from learning the language of drawings and applying this knowledge through practice. The following suggestions will help the trainee quickly understand the language of working drawings:

- Learn the principles of technical projection.
- Learn the meanings of the different types of lines, symbols, abbreviations, and terminology.
- Learn conventional dimensioning methods and the use of scales.
- Review mathematics.
- When looking at the drawings, which are in two dimensions, visualize the actual construction in three dimensions.

The working drawings are the drawings that supervisors use in the field. To understand working drawings, the supervisor must know:

- **Symbols** – Symbols were developed to conserve design time. Standardizing symbols has made it possible for architects in one part of the country to design projects for contractors in another part of the country to build. *Figures 1* through *7* show common symbols found on drawings. These include lines of construction, architectural symbols, civil and structural engineering symbols, mechanical symbols, plumbing symbols, and electrical symbols.
- **Abbreviations** – Abbreviations are commonly used throughout the industry, though they are not as standardized as symbols. *Figure 8* shows a common list of abbreviations. These are used to keep the plans uncluttered, making them easier to read.
- **Dimensions** – The architect is responsible for the dimensions on the drawings, and is required to show all dimensions necessary to construct the project.

continued on page 5.8

 - Elevation or Match Line

 - Window Reference
3 - Window Detail,
B - Window Type

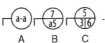 - Section Notations
A   B   C

 - Column Line Symbols

 - Center Line

@ - At

O.C. - On Center

 - Plate

 - Footing Schedule

 - Elevation View Notation

 - Welding Sym.

 - Door I.D. or Column I.D.

 - Door Frame

 - Room Finish Number

 - North Building & True

 - Bench Mark

205F01.EPS

Figure 1 • Common Symbols Found on Drawings

Show Transparency 2 (Figure 1).

Point out an example of dimensions on the drawings, as well as each of the symbols and abbreviations and what they stand for.

Ask trainees to find the lines of construction and architectural symbols of Figures 2 and 3, respectively, on the sample drawings.

Show Transparencies 3 and 4 (Figures 2 and 3).

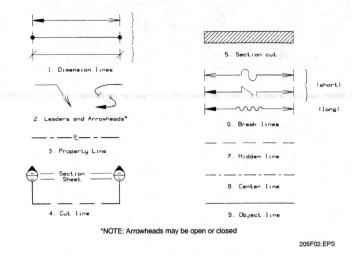

Figure 2 • Lines of Construction

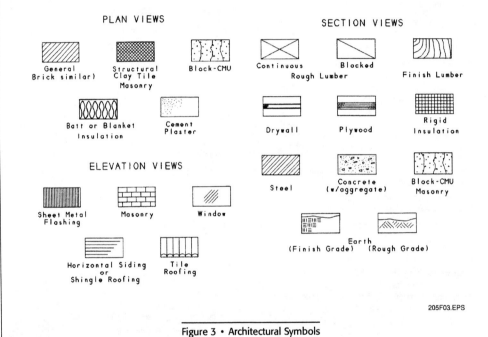

Figure 3 • Architectural Symbols

Instructor's Notes:

Module MT205 ♦ Contract and Construction Documents

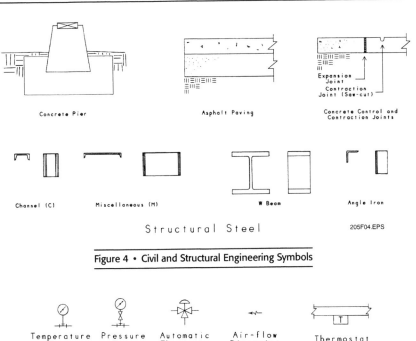

Figure 4 • Civil and Structural Engineering Symbols

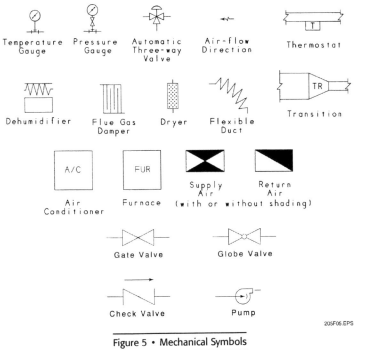

Figure 5 • Mechanical Symbols

Ask trainees to find the engineering and mechanical symbols of Figures 4 and 5, respectively, on the sample drawings.

Show Transparencies 5 and 6 (Figures 4 and 5).

Ask trainees to find the plumbing and electrical symbols of Figures 6 and 7, respectively, on the sample drawings.

Show Transparencies 7 and 8 (Figures 6 and 7).

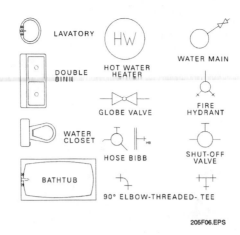

Figure 6 • Plumbing Symbols

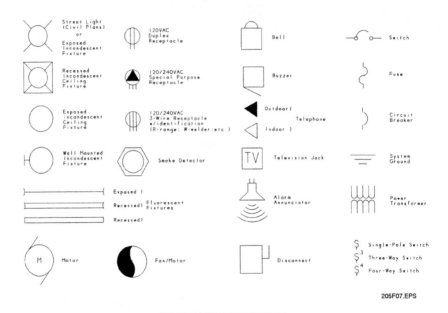

Figure 7 • Electrical Symbols

Instructor's Notes:

ABBREVIATIONS

A.B.	– ANCHOR BOLT	G.L.	– GLU-LAM BEAM
ADD'L	– ADDITIONAL	GR.	– GRADE
ADJ.	– ADJACENT	GR. BM.	– GRADE BEAM
A.I.S.C.	– AMERICAN INSTITUTE OF STEEL CONSTRUCTION	H.A.S.	– HEADED ANCHOR STUD
		HORIZ.	– HORIZONTAL
ALT.	– ALTERNATE	H.S.B.	– HIGH STRENGTH BOLT
ARCH.	– ARCHITECTURAL	I.D.	– INSIDE DIAMETER
A.S.T.M.	– AMERICAN SOCIETY FOR TESTING & MATERIALS	IN.	– INCH
		INT.	– INTERIOR
BLDG.	– BUILDING	JNT.	– JOINT
BM.	– BEAM	LB.	– POUND
B.O.	– BOTTOM OF	LIN. FT.	– LINEAL FEET
BOT.	– BOTTOM	L.L.V.	– LONG LEG VERTICAL
BSMT.	– BASEMENT	MAT'L.	– MATERIAL
BTWN.	– BETWEEN	MAX.	– MAXIMUM
CANT.	– CANTILEVER	MECH.	– MECHANICAL
CB.	– CARDBOARD	MID.	– MIDDLE
CH.	– CHAMFER	MIN.	– MINIMUM
C.J.	– CONTROL/CONSTRUCTION JOINT	MISC.	– MISCELLANEOUS
CLR.	– CLEAR, CLEARANCE	MTL.	– METAL
C.M.U.	– CONCRETE MASONRY UNIT	N.I.C.	– NOT IN CONTRACT
COL.	– COLUMN	NO.	– NUMBER
CONC.	– CONCRETE	NOM	– NOMINAL
CONN.	– CONNECTION	N.T.S.	– NOT TO SCALE
CONST.	– CONSTRUCTION	O.C.	– ON CENTER
CONT.	– CONTINUOUS	O.D.	– OUTSIDE DIAMETER
CONTR.	– CONTRACTOR	O.H.	– OPPOSITE HAND
CTRD.	– CENTERED	OPNG.	– OPENING
DET.	– DETAIL	₧	– PLATE
DIAG.	– DIAGONAL	P.S.F.	– POUND PER SQUARE FOOT
DIAM.	– DIAMETER	P.S.I.	– POUND PER SQUARE INCH
DIM.	– DIMENSION	R.	– RADIUS
DISCONT.	– DISCONTINUOUS	REINF.	– REINFORCEMENT
DWG.	– DRAWING	REQ'D.	– REQUIRED
EA.	– EACH	RM.	– ROOM
E.F.	– EACH FACE	SCHED.	– SCHEDULE
EL.	– ELEVATION	SECT.	– SECTION
ELECT.	– ELECTRICAL	SHT.	– SHEET
ELEV.	– ELEVATOR	SIM.	– SIMILAR
EQ.	– EQUAL	S.L.V.	– SHORT LEG VERTICAL
E.W.B.	– END WALL BARS	SPC.	– SPACE
E.W.	– EACH WAY	SPEC.	– SPECIFICATION
EXIST.	– EXISTING	SQ.	– SQUARE
EXP. JNT.	– EXPANSION JOINT	STD.	– STANDARD
EXT.	– EXTERIOR	STIFF.	– STIFFENER
F.D.	– FLOOR DRAIN	STL.	– STEEL
FDN.	– FOUNDATION	STOR.	– STORAGE
FIN.	– FINISH	SYM.	– SYMMETRICAL
FLR.	– FLOOR	T.&B.	– TOP AND BOTTOM
F.O.B.	– FACE OF BRICK	THK.	– THICKNESS
F.O.CONC.	– FACE OF CONCRETE	T.O.	– TOP OF
F.O.W.	– FACE OF WALL	TYP.	– TYPICAL
FS.	– FLAT SLAB	U.N.O.	– UNLESS NOTED OTHERWISE
FT.	– FOOT	VAR.	– VARIES
FTG.	– FOOTING	VERT.	– VERTICAL
F.W.	– FILLET WELD	V.I.F.	– VERIFY IN FIELD
GA.	– GAUGE	WT.	– WEIGHT
GAL.	– GALVANIZED		

Figure 8 • Abbreviations

Classroom

Ask trainees to find the abbreviations in Figure 8 on the sample drawings.

Ask trainees to find the contour lines on the sample drawings.

Show Transparency 9 (Figure 9).

On drawings, the architect tries to position dimensions to form a continuous series that does not interfere with dimension lines or notations. Supervisors must be able to read dimensions accurately. This is largely a matter of noticing where the arrowheads, dotted or slashed lines, and extension lines are located and knowing what they mean.

- **Scale** – Working drawings are drawn to the scale necessary to show the information to the contractor. When selecting a scale, the architect must consider two things:
  — It must clearly show the information.
  — It is small enough to be manageable on the job site.

Many times an architect resolves the conflict between clarity and size by enlarging only certain portions of the plans.

Sometimes, an architect omits a dimension and the drawing must be "scaled" in order to obtain the information. Scaling is never recommended, because drawings have often been reduced and the results can lead to drastic construction errors. Whenever a dimension is missing from a drawing, the missing dimension should be requested from the architect/engineer in writing.

3.2.1 Interpreting Drawings

A supervisor needs to know other types of information to understand various categories of drawings. A brief review of some of these points is presented here. The best way to learn to read drawings is by experience.

Contour lines are drawn to indicate the slope of the building site before and after construction. The supervisor can tell the difference between finished and existing grades by the way that the lines are drawn. In *Figure 9*, existing grades are dashed, and finished grades are solid.

Center lines are used to divide drawings into equal or symmetrical parts. In *Figure 10*, notice that the center line is a lightweight line with alternate long and short dashes. These lines are helpful in dimensioning and lining up views. Center lines are always used to locate the center of a round hole.

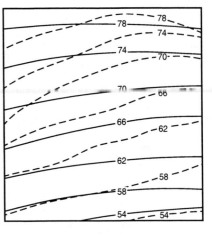

EXISTING GRADE – – – – – – – –
FINISHED GRADE ───────────

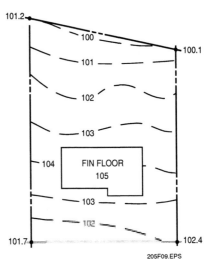

Figure 9 • Contour Lines

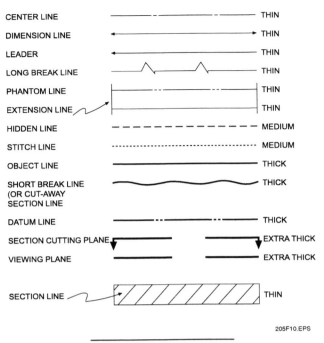

Figure 10 • Drawing Center Lines

Ask trainees to find the center lines on the sample drawings.

Show Transparency 10 (Figure 10).

Dimension lines contain the size, which is placed in a break in the line. Some dimensions are placed between separate arrowhead-tipped lines. In either case, the dimension distance is from the point of one arrowhead to the point of the other arrowhead. To keep dimensions in clear view, they are placed outside the view, although sometimes it may be necessary to put dimensions inside the outline. *Figure 11* shows the use of dimension lines and the other types that are described in this section.

Leader lines tie names to their corresponding lines. Leaders are composed of straight, ruled lines that indicate exactly where dimensions or explanatory notes are to be applied. The note end of the leader should always run either to the beginning or end of the note or dimensions, never to the middle. In general, the leader should be applied on the drawing view that shows the profile of the surface to which the requirement applies. Drawings present a better appearance when all adjacent leaders are drawn parallel. One exception to this is that leaders should not be parallel to adjacent dimension or extension lines. Leaders drawn to symmetrical features should be in line with the center of the feature, and the arrowhead of the leader should terminate exactly on the line representing the profile of the feature.

Many times the draftsperson is cramped for space on the paper. *Long break lines* are space savers. If the whole object were drawn, it would either run off the paper or have undersized views. Long break lines are used to indicate that part has been shortened. The lines do not change the actual length indicated by the dimensions.

Ask trainees to find the dimension lines on the sample drawings.

Show Transparency 11 (Figure 11).

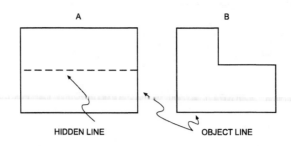

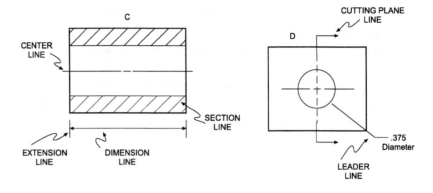

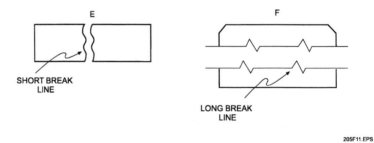

Figure 11 • Dimension Lines

Instructor's Notes:

Phantom lines are used to show an alternate position. They can also indicate a repeated detail or the relative position of an absent part. The draftsperson uses a lightweight stroke, drawing a line consisting of a long dash broken by two short dashes.

Extension lines are lightweight and extend outwards from the outline. They are started about 1/16-inch away from the view outline.

Hidden lines portray what you would see if you could look right through an object and observe the inside of the drawing or the back surface. To represent these invisible lines, the draftsperson uses a medium-weight broken line. A hidden line is shown as a series of short dashes, all of the same length.

Object lines correspond to the boundary lines of the object. These heavy solid lines represent the edges and surface that are visible from the angle at which the view is drawn. Without the outlines or visible lines, identifying the object would be impossible.

Short break lines, or cutaway section lines, indicate that the draftsperson has removed part of an outer surface to reveal the inside structure. Short break lines, like long break lines, indicate that part of an object has been omitted. Usually, these omissions produce a clearer drawing.

3.2.2 Dimensioning Conventions

The method of showing the dimensions will depend on the type of construction, whether the dimensions are for the interior or exterior, and any special conventions used by the architect.

Wood frame – A plan view of a typical construction wall is noted in *Figure 12*. Exterior dimensions are to the outside of the studs, and interior dimensions are to the center line of the wall. Window and door openings are dimensioned to their centers. This convention also holds for frame and masonry veneer.

Solid masonry – Exterior dimensions are to the outside of the wall, while interior dimensions are to wall surfaces. Actual door and window wall openings are dimensional, as noted in *Figure 12*.

3.2.3 Parts of Drawings

The drawings for a building may consist of a few sheets or a considerable number of sheets making up several sets. Generally, drawings consist of the following components:

Site Drawings (Civil Drawings) contain information on existing and final grades, existing conditions, and site development for utilities, paving, sidewalks, landscape, and irrigation.

Architectural Drawings contain information on floor plans, elevations, sections, details, and schedules.

Structural Drawings contain information on foundations, placement of reinforcement steel, floors, and roof construction.

Mechanical Drawings contain information on the plumbing system, HVAC system, mechanical equipment schedule, and fire protection layout.

Electrical Drawings contain information on the electrical system, including wiring diagrams, fixture schedules, and panel locations.

Another important part of drawings is the *title block*. Often overlooked as a source of information, the title block contains the following information:

- Name of the project
- Name of the architect or engineer
- Date the drawings were prepared
- Title of the sheet and the sheet number
- Scale used
- Revisions to the drawings

A plan, such as a floor or framing plan, is a view looking directly down on the subject or object in question. See *Figures 13* and *14* for examples. A site plan is a drawing of the project on its land, looking straight down on the site from a distance, such as from an airplane.

An *elevation* is a view, either inside or outside, of one of the sides of the building or a portion of a side. Elevation drawings are titled by what is shown, such as a front elevation, north elevation, or bookcase elevation. See *Figure 15* for an example.

Ask trainees to find the dimensioning conventions on the sample drawings.

Show Transparency 12 (Figure 12).

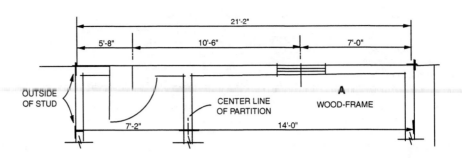

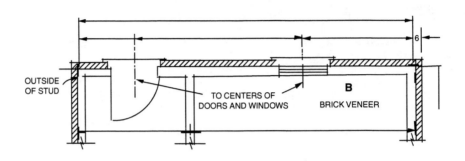

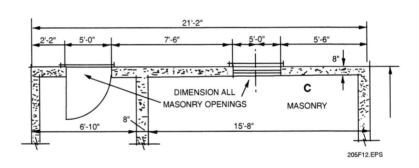

Figure 12 • Dimensioning Conventions

Instructor's Notes:

Module MT205 ♦ Contract and Construction Documents

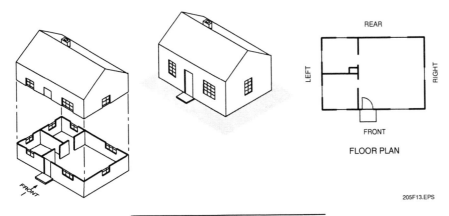

Review Floor or Framing Plan I and II if the trainees have not yet viewed a sample in class.

Show Transparencies 13 and 14 (Figures 13 and 14).

Figure 13 • Typical Floor or Framing Plan One

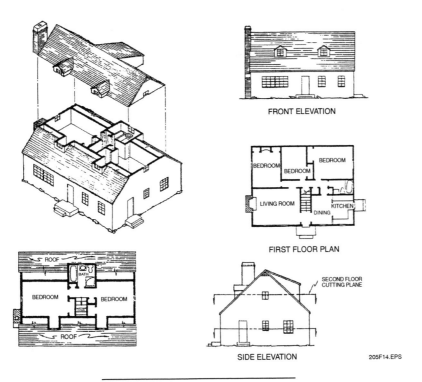

Figure 14 • Typical Floor or Framing Plan Two

Review Elevation Drawing if the trainees have not yet viewed a sample in class.

Show Transparency 15 (Figure 15).

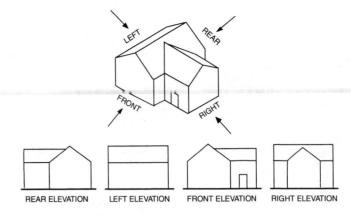

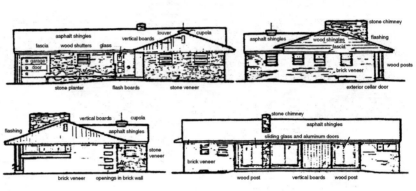

Figure 15 • Typical Elevation Drawings

Sections are drawings showing the construction of walls, stairs, or other details not clearly shown on the elevations or plans. This drawing is very important because it gives the reader valuable information on how construction takes place and what materials are used. See *Figure 16* for an example.

Details are needed where special or unusual construction is to be performed, and this information needs to be communicated to the craftworker doing the work. See *Figure 17* for an example.

Schedules are listings or tables of information needed to build the job that can best be communicated in a table format. Examples include a fixture, finish material, room, or door and window schedule. See *Figure 18* for an example of a fixture schedule.

Note that even though attempts are being made to standardize the meaning of lines, notations, and abbreviations, some architects and engineers do not use conventional notations as presented in this module.

Instructor's Notes:

Module MT205 ◆ Contract and Construction Documents

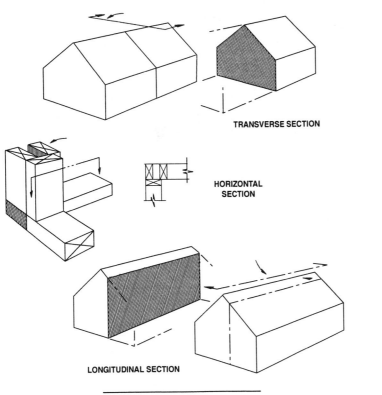

Figure 16 • Typical Section Drawings

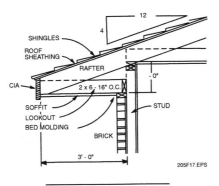

Figure 17 • Detail of Cornice

Review Section Drawing and Detail Drawing if the trainees have not yet viewed a sample in class.

Show Transparencies 16 and 17 (Figures 16 and 17).

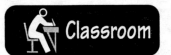

Ask trainees if they use other methods to review drawings besides the suggested method. What are the advantages and disadvantages?

FIXTURE SCHEDULE					
MARK	MANUFACTURE	CAT NO.	WATTAGE	FINISH	REMARKS
A	LITHONIA	26440-A12	4-40	WHITE	GRID CEILING ACRYLIC
B	PRESCOLITE	4020	100	BLACK	
C	PRESCOLITE	4452	4-60	BLACK	
D	PRESCOLITE	74V	75-2-30	WHITE	
E	PRESCOLITE	1252-416	75-R-30	BRONZOTIC	
F	PRESCOLITE	540	100W	WHITE	8" TUBE
G		RECESSED CEILING MT.	100W		SUITABLE FOR USE IN SAUNA
H	LITHONIA	6240-A12	2-40 WHITE	SURFACE	
J	SPAULDING	532	60	BRONZE ANODE	WALLBRACKET
K	PRESCOLITE	37D3	2-60	ALUM	6" A.F.F.
L	PRESCOLITE	93020	175W-MV	BLACK	
N	PRESCOLITE	4220	100	BLACK	
O	ALLOWANCE	OF $250	/OUTLET		OUTLET ONLY
P	---	---	100		PORCELAIN SOCKET
R	PRESCOLITE	1171-900	2-50 R20	BRONZONIC	

Figure 18 • Typical Schedule

3.3.0 Reading Drawings

A supervisor's first step in studying a set of drawings is to become familiar with their general features. This is important even though they may be involved only with a specific section of the project, such as masonry or plumbing. Being familiar with the general layout of the floor plans and the elevations involved is initially more important than spending hours looking at minor construction features.

Later, supervisors can start reviewing the portions they are responsible for and learning how these responsibilities relate to other contractors. For example, a concrete supervisor must coordinate efforts with the contractors responsible for the reinforcing steel, plumbing, and electrical work to ensure that their tasks are completed prior to concrete placement. The drawings indicate where coordinating efforts are required.

The supervisor should review the drawings systematically. One suggested method for reviewing drawings is as follows:

Step 1 Review site drawings for building layout and location, roadways, utilities, site, and building grades.

Step 2 Review structural drawings for sizes of footings, column pads, foundation walls, rebar replacement, areas of excavation, and form work requirements.

Step 3 Review structural drawings for the type of superstructure being used.

Step 4 Review the architectural drawings for:
- Floor plan and layout of individual rooms
- Elevations of height, finish, window, and door openings
- Wall sections, vertical sections cut through the building, dimensions between floors, dimensions of exterior and interior constructions

Step 5 Review the mechanical drawings for plumbing and HVAC layout, location of equipment, and sprinkler layout.

Step 6 Review electrical plans for lighting layout and power distribution.

Instructor's Notes:

Have trainees take time to complete the Participant Activity. Allow 25 minutes.

PARTICIPANT ACTIVITY

The instructor will provide a set of construction drawings. Using the following questions, go through the drawings and answer these questions.

As you do this, if you run into unusual terms or have questions, consult your instructor, supervisor, other crew members, or written materials, such as a dictionary of construction terms.

General Questions for Participant Activity

Site Plan

1. What is the contour interval?
2. What is the finish floor elevation?
3. Is the electrical service underground or overhead?
4. List the elevation at the northwest corner of the lot.
5. In what compass direction does the water flow?
6. What are the dimensions of the site?
7. What direction does the building entrance face?
8. How far is the east side of the building from the east property line?
9. What is the finished grade elevation at the building entrance?
10. What is the highest elevation on the site as shown by the grade lines?

Architectural

1. What are the scales of the drawing?
2. What are the interior dimensions of the _____ room(s)?
3. What types of materials are used in wall construction? Name walls and materials.
4. What materials are used as wall finishes? Name walls and finishes.
5. What wall thickness do you find?
6. How many columns are shown?
7. What is the distance from center of column to center of column?
8. What is the chief exterior material?
9. How high are the windows from sill to head? (Select one side of the building.)
10. What is the finished material used on the ceiling and how thick is it?
11. How thick is the floor? What is the material used?
12. What materials are cabinets and countertops made of?

Structural

1. How thick are the exterior foundation walls? How much reinforcing steel is used in the walls?
2. How thick are the exterior wall footings? How much reinforcing steel is used in the footings?
3. What are footing dimensions for columns?
4. How is the column base plate fastened to the foundation?
5. Where are keyways found? Why?
6. What is the specific compressive strength of concrete?
7. Are there any steel joists indicated? If so, give size.
8. What is the *basic* structural support system?
9. How far apart are the beams or girders spaced?
10. What size of anchor bolts is used, if any?

(continued on following page)

PARTICIPANT ACTIVITY

Have trainees finish the Participant Activity at home. Assign reading of Sections 4.0.0 – 8.13.0 for the next class session.

PARTICIPANT ACTIVITY
(continued)

Plumbing

1. What size and type of pipe is required to bring fresh water into the building?
2. Are there any hose bibbs on the exterior of the building? If so, locate.
3. What size and type of pipe is used to carry wastewater from the building?
4. What types of drainage have been provided on the roof?
5. Where do the main water lines run in the building (floor, ceiling, and/or walls)?
6. Have gutters and downspouts been provided?
7. Are there any natural gas lines? If so, where, and what size and type? What is their purpose?
8. How many different types of toilets are called for?
9. Is there a hot water drain line? If so, locate and give purpose.
10. How many hot water heaters does the building have, and where are they located?

Electrical and Mechanical

1. What type and size duct bank will be used to bring the electrical service to the building?
2. Where are fluorescent light fixtures used?
3. Where are incandescent light fixtures used?
4. Where are the following located?
   - Three-way switch
   - Duplex outlet
   - Telephone outlet
   - Light poles
5. Are there any pieces of mechanical equipment, such as motors, that give the electrical requirements on the drawings?
6. What types of duplex outlet cover plates are specified?
7. Is the main control panel a fuse box or circuit breaker? Where is it located?
8. Are the basic requirements for the building, 110 or 220 volts?
9. Is the building serviced by an intercom system?
10. Are outside (exterior) duplex outlets required?

Heating, Ventilation, Air Conditioning (HVAC)

1. What heating system is in the building?
2. Has air conditioning been provided?
3. Where is the main heating power plant located?
4. Where are the air conditioners located?
5. What controls the heat (such as a thermostat) and where are controls located?
6. Where are the heating elements (such as grilles, ducts, baseboards) located? Wall, ceiling, and/or floor?

PARTICIPANT ACTIVITY

Instructor's Notes:

SECTION 4

4.0.0 TECHNICAL SPECIFICATIONS

The supervisor must work with both the drawings and the specifications during the normal construction process. Many supervisors do not consider the specifications an important part of the project and rely only on the drawings for information. This can lead to serious problems, since the specifications provide important technical information about work to be done and material to be used.

Specifications are necessary because there is simply not enough room on the drawings to include all the required technical information. To illustrate the role of specifications, consider the job of building custom cabinets in a particular building. The plans would show the locations of the units, the elevations, and the arrangement of doors, drawers, and shelving. The specifications would describe the type of wood to be used, the hardware requirements, the finish, and the installation instructions. Obviously, using the plans alone would not produce cabinets that meet job requirements.

Specifications serve several purposes:

- They spell out the provisions of the contract and the responsibilities of the architect, the owner, the contractor, and the subcontractor.
- They supplement the working drawings with detailed technical information about the work to be done and the material to be used.
- They support the drawings and become part of the contract documents.

There are three general types of specifications:

- **Performance specifications** specify the outcome of the construction activity and allow the contractor to decide how to perform the work. For example, if the performance specification for a reinforced concrete foundation is a compressive strength of 3,000 pounds or more per square inch after 28 days, the contractor can select any method to install the foundation as long as the work meets that specification.
- **Descriptive specifications** tell the contractor exactly how to do a particular task. In the example of the reinforced concrete foundation, a descriptive specification might state, among other things, that the concrete must be placed by chute.
- **Standard (Reference) Specifications** are developed by a party other than the project architect, such as installation specifications prepared by the manufacturer of an air-handling unit. Associations, such as the American Concrete Institute, or companies can also formulate such specifications. A copy of all standard specifications for a job must be on the job site where the supervisor can refer to them as needed.

The architect is responsible for developing the specifications, making sure all necessary details are included, and resolving any conflicts between the specifications and the drawings. When conflicts do arise, the architect must decide whether the specifications or the drawings take precedence. Architects differ in their attitudes about specifications and drawings, but generally, the technical specification takes precedence over drawings.

Specifications usually follow the format created by the Construction Specifications Institute (CSI). This format divides the technical data into sixteen divisions, each covering a particular material or product and its installation. *Appendix A* illustrates the CSI format. The supervisor should be familiar with the divisions of the CSI format and the areas they cover.

PARTICIPANT ACTIVITY

Using the set of specifications provided by the instructor, review the parts and format of specifications.

Ensure that you have all the necessary materials to teach the course. Check the Materials and Equipment list at the front of the module.

Ask trainees for examples they have witnessed of the consequences of working with drawings without the specifications during a project.

Distribute and discuss samples of the three general types of specifications.

Ask trainees to relate their experiences with managing submittals and what they would do differently to improve the process.

SECTION 5

5.0.0 SUBMITTALS

Submittals, sometimes referred to as *shops*, is a general term used to describe information provided by subcontractors and vendors that provides the connecting link between design and construction. This information is presented in a variety of forms, but most commonly as shop drawings, catalogs, manufacturers' product data, samples, schedules, and mock-ups. All of this information required by a specific scope of work or scope component is called a *submittal package*.

5.1.0 Managing Submittals

Submittal management involves five basic steps:

Step 1   Establish project submittal requirements.

Step 2   Assign a responsible party to the required submittals.

Step 3   Schedule the prompt receipt of the submittals.

Step 4   Expedite the review and approval of the submittals.

Step 5   Return the submittals and ensure timely release of the material fabrication.

Submittals are typically tracked in two different ways. The *submittal data schedule* is often used on fairly straightforward design/build projects, while a separate *submittal log* is created for use on rather complex projects. Whichever system is used, the intent is the same: to provide meaningful information to all members of the project team.

5.1.1 Submittal Data Schedule

The submittal data schedule lists the submittals required for each division of the project specifications. If this schedule is used, a cursory review of the items listed should be checked against the specifications. This is a precaution to verify that all of the submittal requirements have been indicated. *Figure 19* shows a typical submittal data schedule.

5.1.2 Submittal Log

The submittal log generally provides more specific information than the submittal data schedule. Additional information includes submittal-processing dates such as date received from the subcontractor or vendor; date forwarded to the architect, engineer, and/or owner; the date returned; the date forwarded back to the subcontractor or vendor; and the date issued to the superintendent or field.

5.2.0 Processing a Submittal

Follow these general guidelines to ensure that your submittals are complete. Your company may have additional requirements.

Step 1   Date-stamp all submittals received. These might include shop drawings, bluelines, and manufacturers' data.

Step 2   Review the submittals received for completeness, substitution, and coordination.

Step 3   Enter the submittal information into a submittal log.

Step 4   File one copy of the submittal package in the project files.

Step 5   Transmit the submittal package to the appropriate company representatives for review.

Step 6   Review the returned submittal packages, and respond based upon the review comments.

Step 7   For critical material approved or approved-as-noted submittals, confirm the return of the submittal with the supplier or subcontractor, and obtain specific delivery information.

5.3.0 Reviewing Submittals

Submittals are reviewed to verify conformance with the contract documents including the specifications, drawings, subcontract, and scope. The degree to which project submittals are reviewed will vary from project to project, depending on several factors:

- Definition
- Completeness
- Substitution
- Coordination

Instructor's Notes:

SUBMITTALS REQUIRED						
CERTIFICATES	SHOP DRAWINGS	MFG. DATA	SAMPLE/ COLOR	ITEM	SPEC. SECT	SUPPLIER	DUE DATE

CERT	SHOP	MFG	SAMPLE	ITEM	SPEC. SECT	SUPPLIER	DUE DATE
	X			Flashing at Roof Penetrations	07600		
	X			Cap Flashing	07600		
	X	X		Roof Hatch	07724		
X		X		Joint Sealers	07900		
X		X		Hollow Metal Doors & Frames	08111		
	X		X	Wood Doors	08210		
	X	X		Accordion Folding Grilles	08355		
	X	X		Entrance Doors	08402		
	X	X		Glazing Framing Systems	08411		
		X	X	Finish Hardware	08710		
		X		Glass and Glazing Materials	08800		
			X	Glass Framing Sealants	08851		
	X	X	X	Structural Glazing System	08915		
X		X		Gypsum Drywall Materials	09260		
				Wall & Ceiling Assemblies	09260		
				Test Rpts. - Fire Rated Assemblies	09260		
X			X	Ceramic Tile	09310		
	X	X		Acoustical Material	09511		
			X	Resilient Flooring	09650		

Figure 19 • Submittal Data Schedule

Definition of a submittal or submittal package is important from the outset because submittal requirements differ for each job. It is important to clarify what the package should contain.

Completeness of a submittal is accomplished by answering four questions:

- *Are all of the required items included in the package submitted by the contractor or vendor?*

  Checking the package against the specified requirements and the number of copies included will verify a package's completeness.

- *Is any information missing?*

  Perhaps a drawing is missing or an important detail has been left off. One way to approach this review is to try assembling the work with the information provided in the submittal package. If this cannot be done, then the superintendent or the installer may have the same difficulty.

- *Is the information presented correctly, in an easy-to-understand format?*

  Evaluate the submittal for overall format, clarity, and accuracy. *Format* involves verification that certain submission procedures are observed. *Clarity* deals with the presentation of the information, while *accuracy* considers the correctness of the information. During the review for clarity and accuracy, make sure of the following:

  – The dimensions are obvious and easy to read

  – The manufacturer's product data agrees with the information shown in shop drawings

  – Essential legends or lists of notes are prominent

  – Strings of dimensions add up correctly

  – The details accurately represent the actual field conditions

  – The product warranty includes any conditions pertinent to the project. For example, if a material is not to be used in oceanfront applications and you are building a seaside condominium complex, has this been considered?

- *Have notes been added to the submittal package which would reduce or change the scope of work?*

  Sometimes a vendor or subcontractor will not communicate the contractual requirements of their entire scope to the drafter creating the drawings. Information is either omitted from the drawings or is indicated as "not in contract." Also, vendors or subcontractors will sometimes add "work" to their shop drawings that is not shown on the contract drawings. If the submittal is processed without comment, then the installer may assume that the material will be installed before work begins.

  Substitution generally occurs because the specified product is no longer made, the product was specified incorrectly, or the subcontractor or vendor is trying to save money or time. Substitutions should be taken seriously and evaluated to make sure that they meet or exceed the owner's original requirements.

Coordination between subcontractors is essential. As projects become more and more complex, the interdependence of different trades is on the rise. Review all submittals for multiple trade and discipline conflicts. All material installations should be reviewed.

For example, you may need to ask questions such as "Will the handrail need cast-in-place embeds for attachment?" "Is there an electrical circuit for the overhead doors?" "Where does the in-wall blocking for the millwork go?" For large-scale coordination, you may need to hold a specific coordination meeting. Another method of coordinating the work is to issue a specific set of coordination drawings.

SECTION 6

6.0.0 METHODS OF OBTAINING WORK

To obtain work, contractors must develop a price for the proposed project and submit that price to the owner, the architect, or another contractor. There are four methods for doing this, each based on the way the owner or contractor requests the bid information.

Competitive bidding is the most common method of obtaining and awarding work. In competitive bidding, an owner or general contractor requests bids from several contractors and normally awards the work based on the lowest qualified bid. The competing bids are due on a predetermined date and time.

In **invited bidding,** the owner selects contractors based on their experience, financial strength, present and future workloads, and management systems and asks exclusively these contractors to bid on a project.

In *negotiated bidding*, the owner or contractor selects one or more contractors to prepare and submit bids on a total project or on various phases of the project, and then negotiates the final cost based on the established budget.

In *design and build bidding*, the owner selects a contractor to be responsible for design and construction of a project, within a specific budget and set guidelines.

In all four methods, the contractor must go through the mechanics of compiling estimates

SECTION 7

7.0.0 CONTRACTS

The type of contract awarded by an owner to a contractor or by a contractor to a subcontractor depends upon:

- The type of project
- The time frame of the project
- The accuracy of the drawings and specifications

7.1.0 Types of Contracts

The five most common types of contracts are:

- **Lump sum**
- **Cost of work plus a fee**
- **Unit cost**
- **Construction management** (agent of the owner)
- **Guaranteed maximum price**

7.1.1 Lump Sum Contracts

Lump sum contracts have the total cost of the project established before construction begins. They are used when the drawings and specifications are complete.

7.1.2 Cost of Work Plus a Fee Contracts

Cost of work plus a fee contracts, or cost-plus contracts, are based on the actual cost of work plus a fee to cover the contractor's management cost. This fee may be either a percentage of the cost or a fixed amount.

Cost-plus contracts are usually used when:

- It is impractical for the owner to complete drawings and specifications before construction begins.
- Construction must start immediately in order to take advantage of low interest rates.
- Changes in zoning regulations occur.
- Many changes are anticipated during construction.

For the project. This estimating procedure takes considerable time and resources, including company personnel and support from suppliers and other contractors.

7.1.3 Unit Cost Contracts

Unit cost contracts, or unit price contracts, are based on the price of increments or phases of work within the total project. Unit cost contracts are used when quantities of materials or equipment may have been predetermined or the extent of the work has not been fully established. This type of contract is used primarily in the construction of highways, bridges, and dams, as well as interior work where work cannot be defined until tenants' needs are determined.

7.1.4 Construction Management Contracts

Construction management contracts, or agency contracts, are used when the owner needs a construction manager to review the contract documents and accelerate construction. Owners hire construction managers because of their experience in the technical aspects or administration of a particular type of project. The construction manager usually does not perform any construction, and all purchasing, negotiating, subcontracting, and paying bills is done in the name of the owner.

7.1.5 Guaranteed Maximum Price

In a guaranteed maximum contract, the contractor agrees to perform a specific scope of work for a determined price. If the cost of the work is greater than anticipated, the overrun will be borne by the contractor. If the cost is less than anticipated and savings result, these savings will usually be shared between the contractor and the owner at a percentage defined in the contract. This is normally done within a fixed completion time and may specify liquidated damages and/or bonuses.

7.2.0 Relationships

The type of contract used and the method of obtaining work usually establish the relationship of the parties involved in a construction project. In competitive or invited bidding situations, for example, the owner awards the project to the lowest bidder. The lowest bidder, in turn, awards his subcontracts to suppliers and other contractors who provide the lowest bids. In such a situation, the relationship between the parties is strictly a business relationship, since they were selected only on the basis of their bids and not because of previous experience with the organization.

Discuss the role of relationships in obtaining work in the construction industry.

Distribute and discuss samples of the five most common types of contracts, if available. Try looking for samples in the books listed in the Additional Resources section.

Distribute and discuss samples of closeout documents. Try looking for samples in the books listed in the Additional Resources section.

Here the contract is usually a lump sum contract, with profit being the difference between the contract amount and the actual cost of the work. If the costs exceed the lump sum, the contractor absorbs the loss.

A negotiated bid provides an owner with either a guaranteed maximum price or a cost-plus for the project. The contract award is based on the contractor's qualifications as well as price. In turn, the contractor awards contracts to subcontractors and suppliers based on their qualifications and price. This creates an atmosphere of trust among the parties, allowing changes and disagreements to be worked out in a spirit of cooperation and providing incentives to reduce cost. The owner receives more than just the minimum outlined in the drawings and specifications, and contractual problems are minimized through the negotiation process.

SECTION 8

8.0.0 DOCUMENTATION

Documentation was once a secondary activity for the supervisor, but during the past decade, documentation has grown to become one of the most important tasks. The reason is simple: documentation is vital to a successful business. A study conducted by Arthur Andersen & Company shows that poor documentation is the source of many of the industry's delays (see *Figure 20*).

Documentation is the *who*, *what*, *where*, and *when* of a job — a collection of facts that records the actual history of the project. The extent of documentation required depends on the project, the owner, the architect/engineer, the company, and the project team members. Some form of documentation is required on all projects.

Documentation is primarily an effort to record the progress of a project, control time, and keep abreast of cost. It is also a means to develop strategies throughout the life of the project.

The primary documents a contractor may require a supervisor to maintain are:

- Project communication
- Daily logs or reports
- Meeting minutes
- Photographs
- Schedule updates
- Time cards
- As-built drawings

There are also certain closeout documents that must be completed at the end of a project before the owner accepts the final work. They include:

- **Certificate of Occupancy**
- **Certificate of Substantial Completion**
- **Punch list**
- Operation and maintenance manuals
- Warranties and guarantees

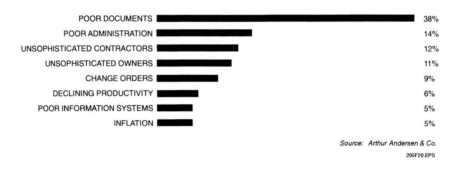

Figure 20 • Study of Industry Delays

Instructor's Notes:

Each of the primary and closeout documents and the supervisor's responsibilities regarding them are discussed further in this module.

8.1.0 Criteria for Good Documentation

Good documentation is at the heart of managing contracts and projects. In order to have good documentation, the following six criteria must be fulfilled:

1. *The documentation should be accurate.* The facts should be told the way they actually happened. If the facts are incorrect in one place, it may be assumed that facts have been incorrect throughout the documentation.

2. *The documentation should be objective.* This means that the facts should be stated in an unbiased way. This may be very difficult since the supervisor may be directly involved in the situation being documented.

3. *The documentation should be complete.* If the company provides the supervisor with a form for documentation, it should be filled out completely. Incomplete forms demonstrate inaccuracy or lack of knowledge of the situation.

4. *The documentation should be uniform.* Information should be recorded in a precise, routine manner. For example, all field orders received from the architect should be numbered and recorded, not just the field orders that involve additional work.

5. *The documentation should be credible.* It should be completed in a manner that indicates timeliness and lack of bias. The supervisor should be certain that all documentation is accurate and carefully done, and not be overly concerned with misspelled or crossed out words.

6. *The documentation should be timely.* All documentation should be completed when events occur, not be put off until later. It is easier to remember what you had for lunch today than what you had for lunch two weeks ago.

8.2.0 Communication

Throughout every project, supervisors have to communicate their thoughts, ideas, and actions to others. Many times, the communication is verbal only. Verbal communication makes for poor documentation because it forces the parties involved to depend on their memories to recall facts and issues that may have been discussed months earlier. Even the best memory is sometimes unreliable. Supervisors should therefore use written correspondence when transmitting information to the owner, architect/engineer, contractors, and vendors, and it should always be transmitted on a timely basis and to the appropriate parties.

Correspondence includes formal letters, speed messages, memos, and other documents authorized by the company. *Figure 21* illustrates one such document — a **Request for Information (RFI)**, which is used to request clarification about a problem. Other types of information that should be in writing are:

- Notification of delays
- Disruption of scheduled work activities
- Changes in project schedule
- Changes to drawings and specifications
- Any problem that may result in a claim

The golden rule of documentation is *always get it in writing!* For example, if the owner or architect instructs the supervisor to perform work that has not been assigned by the manager, the manager should be contacted immediately to make sure the changes were authorized in writing — before starting the work.

8.3.0 Daily Log

Completing a **daily log,** or a daily project report, is very important and cannot be stressed enough. The daily log is the permanent written record of the job as recorded by the supervisor, and the information in it can benefit the company in many ways. For example, no contractor contemplates lawsuits or claims in advance of starting a job. However, if a lawsuit or claim arises, the supervisor's daily logs are the company's primary defense. Testimony based on daily written documentation is more credible and has greater impact than testimony drawn from an individual's memory.

Ask trainees for examples of cases when written correspondence should be used when transmitting information.

DATE ______________ RFI NO. ______________

PROJECT NAME ______________________________ PROJECT NO. ____________________

REQUEST: REF D.W.G.NO. REV. OTHER

BY: ______________________________ REPLY BY (DATE): ____________________

REPLY:

DATE: ____________________

205F21.EPS

Figure 21 • Request for Information

Instructor's Notes:

The daily log provides the following information on a daily basis:

- Facts about the work
- Work progress
- Special requirements or problems
- Notice of equipment and material deliveries
- Company and subcontractor personnel on site
- Visitors to the job, including the owner, architects, inspectors, testing agents, and anyone else who is not employed by the company

The format of daily logs varies with the type of information each company requires from its supervisor, but the basic intent of a daily log remains the same: to document what takes place at the job site and to provide information to management on a daily basis.

Most companies have their own daily log forms. *Figure 22* shows a typical daily log form and the type of information contained.

The supervisor must complete a log form every day when the information is fresh. It should never be done at the end of the week or the end of the month since this forces you to rely on your memory and inevitably leads to serious errors or omissions.

Teaching Tip

Distribute and discuss samples of daily log forms or discuss Figure 22.

DAILY PROJECT REPORT

Project: Midtown Office Bldg. Job #63   **Date:** October 13, 20 – –
Report By: Joe Green   **Report No.** 56
Weather: Avg. Temp: 78   Sunny   Cloudy   Rainy   Clear

Crew: (Subcontractors)

Plumbers	6 incl. 1 foreman	Sheet Metal	4	Cooling Tower	
Fitters	7 incl. 1 foreman	Insulation	1	Refrigeration	
Laborers	1	Temp. Controls	1	Others	1 Project Engineer
Operators		Painter			

Job Progress Report:
3 plumbers working in basement pump room
2 plumbers roughing in 3rd floor
3 pipefitters in main equipment room–chilled and condenser water lines
2 fitters working on HVAC risers–4th floor
1 fitter fabricating coil connections
4 chilled and condenser water pumps delivered today

Special Decisions or Instructions: *General Contractor requested that we pull off 3rd floor so drywall contractor can stock floor with sheet rock. (This will affect sheet metal sub and will cause us to shift crew of plumbers to another work area.)*

Items Needing Expediting: *Air handling units were scheduled to ship October 6, but I've heard no word. Please check with supplier and expedite. Need badly or will have to cut back on pipefitter crew and will be covered up by sheetrocker.*

Figure 22 • Typical Daily Log Form

Have trainees give suggestions for other items they might place on a meeting agenda and why.

8.4.0 Meetings

The supervisor may be asked to attend job meetings. Job meetings can be very helpful if their purpose is to bring people together who have different areas of responsibility and different views of the project progress in order to resolve problems, take corrective action, or make decisions. As a result, every meeting should be designed to exchange ideas, not just to pass out information.

Generally, a well-planned meeting has an agenda that advises attendees in advance of the objective of the meeting. *Figure 23* is a typical job-site meeting agenda. When preparing for a meeting, the supervisor should make notes on the topics that will affect the company. The better prepared the supervisor is, the more quickly the meeting moves.

During the meeting, the supervisor is responsible for seeing that notes or minutes are taken. Minutes and notes are part of the project documentation and become a source of information in any potential legal actions. Therefore, their accuracy is extremely important. *Figure 24* illustrates the type of information that should be included in job minutes.

8.5.0 Project Photographs

Photographs are one of the best forms of documentation. One picture can show what twenty pages of words cannot describe. Photographs are useful for documenting the project. Photographic logs of projects are useful for reference.

A photograph can:

- Show daily progress and quality of workmanship
- Communicate problems to the architect/engineer or to the home office
- Show construction techniques to the estimating department

The number and subjects of photographs that the supervisor takes will vary from project to project, but generally, photographs should be taken when:

- Existing conditions vary from the contract documents
- Materials are stored on site
- Others damage your work
- Unsafe conditions or accidents occur
- The supervisor wants to record something for reference later

Photographs are also useful for:

- Showing project progress in a local newspaper or the company newsletter
- Demonstrating new construction techniques
- Documenting job-site pickets or other labor unrest

MEETING AGENDA

*1. Review last meeting's minutes.
*2. Review work progress since last meeting.
*3. Note field observation, problems, and decisions.
*4. Identify problems that impede planned progress.
 5. Review off-site fabrication problems.
*6. Develop corrective measures and procedures to regain planned schedule.
*7. Revise Construction Schedule, as necessary.
*8. Plan progress during next work period.
*9. Coordinate schedule of trades.
 10. Review submittal schedules and means of expediting them to maintain schedule.
*11. Discuss maintaining quality and work standards.
 12. Review changes proposed by owner for their
   - Effect on the Construction Schedule
   - Effect on the Completion Date
 13. Complete other current business.

* Indicates items for superintendent's weekly meetings; all items are included in project manager's meetings.

Figure 23 • Typical Job-Site Meeting Agenda

Instructor's Notes:

Module MT205 ◆ Contract and Construction Documents

Job Meeting Minutes
XYZ Company, Inc.
Minutes May 27, 2001
Subject: Subcontractors' Meeting
Time: 9:00 a.m.
Attendees:

Roger Jones	XYZ Mechanical Company
Philip Smith	XYZ Formwork Company
Dick White	XYZ Electric Company
Don Helms	XYZ Drywall & Painting Company
Mark Lord	XYZ Flooring Company
Dick Taylor	XYZ Company, Inc.
Susan Johnson	XYZ Glass Company

Safety
All subcontractors were reminded that this project is a hard hat job. Hard hats must be worn at all times.

Schedule
The total project is nine days ahead of schedule. All subcontractors must maintain their present crew level and productivity in order to remain ahead of schedule.

General Notes
- Next window delivery is scheduled for June 2nd.
- Metal studs on the 5th floor will be completed by June 5th.
- Domestic water rough in on 5th floor will be completed on June 9th.
- Electric rough in on 5th floor will start June 9th.
- Flooring for top floors will start next week.

1. **Change Order.** The owner has revised the first floor layout. All work on the first floor is on "hold" until the architect releases new drawings.

2. **Clean up.** Subcontractors must clean up their own trash. A dumpster is located at the south end of the property for all contractors to use.

Meeting adjourned at 9:32 am.

The meeting minutes as transcribed will become part of the project record. If you disagree with the minutes, contact this office within seven days with your corrections or comments.

Minutes recorded and transcribed by: _____

Peggy Worton, Project Secretary

Figure 24 • Typical Job-Site Meeting Minutes

Classroom

Have trainees give suggestions for other items they might see in typical job-site meeting minutes.

Ask trainees to suggest other reasons photographs might be useful in the workplace.

To make each photograph a document, the following information should be recorded on its reverse side:

- Job name and number
- Date of the photograph
- Name of the person taking the photograph
- Subject or purpose of the photograph
- Other comments describing the photograph, including location, orientation, and time of day

Figure 25 illustrates job photos and the type of information needed to describe them properly.

Every job site should have a Polaroid or a 35mm digital camera, and the supervisors should use it on a regular basis. Digital cameras allow the images to be stored and easily manipulated in project logs or reports that are transmitted or archived digitally. Another good alternative is a portable video camera. The miniaturization of video technology has made cameras small and easy to operate, and the fact that videotape not only shows action, but also can be used to record verbal notes as scenes are being recorded gives them certain advantages over still photos. The supervisor must be certain that the label of each videocassette has the same information that would be recorded on the back of a photograph.

Job: Zenith Office Complex – Office #1986
Date: March 14, 2001
Form Work

Job: Zenith Office Complex – Office #1986
Date: April 22, 2001
Exterior Wall

205F25.EPS

Figure 25 • Job Photos

Instructor's Notes:

8.6.0 Schedule

The **project schedule** is an estimate of the time and the sequence of events necessary to get the job done and is critical to the timely completion of work. The supervisor should review the schedule on a regular basis and make sure the completion date can be met. If the completion date *cannot* be met, the superintendent or project manager must be notified immediately. In addition, all events that affect the schedule must be documented, whether they are events the supervisor causes or controls or others cause or control.

The supervisor should notify management of events that influence the schedule and be prepared to suggest methods to resolve the scheduling problem.

8.7.0 Daily Time Reports

Time reports are the one source of documentation readily available to the supervisor on a daily basis. Daily time reports should be completed every day and should contain information such as:

- Hours worked
- Areas or activities worked on
- Production rates

Time reports also reflect additional work done because of a change order or directive from the architect or owner, and are therefore a valuable record of the cost of such work. As a result, the supervisor should treat time reports as documents and make sure they are accurate. *Figure 26* is an example of a supervisor's daily time report.

8.8.0 As-Built Drawings

As-built drawings are developed to show changes to the project as it is being built. When changes are made, they must be recorded in order to avoid conflicts with crews. It is usually the supervisor's responsibility to maintain and record as-built drawings.

Normally, the architect/engineer will furnish a set of drawings (either prints or sepias) for recording changes. These drawings must be clearly identified, kept at the job site, and kept separate from other job drawings.

The general contractor is responsible for maintaining a complete set of as-built drawings and for making sure that each subcontractor keeps the as-builts updated.

At the completion of the project, the as-built drawings are submitted to the architect/engineer as part of the closeout documents. Many times the architect/engineer or the owner will withhold monthly or final payment until as-built drawings have been submitted.

8.9.0 Closeout Documents

After construction has been completed, the project closeout documents still must be completed. Among the closeout documents usually required are:

- Certificate of Occupancy
- Certificate of Substantial Completion
- Punch list
- Operation and maintenance manuals
- Warranties and guarantees

8.9.1 Certificate of Occupancy

The certificate of occupancy is the document provided by the building inspector certifying that all items meet local codes and that the building is now safe for occupancy.

Several types of inspections such as electrical and environmental are usually required before the certificate of occupancy is issued.

8.9.2 Certificate of Substantial Completion

When the contractor and the owner agree that the substantial portions of the contract have been honored and the majority of the work has been completed and approved, a certificate of substantial completion can be issued. This allows the owner to begin using the project for its intended purpose. In order for this certificate to be issued, the contractor's remaining work must not interfere with this use. Only minor work must remain, such as finishing the landscaping, detail work, or the punch list, discussed next.

Ask trainees to give examples of types of events that might affect the project schedule and how the events should be documented.

Description of Work & Activity Code Number

Contract No. Location

Date Day of the Week

Supervisor Signature

Approved By

Name	Social Security No.	R O	R O	R O	R O	R O	R O	R O

205F26.EPS

Figure 26 • Typical Supervisor's Daily Time Report

Instructor's Notes:

8.9.3 Punch List

The punch list is the result of a formal detailed inspection of completed work by the architect/engineer before final acceptance. It identifies items that are not acceptable or not constructed according to the contract requirements. Some items on the punch list, such as repainting a wall, repairing a damaged item, or adjusting doors, can be relatively easy to correct. Others, such as fan noise or irregular brickwork are more difficult to correct. In either case, the supervisor should take each item seriously because final payment for work will not be made until all punch list items have been completed.

8.9.4 Operation and Maintenance Manuals

The equipment provided under the project contract should have manuals containing all information necessary for:

- Operating
- Maintaining
- Repairing
- Dismantling
- Assembling
- Identifying parts for ordering replacements

Operation and maintenance manuals must be furnished to the owner upon completion of the project.

8.9.5 Warranties and Guarantees

Contractors and suppliers who guarantee workmanship and material integrity for a specified length of time furnish warranties and guarantees.

8.10.0 Change Orders

Many times during the construction process, the owner or the architect/engineer will request changes to the project. It is the supervisor's responsibility to carry out those changes when they appear in the form of **change orders**.

A change order occurs because of:

- Budget revisions
- Design defects and deficiencies
- Revised project needs

8.10.1 Budget Revisions

The first type of change order occurs when an owner's project budget changes, often because of adding or deleting work. Such changes are usually done before the actual construction, and the supervisor's involvement is minimal.

8.10.2 Design Defects and Deficiencies

Changes due to design defects and deficiencies affect supervisors more directly because they are often the ones who discover the defect. The home office should handle this type of change order, as it probably will result in disagreement among the contractor, the owner, and the design professional as to where the responsibility lies for the deficiency. This type of change order is often the most difficult to resolve.

8.10.3 Revised Project Needs

Change orders resulting from revised project needs come from the architect in response to a request by the owner. In issuing the change order, the architect makes it clear that the contractor will be duly compensated for the work involved. Generally, change orders resulting from revised project needs are the easiest to resolve.

The supervisor often assists in the documentation of change orders. The supervisor must be aware of the changes and be prepared to assist the project manager or home office in preparing estimates of the expected costs of doing the change work requested.

The supervisor may be involved in two types of change orders: minor change orders or time and material change orders

8.10.4 Minor Change Orders

Minor change orders are often the result of interference or obstructions to work. They are situations discovered in the field, and the company is entitled to compensation for the additional work involved. Although they do not often involve large sums of money, minor change orders can accumulate, and the cumulative costs can become a hardship for the company if proper compensation is not received. It is therefore the supervisor's responsibility to notify the home office of these types of changes using whatever forms the company specifies for this purpose.

Review the seven basic types of insurance. Discuss when each type of insurance is used and what is covered in different situations such as a tree falling on a rental forklift, on a company-owned backhoe, or an employee.

Show Transparency 18 (Basic Types of Insurance).

8.10.5 Time and Material Change Orders

Time and material change orders can be very costly to the company and its subcontractors if not properly handled. It is usually the supervisor's responsibility to properly track the actual cost of this type of change order, and *Figure 27* shows an example of a form for doing this. A supervisor should never proceed with a change until written authorization from a superior and/or the architect/engineer has been received.

8.11.0 Insurance

In order for a company to operate, it must acquire and maintain an extensive amount of insurance of various types. Insurance is necessary to protect the employees, the owners, the public, and the company itself from losses that the company cannot afford. When a company has this kind of insurance coverage, it means they are paying high annual premiums.

Insurance does not protect the company from all losses, since a deductible is often stipulated in each policy. In addition, when an accident or loss occurs, the company's premiums increase. The supervisor should understand this and assist management by being aware of risk exposure and helping to reduce premiums.

The basic types of insurance a company may carry are:

- Worker's compensation
- General liability
- Vehicle
- Builder's risk/installation floater
- Equipment floater
- Property
- Medical/dental/life/disability

8.11.1 Workers' Compensation Insurance

Workers' compensation insurance pays employees injured on company time. Most states require companies to maintain coverage in each state in which they work or risk financial penalty. Supervisors must report all injuries to their immediate supervisors. Failure to do so may result in the injured employee losing benefits and the company being penalized.

8.11.2 General Liability Insurance

General liability insurance provides coverage from claims that the company is obligated to pay because of bodily injury or property damages. Many general liability policies do not cover all possible situations, so the supervisor should be aware of what is covered under the company's policy and alert the home office if any situation arises that may be covered by the policy.

Premiums for general liability insurance can be very high, and the cost is included in the company's overhead or job cost. Premiums increase due to a loss or claim, and consequently the company's costs rise, making the company less competitive.

8.11.3 Vehicle Insurance

Vehicle insurance covers losses the company incurs from ownership or usage of automobiles or trucks and usually includes physical damage to company vehicles due to collision, fire, theft, and vandalism. Many companies carry insurance on leased or employee-owned vehicles used for company business.

If an accident involving a company-insured vehicle occurs, the supervisor should notify the home office. The employee involved in the accident must complete the accident report with:

- Information on the cause of the accident
- Names of witnesses
- Names of parties involved
- Photographs of the accident
- Sheriff or police reports

All pertinent documents should be sent to the home office as soon as possible so that the insurance company can be informed and involved in any potential legal action.

8.11.4 Builder's Risk/Installation Floater Insurance

Builder's risk/installation floater insurance only covers installed work, such as losses of materials and labor due to fire, theft, vandalism, and windstorms. Unless contractually provided by the owner, the contractor maintains all-risk builder's risk insurance in an amount equal to the cost of the project.

Instructor's Notes:

Review the process of completing a time and material change order and discuss when and why this form is used.

Figure 27 • Time and Material Change Order

An installation floater policy is usually written in the amount of the subcontractor's contract or the general contract. The policy is in effect when the contractor gains possession of the materials, during temporary storage on or off site, while in transit, and during installation. Coverage starts upon acceptance of the work performed.

Should a loss occur on the project, the supervisor should notify the home office immediately so that the insurance company can be advised. The supervisor may be required to prepare a report that contains a list of the material lost or damaged, photographs, names of any witnesses, and a copy of the sheriff or police report.

8.11.5 Equipment Floater Insurance

Equipment floater insurance provides coverage against loss or damage of company-owned or leased equipment and tools that are not covered under the builder's risk insurance policy. The policy covers only pieces of equipment that are listed on the insurance policy schedule of equipment. It is up to the supervisor to advise the home office of any additions of equipment on the job so that the policy can be adjusted to reflect the changes. The coverage includes damage or loss of equipment due to fire, theft, or accidents. Supervisors should understand company policies and avoid paying for additional insurance that may be charged by equipment rental companies.

8.11.6 Property Insurance

Property insurance provides coverage of all company-owned and leased property, including the contents of main office buildings and jobsite offices.

8.11.7 Medical/Dental/Life/Disability Insurance

Medical/dental/life/disability insurance provides coverage to all eligible employees and is considered an employee benefit.

Insurance is an extremely complex subject, but the supervisor must be able to recognize potential problems and report them properly to the home office. The supervisor must also assist the company by avoiding accidents and losses that lead to insurance claims.

8.12.0 General and Special Conditions

The general and special conditions in the contract documents define the relationships and duties of the owner, the architect/engineer, and the contractor. Generally, the architect/engineer uses a standard set of General Conditions developed and published by the American Institute of Architects. These standards are known as AIA Document A201.

When it is necessary to modify the standard general conditions, the architect/engineer issues special or supplemental conditions.

The supervisor must carefully read and understand the clauses in both the general and special conditions. The following represent simple clauses typically found in general and special conditions.

Change clause gives both the contractor and owner the right to make changes within the scope of the contract without having to negotiate a new contract each time a change is requested. The clause outlines the procedure to be followed when a change is being considered.

Concealed site condition clause covers soil conditions, water problems, or other hazardous conditions such as abandoned fuel tanks, asbestos, mercury, or lead that differ from those anticipated at the time of bidding and contract acceptance. The clause allows the contractor to charge the owner for extra costs incurred because of unforeseen conditions. When unforeseen site conditions are discovered, supervisors must notify their superiors immediately and take the following steps to document the condition:

- Note the condition of the site prior to uncovering the unforeseen condition.
- Describe the conditions that were discovered.
- Note the difference between what was anticipated and what was actually found.
- Document all facts in writing and take photographs of the condition.

Hold harmless clause states that if the owner is sued regarding a situation involving the job, and the owner loses the lawsuit due to the fault of the contractor, the contractor will reimburse the owner for all costs.

Instructor's Notes:

Liquidated damage clause states that if a contractor does not complete work within the time specified in the contract, a penalty—usually a dollar amount per day—will be charged to the contractor. It is important that the supervisor maintains records for excusable delays such as those from unanticipated weather conditions or changes made by the owner or architect/engineer in order to avoid these charges. The supervisor should also document all requests for time extensions and all events and situations that might prevent completing work on time, such as a delay in the return of shop drawings or owner-furnished equipment not being delivered on schedule.

8.13.0 Back Charges

During the course of a project, the supervisor may be requested to perform work that is outside the terms of the contract documents or the contractor's normal scope of work, such as furnishing labor, material, or equipment in assisting another trade or repairing work damaged by another contractor.

These extra costs incurred between contractors on a job site are called *back charges*. The supervisor must recognize that back charges are between contractors and do not affect the contract with the owner; nonetheless, the accumulation of back charges can cause the company's costs to increase or decrease.

Among the most common back charges a supervisor may be involved with are:

- One contractor furnishing scaffolding to another contractor
- One contractor providing saw cutting and patching for another contractor
- One contractor furnishing a pump to another contractor to remove water from a trench
- One contractor providing labor and equipment to remove another contractor's trash

Any time a supervisor is asked to approve a back charge, it should be checked with their manager or home office to verify that the back charge is justified and signing it is authorized. When a back charge is signed, a copy should be sent to the home office for action; the reasons for the back charge should be noted on the daily log report.

SUMMARY

Understanding and using contract and construction documents is an essential part of construction project supervision. In order to be effective at supervising construction projects, you need to be able to understand what documents are used in construction, from pre-planning documents to actual plans and project tracking documents. This involves being able to read basic plans and understand and recognize standard construction symbols. It means having a working knowledge of industry specification standards and formats and knowing how they are used. Understanding submittals and how to manage them is one of the most important pieces of project communication and follow-though.

Supervisors need to understand the different type of contracts that exist in the construction industry as well as how companies get work. Contract and construction documents and their maintenance directly relate to the success of the business. Being a good manager involves putting proper systems in place for understanding comprehensive contracts and managing the project throughout with meetings and other forms of communication.

Project closeout requires a great deal of review and documentation. Supervisors must recognize that projects are extremely complex, and take into consideration legal considerations as well as regulatory constraints to be able to understand what constitutes a finished project.

Ask trainees to share examples of back charges they have experienced in their workplace and what the company did to prevent back charges in the future.

Instructor's Notes:

Have trainees complete the Review Questions. Discuss the correct answers, located at the end of this module.

Review Questions

1. The spec book is sometimes called a _____.
   a. set of drawings
   b. general condition
   c. form of agreement
   d. project manual

2. The term *contract* is often referred to as a _____.
   a. general condition
   b. form of agreement
   c. bid form
   d. payment of performance bond

3. The following are all types of drawings except _____.
   a. competitive
   b. presentation
   c. preliminary
   d. working

4. Project photographs are useful for all of the following purposes except _____.
   a. documenting vandalism
   b. uncovering estimation errors
   c. showing daily progress
   d. illustrating new construction techniques

5. All of these statements apply to specifications except _____.
   a. they support the drawings and become part of the contract documents
   b. they supplement the working drawings with detailed technical information
   c. Their creation is the responsibility of the architect
   d. their creation is the responsibility of the general contractor

6. All of the following are true about submittals except _____.
   a. they include a schedule of the submittal items
   b. they are referred to as *mocks*
   c. they include such items as shop drawings, product data, and samples
   d. they improve coordination between subcontractors

7. The most common way to obtain and award work is _____.
   a. negotiated bidding
   b. competitive bidding
   c. invited bidding
   d. design and build bidding

8. The estimating process includes all of the following except _____.
   a. using information from subcontractors
   b. using company personnel
   c. using information from suppliers
   d. using estimates from competing general contractors

9. The following are examples of the most common type of contract *except* _____.
   a. lump sum
   b. guaranteed minimum
   c. unit cost
   d. guaranteed maximum price

10. *Cost of work plus a fee* contracts are used for _____.
    a. bridges
    b. changes in zoning
    c. undefined interior work
    d. accelerated construction requests

11. Many of the completion delays in the construction industry are caused by _____.
    a. poor safety programs
    b. inadequate training
    c. high wages
    d. poor documentation

12. All of the following documents must be completed before the owner accepts the final work *except* _____.
    a. statement of errors and omissions
    b. certificate of occupancy
    c. certificates of substantial completion
    d. warranties and guarantees

13. The daily log provides all of the following *except* _____.
    a. work progress
    b. special requirements or problems
    c. safety procedures
    d. notice of deliveries

14. As-built drawings must meet all of the following requirements *except* _____.
    a. clearly identified
    b. created by the subcontractors
    c. kept at the job site
    d. kept separate from the other drawings

15. A change order occurs for all of the following reasons *except* _____.
    a. budget revisions
    b. design defects
    c. documentation requests
    d. revised project needs

Instructor's Notes:

Trade Terms Introduced in This Module

Addenda: A written comment used to modify the contract documents before the bidding process.

Architectural drawings: Drawings that contain information on floor plans, elevations, sections, details, and schedules.

As-built drawings: Drawings are developed to show changes to the project as it is being built.

Back charges: Extra costs incurred when companies bill each other for work provided at the job site.

Bid bond: A legal document posted by the bidder as an assurance that the bidder will sign the contract if it is awarded to them.

Certificate of Occupancy: A document that states the building has met inspection standards and is ready to be occupied.

Certificate of Substantial Completion: A document indicating that the majority of a project has been completed and that the project may be used or occupied while the contractor completes the remaining items. It must be approved by both the owner and contractor.

Change clauses: Statements that give both the contractor and owner the right to make changes within the scope of the contract without having to negotiate a new contract.

Change order: Documents to modify a project when budget revisions or design defects and deficiencies occur. They are also created to deal with revised project needs.

Competitive bidding: An owner or general contractor requests bids from several contractors and normally awards the work based on the lowest qualified bid. This is the most common method of obtaining and awarding work.

Concealed site condition clauses: Legal clauses that cover conditions that differ from those anticipated at the time of bidding and contract acceptance.

Construction management contracts: Agreements used when the owner needs a construction manager to review the contract documents and accelerate construction. Also called *agency contracts*.

Cost of work plus a fee contracts: Contracts based on the actual cost of work plus a fee to cover the contractor's management cost. Also called *cost-plus contracts*.

Daily log: Document that is the permanent written record of the job as recorded by the supervisor.

Descriptive specifications: Instruct the contractor exactly how to do a particular task.

Design and build bidding: The owner selects a contractor to be responsible for design and construction of a project, within a specific budget and set guidelines.

Guaranteed maximum contract: A contract in which the contractor agrees to perform a specific scope of work for a determined maximum price.

Ask trainees to review the Glossary for unfamiliar terms.

Review the objectives of the module and then answer any questions the trainees may have.

Have trainees prepare for the Module Examination.

Administer the Module Examination. Be sure to record the results of the Exam on Craft Training Report Form 200 and submit the results to the Training Program Sponsor.

Hold harmless clause: Clauses that state that if the owner is sued regarding a situation involving the job, and the owner loses the lawsuit due to the fault of the contractor, the contractor will reimburse the owner for all costs.

Invited bidding: Type of bidding very similar to competitive bidding except that only pre-qualified or selected contractors are invited to submit bids.

Lump sum contracts: A contract in which the total cost of the project is established before construction begins. They are used when the drawings and specifications are complete.

Negotiated bidding: The owner or contractor selects one or more contractors to prepare and submit bids on a total project or on various phases of the project, and then negotiates the final cost based on the established budget.

Payment and performance bond: Bond posted by the contractor that guarantees the owner that the contractor will perform the work in accordance with the terms of the contract and pay all bills and obligations incurred under the contract.

Performance specification: Specify the outcome of the construction activity and allow the contractor to decide how to perform the work.

Preliminary drawings: Beginning drawings prepared by the architect/engineer during the promotion stage of the building's development.

Presentation drawings: Drawings that include architectural perspective views of the building used to make formal presentations

Project manual: See *specifications*.

Project schedule: An estimate of the time and the sequence of events necessary to get the job done; it is critical to the timely completion of work.

Punch list: List of items that need to be fixed before the project receives final acceptance.

Request for Information (RFI): Document used to request clarification about a problem.

Schedules: Listing or table of information needed to build the job.

Site drawings (civil drawings): Drawings that contain information on existing and final grades, existing conditions, and site development for utilities, paving, sidewalks, landscape, and irrigation.

Specifications: Spell out the provisions of the contract and supplement the working drawings with detailed technical information.

Standard (reference) specifications: Specifications that are developed by a party other than the project architect such as by an equipment manufacturer or a trade association.

Structural drawings: Drawings that contain information on foundations, placement of reinforcement steel, floors, and roof construction.

Unit cost/price contracts: Agreements based on the price of increments or phases of work within the total project. Unit cost contracts are used when quantities of materials or equipment may have been predetermined or the extent of the work has not been fully established.

Working drawings: Drawings that are needed by the various trades to complete a project.

Instructor's Notes:

Divisions of the Construction Specification Institute Format

Division 1 General Requirements
- 01100 Summary
- 01200 Price and Payment Procedures
- 01300 Administrative Requirements
- 01400 Quality Requirements
- 01500 Temporary Facilities and Controls
- 01600 Product Requirements
- 01700 Execution Requirements
- 01800 Facility Operation
- 01900 Facility Decommissioning

Division 2 Site Work
- 02050 Basic Site Materials and Methods
- 02100 Site Remediation
- 02200 Site Preparation
- 02300 Earthwork
- 02400 Tunneling, Boring, and Jacking
- 02450 Foundation and Loadbearing Elements
- 02500 Utility Services
- 02600 Drainage and Containment
- 02700 Bases, Ballasts, Pavements, and Appurtenances
- 02800 Site Improvements and Amenities
- 02900 Planting
- 02950 Site Restoration and Rehabilitation

Division 3 Concrete
- 03050 Basic Concrete Materials and Methods
- 03100 Concrete Forms and Accessories
- 03200 Concrete Reinforcement
- 03300 Cast-In-Place Concrete
- 03400 Precast Concrete
- 03500 Cementitious Decks and Underlayment
- 03600 Grouts
- 03700 Mass Concrete
- 03900 Concrete Restoration and Cleaning

Division 4 Masonry
- 04050 Basic Masonry Materials and Methods
- 04200 Masonry Units
- 04400 Stone
- 04500 Refractories
- 04600 Corrosion-Resistant Masonry
- 04700 Simulated Masonry
- 04800 Masonry Assemblies
- 04900 Masonry Restoration and Cleaning

Division 5 Metals
- 05050 Basic Metal Materials and Methods
- 05100 Structural Metal Framing
- 05200 Metal Joists
- 05300 Metal Deck
- 05400 Cold-Formed Metal Framing
- 05500 Metal Fabrications
- 05600 Hydraulic Fabrications
- 05700 Ornamental Metal
- 05800 Expansion Control
- 05900 Metal Restoration and Cleaning

Division 6 Wood and Plastics
- 06050 Basic Wood and Plastic Materials and Methods
- 06100 Rough Carpentry
- 06200 Finish Carpentry
- 06400 Architectural Woodwork
- 06500 Structural Plastics
- 06600 Plastic Fabrications
- 06900 Wood and Plastic Restoration and Cleaning

Division 7 Thermal and Moisture Protection

07050	Basic Thermal and Moisture Protection Materials and Methods
07100	Damproofing and Waterproofing
07200	Thermal Protection
07300	Shingles, Roof Tiles, and Roof Coverings
07400	Roofing and Siding Panels
07500	Membrane Roofing
07600	Flashing and Sheet Metal
07700	Roof Specialties and Accessories
07800	Fire and Smoke Protection
07900	Joint Sealers

Division 8 Doors and Windows

08050	Basic Door and Window Materials and Methods
08100	Metal Doors and Frames
08200	Wood and Plastic Doors
08300	Specialty Doors
08400	Entrances and Storefronts
08500	Windows
08600	Skylights
08700	Hardware
08800	Glazing
08900	Glazed Curtain Wall

Division 9 Finishes

09050	Basic Finish Materials and Methods
09100	Metal Support Assemblies
09200	Plaster and Gypsum Board
09300	Tile
09400	Terrazzo
09500	Ceilings
09600	Flooring
09700	Wall Finishes
09800	Acoustical Treatment
09900	Paints and Coatings

Division 10 Specialties

10100	Visual Display Boards
10150	Compartments and Cubicles
10200	Louvers and Vents
10240	Grilles and Screens
10250	Service Walls
10260	Wall and Corner Guards
10270	Access Flooring
10290	Pest Control
10300	Fireplaces and Stoves
10340	Manufactured Exterior Specialties
10350	Flagpoles
10400	Identification Devices
10450	Pedestrian Control Devices
10500	Lockers
10520	Fire Protection Specialties
10530	Protective Covers
10550	Postal Specialties
10600	Partitions
10670	Storage Shelving
10700	Exterior Protection
10750	Telephone Specialties
10800	Toilet, Bath, and Laundry Specialties
10880	Scales
10900	Wardrobe and Closet Specialties

Division 11 Equipment

11010	Maintenance Equipment
11020	Security and Vault Equipment
11030	Teller and Service Equipment
11040	Ecclesiastical Equipment
11050	Library Equipment
11060	Theater and Stage Equipment
11070	Instrumental Equipment
11080	Registration Equipment
11090	Checkroom Equipment
11100	Mercantile Equipment
11110	Commercial Laundry and Dry Cleaning Equipment
11120	Vending Equipment
11130	Audio-Visual Equipment
11140	Vehicle Service Equipment
11150	Parking Control Equipment
11160	Loading Dock Equipment
11170	Solid Waste Handling Equipment
11190	Detention Equipment
11200	Water Supply and Treatment Equipment
11280	Hydraulic Gates and Valves
11300	Fluid Waste Treatment and Disposal Equipment
11400	Food Service Equipment
11450	Residential Equipment
11460	Unit Kitchens
11470	Darkroom Equipment

Instructor's Notes:

11480	Athletic, Recreational, and Therapeutic Equipment		13400	Measurement and Control Instrumentation
11500	Industrial and Process Equipment		13500	Recording Instrumentation
11600	Laboratory Equipment		13550	Transportation Control Instrumentation
11650	Planetarium Equipment		13600	Solar and Wind Energy Equipment
11660	Observatory Equipment		13700	Security Access and Surveillance
11680	Office Equipment		13800	Building Automation and Control
11700	Medical Equipment		13850	Detection and Alarm
11780	Mortuary Equipment		13900	Fire Suppression
11850	Navigation Equipment			
11870	Agricultural Equipment			
11900	Exhibit Equipment			

Division 12 Furnishings

12050	Fabrics
12100	Art
12300	Manufactured Casework
12400	Furnishings and Accessories
12500	Furniture
12600	Multiple Seating
12700	Systems Furniture
12800	Interior Plants and Planters
12900	Furnishings Restoration and Repair

Division 13 Special Construction

13010	Air-Supported Structures
13020	Building Modules
13030	Special Purpose Rooms
13080	Sound, Vibration, and Seismic Control
13090	Radiation Protection
13100	Lightning Protection
13110	Cathodic Protection
13120	Pre-Engineered Structures
13150	Swimming Pools
13160	Aquariums
13165	Aquatic Park Facilities
13170	Tubs and Pools
13175	Ice Rinks
13185	Kennels and Animal Shelters
13190	Site-Constructed Incinerators
13200	Storage Tanks
13220	Filter Underdrains and Media
13230	Digester Covers and Appurtenances
13240	Oxygenation Systems
13260	Sludge Conditioning Systems
13280	Hazardous Material Remediation

Division 14 Conveying Systems

14100	Dumbwaiters
14200	Elevators
14300	Escalators and Moving Walks
14400	Lifts
14500	Material Handling
14600	Hoists and Cables
14700	Turntables
14800	Scaffolding
14900	Transportation

Division 15 Mechanical

15050	Basic Mechanical Materials and Methods
15100	Building Service Piping
15200	Process Piping
15300	Fire Protection Piping
15400	Plumbing Fixtures and Equipment
15500	Heat-Generation Equipment
15600	Refrigeration Equipment
15700	Heating, Ventilating, and Air Conditioning Equipment
15800	Air Distribution
15900	HVAC Instrumentation and Controls
15950	Testing, Adjusting, and Balancing

Division 16 Electrical

16050	Basic Electrical Materials and Methods
16100	Wiring Methods
16200	Electrical Power
16300	Transmission and Distribution
16400	Low-Voltage Distribution
16500	Lighting
16700	Communications
16800	Sound and Video

NATIONAL
CENTER FOR
CONSTRUCTION
EDUCATION AND
RESEARCH

MODULE MT205

Answers to Review Questions

	Answer	Section Reference
1.	d	2.0.0
2.	b	2.0.0
3.	a	3.1.0
4.	b	8.5.0
5.	d	4.0.0
6.	b	5.0.0
7.	b	6.0.0
8.	d	6.0.0
9.	b	7.1.0
10.	b	7.1.2
11.	d	8.0.0
12.	a	8.0.0
13.	c	8.3.0
14.	b	8.8.0
15.	c	8.10.0

CONTREN™ LEARNING SERIES — USER UPDATES

The NCCER makes every effort to keep these textbooks up-to-date and free of technical errors. We appreciate your help in this process. If you have an idea for improving this textbook, or if you find an error, a typographical mistake, or an inaccuracy in NCCER's Contren™ textbooks, please write us, using this form or a photocopy. Be sure to include the exact module number, page number, a detailed description, and the correction, if applicable. Your input will be brought to the attention of the Technical Review Committee. Thank you for your assistance.

Instructors – If you found that additional materials were necessary in order to teach this module effectively, please let us know so that we may include them in the Equipment/Materials list in the Instructor's Guide.

Write: Curriculum Revision and Development Department
National Center for Construction Education and Research
P.O. Box 141104, Gainesville, FL 32614-1104

Fax: 352-334-0932

E-mail: curriculum@nccer.org

Craft _____ Module Name _____

Copyright Date _____ Module Number _____ Page Number(s) _____

Description _____

_____

_____

_____

_____

(Optional) Correction _____

_____

_____

_____

(Optional) Your Name and Address _____

_____

_____

_____

Project Supervisor

Module MT206-01

Document Control and Estimating

Document Control and Estimating
Instructor's Guide

Module MT206

MODULE OVERVIEW

This module introduces the project supervisor trainee to document control and estimating. This module teaches the trainee the components of and the importance of an effective document control system. The trainee will also learn about the kinds of estimates and how to complete a simple material estimate.

PREREQUISITES

There are no prerequisites for this module; however, prior to training with this module, it is recommended that the trainee complete the following modules:
Project Supervision, Modules MT201 through MT205

LEARNING OBJECTIVES

Upon completion of this module, the trainee will learn and be able to:

1. Describe the importance of document control.
2. Discuss your role as a project supervisor in document control.
3. Explain the estimating process.
4. Complete a simple material estimate.

PERFORMANCE OBJECTIVES

This is a knowledge-based module – there is no performance profile examination.

NCCER STANDARDIZED TRAINING PROGRAM

The National Center for Construction Education and Research (NCCER) provides a standardized national program of accredited craft training. Key features of the program include instructor certification, competency-based training, and performance testing. The program provides trainees, instructors, and companies with a standard form of recognition through a National Craft Training Registry. The program is described in full in the Guidelines for Accreditation, published by the NCCER. For more information on standardized craft training, contact the NCCER by writing us at P.O. Box 141104, Gainesville, FL 32614-1104; calling 352-334-0911; or e-mailing info@nccer.org. More information may be found at our Web site at www.nccer.org.

HOW TO USE THIS ANNOTATED INSTRUCTOR'S GUIDE

Each page presents two sections of information. The larger section displays each page exactly as it appears in the Trainee Module. The narrow column ties suggested trainee and instructor actions to each page and provides icons to call your attention to material, safety, audiovisual, or testing requirements. The bottom of each page includes space for your notes.

If you see the Teaching Tip icon, that means there is a teaching tip associated with this section. Also refer to the suggested teaching tips at the end of the module.

PREPARATION

Before teaching this module, you should review the Module Outline, Learning Objectives, and the Materials and Equipment List. Be sure to allow ample time to prepare your own training or lesson plan and gather all required equipment and materials.

MATERIALS AND EQUIPMENT LIST

Materials:

Transparencies

Markers/chalk

Module Examinations*

Samples of a document control system**

Samples of labor, material, and equipment estimates**

Equipment:

Overhead projector and screen

Whiteboard/chalkboard

*Located in the Test Booklet packaged with this Annotated Instructor's Guide.
**If avilable on loan from your workplace or other resource.

ADDITIONAL RESOURCES

This module is intended to present thorough resources for task training. The following reference works are suggested for both instructors and motivated trainees interested in further study. These are optional materials for continued education rather than for task training.

Construction Materials and Building Publication No. 91, International Organization for Standardization (ISO), ISO Online Catalog at www.iso.ch.

Professional Construction Management: Including Contracting C M, Design-Construct, and General Contracting, 1991. Donald S. Barrie and Boyd C. Paulson (Contributor). Upper Saddle River, NJ: McGraw-Hill Higher Education.

Construction Management, 1997. Daniel W. Halpin and Ronald W. Woodhead. New York: John Wiley & Sons.

Construction Operations Manual of Policies and Procedures, 2000. Andrew Civitello, Jr. Upper Saddle River, NJ: McGraw-Hill Professional Book Group.

TEACHING TIME FOR THIS MODULE

An outline for use in developing your lesson plan is presented below. Note that each Roman numeral in the outline equates to one session of instruction. Each session has a suggested time of 2 1/2 hours. This includes 10 minutes at the beginning of each session for administrative tasks and one 10-minute break during the session. Approximately 10 hours are suggested to cover *Document Control and Estimating*.

Topic	Planned Time
Session I. Introduction to Document Control	
A. Introduction	_____
B. Document Control	_____
1. Operational Efficiency	_____
2. Communications Efficiency	_____
3. Maintaining Document Control	_____
4. Participant Activity	_____
Session II. Introduction to Estimating	
A. Estimating	_____
1. The Material Estimate	_____
2. Participant Activity	_____
Session III. Estimating, Continued	
A. Estimating	_____
1. The Labor Estimate	_____
2. Participant Activity	_____
3. Equipment Estimates	_____
4. Participant Activity	_____
5. Developing the Estimate – Uses, Organization, and Errors	_____
6. Participant Activity	_____
Session IV. Material Estimate Example	
A. Example of a Material Estimate	_____
1. Participant Activity	_____
B. Summary	_____
1. Summarize module	_____
2. Answer trainee questions	_____
C. Module Examination	_____
1. Trainees must score 70% or higher to receive recognition from the NCCER.	_____
2. Record the testing results on Craft Training Report Form 200 and submit the results to the Training Program Sponsor.	_____

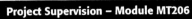

Document Control and Estimating

Instructor's Notes:

Instructor's Notes:
_____
_____
_____
_____
_____
_____

ACKNOWLEDGMENTS

The NCCER wishes to acknowledge the dedication and expertise of Phil Copare and Roger Liska, the original authors and mentors for this module on document control and estimating.

Philip B. Copare, MBA
President, Construction Services Enterprises
Education and Safety Consultant
Zellwood, FL

Roger W. Liska, Ed.D., FAIC, CPC, FCIOB, PE
Clemson University
Chair & Professor
Department of Construction Science & Management

We would also like to thank the following reviewers for contributing their time and expertise to this endeavor:

J.R. Blair
Tri-City Electrical Contractors
An Encompass Company

Mike Cornelius
Tri-City Electrical Contractors
An Encompass Company

Dan Faulkner
Wolverine Building Group

David Goodloe
Clemson University

Kevin Kett
The Haskell Company

Danny Parmenter
The Haskell Company

Course Map

This course map shows all of the modules of the *Project Supervision* curriculum. The suggested training order begins at the bottom and proceeds up. Skill levels increase as you advance on the course map. The local Training Program Sponsor may adjust the training order.

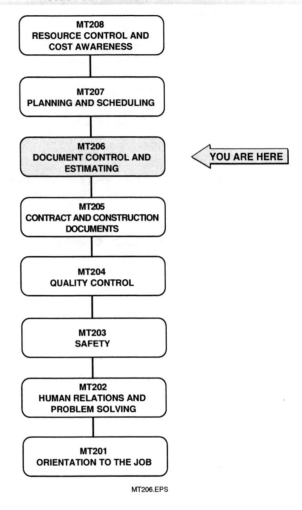

Instructor's Notes:

MODULE MT206

TABLE OF CONTENTS

1.0.0	**INTRODUCTION**	6.1
2.0.0	**DOCUMENT CONTROL**	6.2
2.1.0	Operational Efficiency	6.2
2.2.0	Communications Efficiency	6.2
2.3.0	Maintaining Document Control	6.2
2.3.1	*How to Use Document Control*	6.3
2.3.2	*Documents Needing Maintenance*	6.3
2.3.3	*PDAs in Document Control*	6.5
3.0.0	**ESTIMATING**	6.6
3.1.0	The Estimate	6.6
3.1.1	*The Material Estimate*	6.7
3.1.2	*The Labor Estimate*	6.12
3.1.3	*Equipment Estimates*	6.14
3.2.0	Developing the Estimate	6.14
3.2.1	*Uses for the Estimate*	6.16
3.2.2	*Organization of the Estimate*	6.17
3.2.3	*Errors in Estimating*	6.17
3.3.0	Example of a Material Estimate	6.17
	SUMMARY	6.21
	GLOSSARY	6.23

LIST OF FIGURES

Figure 1	•	Superintendent's Pre-Planning Checklist 6.4
Figure 2	•	Worksheet. 6.8
Figure 3	•	Completed Worksheet. 6.9
Figure 4	•	Summary Sheet 6.10
Figure 5	•	Completed Summary Sheet 6.11
Figure 6	•	Labor Summary Sheet. 6.13
Figure 7	•	Recap Sheet 6.15
Figure 8	•	Completed Recap Sheet. 6.16
Figure 9	•	Rectangular Foundation Plan 6.19
Figure 10	•	L-Shaped Foundation Plan 6.20

Instructor's Notes:

NATIONAL CENTER FOR CONSTRUCTION EDUCATION AND RESEARCH

MODULE MT206

Document Control and Estimating

OBJECTIVES

Upon the completion of this module, you will be able to do the following:

1. Describe the importance of document control.
2. Discuss your role as a project supervisor in document control.
3. Explain the estimating process.
4. Complete a simple material estimate.

SECTION 1

1.0.0 INTRODUCTION

Communication during a project is extremely important. One of the ways to increase and enhance project communication is through the establishment of an effective document control system. Construction documents need to be controlled, maintained, and archived in all phases of a project. Your understanding of this process in the company is very important to your success as a project supervisor.

Estimating is the process of determining the cost of work before it is done. Estimating requires extreme care and precision because an error in the estimate can cause problems throughout the construction process.

How a construction company is organized has a great deal to do with how estimating is handled. In companies organized along strict departmental lines, the estimating department staff is responsible for all estimating functions. In companies organized along functional lines, a single person, usually the project manager, may be responsible for all functions related to the project — estimating, bidding, scheduling, purchasing, and management.

Ensure that you have all the necessary materials to teach the course. Check the Materials and Equipment list at the front of the module.

Ask trainees to discuss the types of document control in place at their job site.

Show Transparency 1 (Course Objectives).

Assign reading of Module MT206, Sections 1.0.0 – 2.3.3.

Copyright © 2003 National Center for Construction Education and Research, Gainesville, FL 32614-1104. All rights reserved. No part of this work may be reproduced in any form or by any means, including photocopying, without written permission of the publisher.

Classroom

Ask trainees to give examples of communications efficiencies they use that work.

Field supervisors must work within the guidelines of the project estimate and are responsible for monitoring and reporting differences between the estimate and the actual cost of doing the work. They may also be required occasionally to complete material and labor estimates in order to plan the work.

SECTION 2

2.0.0 DOCUMENT CONTROL

An important challenge in the construction industry today is timely and proper communication. A large part of communication is paper and electronic documentation. How your staff manages the processing, distribution, and filing of these documents is critical to the project's success. You should consider document control an important component of being a supervisor.

Documents need to be processed and tracked from the beginning to the end of a construction project. Applying document control creates two types of efficiencies: operational and communications.

2.1.0 Operational Efficiency

To promote accuracy and productivity, a document is tracked through the following phases:

- Design
- Estimating
- Materials
- Construction
- Project planning

The process ensures that documents are correct, current, and available when needed. The engineering or commercial documents will be secure and accessible to authorized personnel. Documents affected may contain the following elements:

- Design revisions
- Design standards
- Codes/regulations
- Inspection and test reports
- Drawings
- Bills of material
- Performance calculations
- Supplier invoices and forms
- Customer specification requirements

2.2.0 Communications Efficiency

In all areas of the company, communication needs to be properly documented to ensure efficiency. The entire package of documents required to support a project must be available before the work begins in the field, shop, or office. A supervisor can ensure that these documents are scheduled to support work completed accurately and on time.

For example, when scheduling a project, a supervisor needs to know what documents are required, who is responsible for obtaining each document, and when each document is needed in order to meet the deadline.

A complete document control system also tracks documents during the construction phase. For example, when a design change occurs, it is important that both the affected project managers or crew leaders and the company administration are notified to ensure that documents are properly revised. Automatic notice of any change promotes communications on all levels of the company.

In the past, companies have focused on document control by using small packets of paper to communicate. Today, the design, management, and supervisory groups of many construction companies are exchanging information online. Many groups within the company can check the status of a project by reviewing documents online and updating them, if necessary.

2.3.0 Maintaining Document Control

Regardless of size, most companies have a system in place to control all information created in the design and construction process. A company needs to decide what software or manual record system to use and how to include and update the data. It also needs to set standards for document design and data input. Whether your company has a manual or electronic method of document control, the process required is the same.

Instructor's Notes:

A supervisor has two basic responsibilities when managing project documents:

- *Knowing acronyms and terms used in each document type.* Most companies have shortened versions of a name for or specific to a document or process. Learning what these acronyms are and other terms commonly used is crucial to completing the document or process accurately and successfully.
- *Seeing the big picture of how each type of communication affects the project.* Understanding how informal and formal documents fit into the overall project is key to document control. Some documents are informal, such as a to-do list or a checklist for preparing a large document; other documents are formal, such as a soil test or a request for tender.

Supervisors use documents to stay organized on the job. If document control is electronic, computer file management is used. If document control is paper-based, then manual maintenance and filing systems update and store the documents. Both types of document control require training. The extent of training needed depends on the complexity of the system.

To understand document control, a supervisor will need to learn:

- How to use document control
- What documents need maintenance
- How to update a document

2.3.1 How to Use Document Control

An effective way to use document control is to be organized, communicate clearly, and resolve issues quickly. Be organized and communicate clearly by:

- Distributing project documents
- Creating customized forms and workflows
- Receiving notification via e-mail or fax when documents are approved or changed
- Managing team contact information
- Broadcasting files or memos to the entire team
- Routing documents to team members via e-mail or fax
- Creating a project homepage or message board for important team information

Resolve issues quickly by:

- Viewing real-time, up-to-date project information logs
- Routing and tracking files, and assigning individual responsibility for documents
- Confirming that team members have received materials
- Monitoring document approvals via e-mail or fax
- Creating an audit trail of your project history.

2.3.2 Documents Needing Maintenance

Which documents need to be maintained and how frequently to update them varies with the type of document, company policy, regulatory and legal requirements, and project type. *Figure 1* shows a checklist that some companies might consider to be a formal document.

The *project checklist* is a convenient listing of tasks a craftworker may perform on a given project. This checklist will assist the manager in recognizing required tasks and in locating data necessary to fulfill assigned responsibilities. By providing space for notes on actions taken, assignment of tasks, and time frames for completion, checklists may also serve as a permanent record of the actions and decisions of the owner, contractor, and architect.

Usually, only formal documents are maintained as part of document control. Formal documents need to be recorded, updated, and stored securely. You may use all or some of the documents listed below. Refer to your company's policies for instructions on how these documents should be handled. See the module on *Contract and Construction Documents* for additional information.

Requests for Information (RFI) are used to request information about a problem or variance.

Requests for Proposal (RFP) are used in the pre-bid phase by the owner to gather information about how to approach a project and to identify firms that are able to provide the solution.

Transmittals include all information sent about a project.

Ask trainees for items they feel should be added to the pre-planning checklist.

Show Transparency 2 (Types of Formal Documents).

SUPERINTENDENT'S PRE-PLANNING CHECKLIST

Implement Quality Assurance Program

Identify methods to be used to maintain the following items throughout project duration:

_____	Job folder	_____	Construction drawings
_____	Specifications	_____	Job schedule
_____	Scope of work	_____	Electrical code
_____	Schedule of drawings	_____	Investigation/coordination
_____	Submittals		

Completed on: _____

Temporary Service & Mobilization to Site

Calculate service size. Meet with power company engineer to discuss location of temporary service. Discuss scope of work concerning utility conduits and metering requirements.

Review on site needs for:

_____	Telephone	_____	Filing system in place
_____	Copier	_____	Establish method of documentation
_____	Facsimile machine	_____	Security plan
_____	Needed storage space for materials and equipment		

Completed on: _____

Safety & Training

_____ Review
_____ Safety requirements
_____ OSHA board
_____ MSDS sheets
_____ Schedule crew to watch training videos for specific phase of work
Completed on: _____

Personnel & Materials

_____ Personnel requirements/job structure
_____ Review material flow process
_____ Tools/equipment
_____ Verification of fixtures and gear
_____ Review job schedule; discuss target dates and ship dates of materials and equipment
Completed on: _____

Executing the Work

_____ Confirm layout
_____ Prepare field drawings
_____ Review subcontractors work
Completed on: _____

Figure 1 • Superintendent's Pre-Planning Checklist

Instructor's Notes:

Submittals include information provided by subcontractors and vendors that provides the connecting link between design and construction. This information is presented in a variety of forms, but most commonly as shop drawings, catalogs, manufacturers' product data, samples, schedules, and mock-ups.

Correspondence includes all written communication about the project, including submittals, transmittals, letters, memos, and e-mails.

Contract documents include all documents that pertain to any aspect of specifications, drawings, subcontracting, and scope.

Change orders may be used as written documentation of changes in the work, contract sum, or contract time that are mutually agreed to by the owner and contractor. The form provides space for the signatures of the owner, architect, and contractor and for a complete description of the change.

A *change order request* is used to obtain price quotations required in the negotiation of change orders. A change order request is not a change order or a direction to proceed with the work; it is simply a request to the contractor for information related to a proposed change in the construction contract.

A *change order proposal* is the response to the change order request, showing what the implications of the proposed change are.

Subcontractor change orders are change orders initiated by subcontractors. The content relates only to their specific scope of work.

Document logs are computerized or manual logs that track detailed information about submittal documents, such as date received from the subcontractor or vendor, date forwarded to the architect or engineer and/or owner, the date returned, the date forwarded back to the subcontractor or vendor, and the date issued to the superintendent or field.

Architect's Supplemental Instructions (ASIs) are used by the architect to issue additional instructions or interpretations or to order minor changes in the work. The form is intended to assist the architect in performing obligations as interpreter of the contract document requirements in accordance with the owner-architect agreement and the general conditions. This form should not be used to change the contract sum or contract time. If a change in the contract sum or contract time is involved, a change order should be used.

Field orders are standard forms for a project representative to use in maintaining a concise record of site visits or, in the case of a full-time project representative, a daily log of construction activities.

Architects, owners, and contractors must be able to track the development of drawings and specifications. These documents are tracked using the submittal process. Since this process tends to be complex, submittal logs track the progress of a submittal, which in turn contributes to the orderly processing of work.

Document control may also include distribution as maintenance. A simple method places all formal documents on a distribution list and each person on the list records that they have read the document. A more sophisticated method records the date the document was received, places the document on a distribution list, and then logs the document in a folder, either electronically or manually.

2.3.3 PDAs in Document Control

Updating a document involves time and effort regardless of whether document control is by paper or computer. Using a computer may seem like a disadvantage to field personnel who don't have time to go to a computer to enter data during the day to keep the system updated. However, **personal digital assistants,** or PDAs, offer a unique solution. A PDA is a hand-held computer that manages various daily functions, such as contact lists, schedules, things-to-do lists, expenses, and e-mail. PDAs fit into a shirt pocket and are inexpensive, fast, and simple.

A supervisor issued a PDA can note the actual start and completion dates for every scheduled activity. At the end of each work day, the supervisor can place the PDA in its cradle and, with the push of a button, send the updated information to the main office, updating the project schedule. The credibility of a daily-maintained schedule in an insurance claim case is invaluable.

Ensure that you have all the necessary materials to teach the course. Check the Materials and Equipment list at the front of the module.

Have the trainees complete the Participant Activity.

Assign reading of Sections 3.0.0 – 3.1.1 for the next class session.

PARTICIPANT ACTIVITY

Discuss the following questions.

1. What form of document control is in place in your company?
2. List suggestions for improving document control in your company.
3. Name the two types of efficiencies created by applying technology to document control.
4. Discuss the importance of tracking a document through all phases of a project.
5. What is a PDA?
6. Do you use a PDA? If so, how do you use it in your work?
7. Discuss your responsibilities for document control as a project supervisor.

PARTICIPANT ACTIVITY

SECTION 3

3.0.0 ESTIMATING

The estimator is among the most important employees in the contractor's organization because the company's reputation and financial well-being are in this individual's hands. For this reason, the supervisor must be very careful when performing any estimating function.

The following qualifications are essential in a successful estimator:

- Ability to read and measure plans
- Knowledge of basic mathematics
- Patience
- Ability to do careful, precise, thorough work
- Ability to visualize the project from the drawings
- Enough construction experience to understand the effect of job conditions on work
- Sufficient knowledge of labor performance and operations so that labor costs on a project can be estimated
- Ability to collect, organize, and utilize a large pool of information on costs of all kinds, including those of labor, material, overhead, and equipment
- Ability to meet tight deadlines calmly
- Creativity in discovering competitive edges
- Knowledge of the use of the computer for performing the estimating function

A person can be taught the mechanics of preparing an estimate and can be alerted to possible errors, problems, and dangers. However, the good judgment required to apply what has been learned comes only with experience, and experience is what makes a good estimator.

3.1.0 The Estimate

Supervisors are not, in most cases, required to prepare complete estimates. However, there may be a requirement to perform estimating functions in order to effectively carry out their duties and responsibilities. For this reason, the supervisor must understand the estimating process.

The steps for developing a project estimate are as follows:

Step 1 Determine the material and equipment requirements for the project.

Step 2 Determine the labor requirements.

Step 3 Price the materials, equipment, and labor.

Step 4 Determine job overhead costs.

Step 5 Obtain subcontract quotations.

Step 6 Summarize the final costs of all items.

Step 7 Apply an adequate mark-up factor to cover all general overhead items and an appropriate profit margin.

Step 8 Tally the final amount of the estimate.

Step 9 Review the entire estimate for thoroughness and accuracy.

Instructor's Notes:
_____
_____
_____
_____
_____
_____

All estimate information should be recorded on a standard estimate format sheet, and separate sheets should be used for each part or phase of the estimate. The most effective format is developed by the company to meet its own specific needs, but there are two basic types of format sheets: worksheets and summary sheets.

Worksheets are used for recording all material requirements during the construction of the project and are sometimes referred to as **quantity takeoff sheets**. A typical worksheet is shown in *Figure 2*. *Figure 3* shows a completed worksheet.

Summary sheets are used to bring together the quantities and prices for labor, material, and equipment and are also known as *pricing sheets*. A typical summary sheet is shown in *Figure 4* and *Figure 5* is a completed summary sheet.

3.1.1 The Material Estimate

The **material estimate,** sometimes called the *quantity takeoff,* determines the amount of material required for the construction of the project. In preparing it, the estimator "takes off" the various materials from the drawings and specifications and tabulates them on a standard work sheet, as described below:

- Review material specifications, such as the specifications of the bricks required for the exterior facing. Also review any addenda which may exist in the contract documents.
- Arrange the material takeoff so that it closely follows the sequence in which the job will be constructed and the order of the material descriptions in the specifications.
- Take off every item as it is shown or stated on the drawings or in the specifications.
- Avoid approximations, radical roundoffs, and changes, even if based on sound judgment.
- Avoid scaling whenever possible. If scaling is required, check to be absolutely certain that the scale used is accurate.
- Keep different items separated. For example, when doing a takeoff of paint, list each type of paint separately.
- Start the takeoff in the same place on each drawing and follow the same pattern through each drawing. This ensures consistency and helps the estimator remember what has been done and what is left to do if it becomes necessary to stop and then return to the drawing later. Many estimators make it a habit to begin at the top left corner of each drawing and proceed clockwise.
- Mark up each portion of each drawing as it is completed.
- Take advantage of duplication of design. If ten classrooms in a school building are identical, the estimator can take off the interior finishes for one classroom and multiply the answer by ten. Before doing this, however, check all figures and calculations. A small error could prove to be very costly.
- Allow for waste and breakage on all materials. Be sure to note how much is included as the allowance.
- When totaling, round up to a standard order quantity. For example, 1800 square feet of ½-inch CDX plywood (1800 square feet ÷ 32 square feet per sheet = 56.25 sheets) should be recorded as 57 sheets.

continued on page 6.12

Ask trainees to give examples of both good and bad estimating and state their reasons.

Review both the worksheet and the sample summary sheet and determine if there is other information the trainees deem relevant for calculating the estimate.

PARTICIPANT ACTIVITY

Discuss the following:

1. Describe how you might use construction estimates in your job.
2. Describe estimates that you might be required to perform.
3. List the qualifications of an estimator that you feel you possess.
4. List the qualification of an estimator that you need to develop and practice before you would feel comfortable doing an estimate.

PARTICIPANT ACTIVITY

Show Transparencies 3 and 4 (Figures 3 and 5).

Figure 2 • Worksheet

Instructor's Notes:

WORKSHEET

Takeoff By: _____ Date _____ Project _____
Checked By: _____ Sheet ___ of ___ Architect _____
Page _____

Ref.	Description	No.	Ref. Length	Width	Height	Extension	Subtotal Quantity	Unit	Total Quantity	Unit	Remark
A	1" CONDUIT	1	1000	---			1000	LF			
							50	LF	1050	LF	+ 5% WASTE
B	CARPET	1	15	24			1050	SF			
							360	SF			
							29	CF	43	SY	+8% WASTE
C	CONCRETE	1	120	80	0 6"		389				
							4800	CF			
							240	CF			
							5040	CF	187	CY	+5% WASTE

Figure 3 • Completed Worksheet

Summary Sheet

Takeoff By: _____ Date _____
Checked By: _____ Sheet _____ of _____

Project: _____
Work Order No.: _____
Title: _____
Page No.: _____

Ref.	Description	Quantity Total	UT	Material Costs Per Unit	Material Costs Total	Labor Man Hours Craft	Labor Man Hours Pr Unit	Labor Man Hours Total	Factors Rate	Factors Cost	Labor Cost Per	Labor Cost Total	Item Cost Per Unit	Item Cost Total
	3000 per footing concrete	272	cy	50 00	13,600 00	3CW[1]	0.1[2]	27.2[3]	18 00	489 60[4]	489 60	1468 80[5]	55 40	1538 80 cy
		13600 00										1468 80		1538 80
		Material										*Labor*		*Total*

[1] Concrete workers
[2] .1 hr/cy or 10 cy/hr (crew output)
[3] .1 hr/cy x 272 cy = 27.2 hrs.
[4] Cost per CW = $18/hr x 27.2 hr. = $489.60
[5] Total Labor Cost = # of workers x cost per CW

Figure 4 • Summary Sheet

Instructor's Notes:
_____
_____
_____
_____
_____
_____

Module MT206 ◆ Document Control and Estimating

Summary Sheet

Computer By: _____ Date _____
Checked By: _____ Sheet ___ of ___ Page No. _____

Project _____
Architect _____

Ref.	Description	Quantity Total	UT	Material Cost Unit Cost	Material Subtotal	Labor Cost Crew Daily	Crew Days	Crew Hrs.	Craft	#	Pay Rate	Labor Subtotal	Equipment Cost Daily Cost	Equipment Subtotal	Cost Per Unit	UT	Subtotal
A	1" CONDUIT	1050	LF	0 80	840.00	AL 2400	2	16	Carp.	3	12 00	576.00					
		150	SH					16	C.H.	2	6 00	192.00					
B	CARPET	42	SY	16 00	688.00							768.00					
C	CONCRETE 3000 psi	187	CY	62 00	11,594.00												
↙ Totals This Page Transferred To				Material → Cost $		Crew → Days					Labor → Costs	$ 768.00	Equip. → Cost $		↙ Total Cost		$

Figure 5 • Completed Summary Sheet

Ensure that you have all the necessary materials to teach the course. Check the Materials and Equipment list at the front of the module.

Have the trainees complete the Participant Activity.

(Optional) Ask trainees what documents they would need to complete the labor estimate.

Assign reading of Sections 3.1.2 – 3.2.3 for the next class session.

- Work neatly. Record all information in such a manner that others are able to read and understand it. Remember, someone else should be able to duplicate the work within a reasonable amount of time.
- Recheck all calculations.
- Review specifications a final time for items overlooked. Omission is the easiest mistake to make and the hardest to find.

Worksheets organize the material takeoff in an orderly fashion that allows the quantity of each item to be totaled and brought forward in condensed form for recording on a summary sheet. Once the total quantities of each type and grade of material are listed on summary sheets, the cost of each is estimated.

In determining costs, the estimator must decide whether the materials are to be purchased from outside the organization, are already available in the inventory, or will be manufactured by the company itself. If the material is to be purchased from suppliers, the cost should be obtained either through a formal bidding process or by acquiring quotes from acceptable suppliers, a list of which should be available to the estimator. An acceptable supplier is one who can supply the needed quantity and quality of material within the schedule. The estimator should not forget to include the costs of shipping, unloading, storage, inspection, testing, and insurance in the estimate.

3.1.2 The Labor Estimate

An estimate of the labor required to complete the project must be completed after the materials takeoff and must include the number of work-hours (or crew-days) required and the cost of those hours or days. The **labor estimate** should be itemized by craft for each of the work items so that the supervisor can use this information to plan the work involved.

Consider the task of placing wood siding on a building. The material estimate shows that 150 sheets (4800 sqare feet) of siding are required. The estimator has determined that a crew of three carpenters and two carpenter helpers can complete the work at a rate of 300 square feet per hour. It will take the crew 18 hours to complete the work, or 48 carpenter work-hours and 32 helper work-hours. If carpenters are paid $12.00 per hour and carpenter helpers $6.00 per hour, the total cost of the work will be $768.00. (See Labor Summary Sheet, *Figure 6*).

Initially, the labor required to perform a specific job is most likely determined from previous experience on similar jobs. This information should be available from the company's records in the form of labor production rates that tell the estimator how many work-hours or crew-hours certain tasks should take. Some factors which affect productivity are working conditions, available equipment and tools, weather, and other unexpected job-site conditions.

PARTICIPANT ACTIVITY

Task 1

Carefully look over the worksheet and the summary sheet (See *Figures 2* and *4*). Note where all descriptions and dimensions are entered.

Task 2

Note how the following information is recorded on the completed worksheet (*Figure 3*) and the completed summary sheet (*Figure 5*):

— 1000 linear foot of 1-inch conduit at $0.80/linear foot

— Carpet for a room measuring 15 feet by 24 feet and priced at $16.00/square yard

— Concrete (3000 psi) for a slab measuring 120 feet by 80 feet and 6 inches thick, priced at $62.00/cubic yard

Task 3

Explain how the project estimate is prepared.

PARTICIPANT ACTIVITY

Instructor's Notes:

Show Transparencies 5 and 6 (Figures 6 and 8).

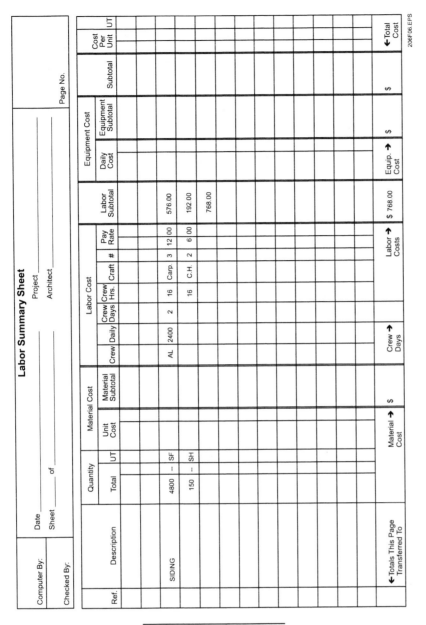

Figure 6 • Labor Summary Sheet

Ask trainees to list all the indirect costs that would affect an estimate and how they would determine these costs as a percentage of the company's general overhead.

Have trainees complete the Participant Activities.

These records should be reviewed periodically and updated to maintain the accuracy of the production rates, because reliable estimates can only be produced if they are accurate.

Once the labor requirements have been determined and production rates assigned, the total number of work-hours required to perform the activity is calculated. Then, using the wage rates of the various trade categories, the total labor cost can be derived, and all information recorded on the summary sheet.

Labor estimates must allow for variations in wages, time required to perform unique job tasks, working conditions, available equipment and tools, and required worker skills. In addition the estimator must consider other items which may require changes in the standard production rate. For example, the standard crew size used by the company may not be the best for all projects.

The following should be considered in determining crew size:

- Space available in which to handle materials and equipment
- Whether or not a worker needs a helper to perform the job safely
- Urgency of completing the job
- Demands of the job

PARTICIPANT ACTIVITY

Discuss the following:

1. Describe the type of labor estimates that you might be required to perform.
2. Where does an estimator find the production rates of labor needed to do a job?

PARTICIPANT ACTIVITY

3.1.3 Equipment Estimates

Equipment estimates are completed in much the same manner as labor estimates. Estimators determine the types of equipment and tools needed to construct the job using the company's historical records and their own experience as references.

The difference between equipment estimates and labor estimates lies in how the costs are estimated. If the equipment is owned by the company, the hourly cost of owning and operating the equipment should be available in the company records. If the equipment is to be rented, cost information is obtained from the renting or leasing agency. If the company must use a piece of equipment for a long period of time, it should consider buying it.

PARTICIPANT ACTIVITY

Discuss the following:

1. What kinds of projects are you likely to perform equipment estimates for?
2. Describe the types of equipment and tools that would be included in your equipment estimate.

PARTICIPANT ACTIVITY

3.2.0 Developing the Estimate

Estimated costs for material, labor, and equipment are summarized on a **recapitulation (recap) sheet,** as shown in *Figure 7*.

The estimator then adds the costs of work to be performed by subcontractors or other contractors, based on quotes supplied by those contractors. All of the figures are then tallied and the appropriate taxes, such as sales tax and labor taxes, are added to the total.

Based on this total, the estimator calculates job overhead costs, such as the field office staff, job trailer, job telephone, and job utilities and adds them to the prior total. The result is the total cost of all materials, labor, equipment, subcontractors, taxes, and job overhead. This is known as the total *direct cost* of the job. These are costs that the company incurs directly as a result of doing work on the project.

There are also *indirect costs* to consider. These are general overhead costs that take into account the expense of maintaining and operating the home office such as office equipment and supplies, office utilities, and office salaries. A percentage of the company's annual general overhead is added to each project estimate to help finance the company's general operations.

Instructor's Notes:
_____
_____
_____
_____
_____
_____

Listed By:	RECAP SHEET	Page
Checked By:	Project ______ Date ______ Architect ______ $/SF ______	Of

PAGE REF.	ITEM	MATERIAL	LABOR	SUB	EQUIPMENT	TOTAL

206F07.EPS

Figure 7 • Recap Sheet

The final item that must be considered is the amount of profit to include. Profit is often calculated as a percentage of the total costs. It varies from one job to the next, based upon factors such as where the job is located, how easy or difficult the work is, who the other contractors on the job are, how risky the job is, and what other contractors are bidding on the job.

Totaling all of these items gives the estimator a final estimate of the cost of doing the project, as shown on the completed recap sheet in *Figure 8*.

RECAP SHEET

DATE: July 8, 20XX
PREPARED BY: Kyle A. Reaves
TITLE: Project Manager
PROJECT: Washington County Nursing Home
ARCHITECT: Charles McCauley Associates
SHEET 1 OF 1

PAGE REF.	ITEM	MATERIAL	LABOR	SUB	EQUIPMENT	TOTAL
	Bulk Excavation			8,324		8,382
	Structural Excavation	495	609		308	1,412
	Soil Treatment			1,487		1,487
	Seeding and Sodding			3,901		3,901
	Concrete	34,614	32,106		2,520	69,240
	Masonry	38,938		91,290		130,228
	Structural Steel & Joists			73,828		73,828
	Metal Steel Partitions	1,843	3,326			5,169
	Carpentry	16,428	5,128		224	21,780
	Membrane Roofing			20,947		20,947
	Misc. Moisture Protection	12,120	9,218			21,338
	Coors and Metal Frames	23,009	2,496		143	25,648
	Windows and Glazing	8,937	3,328			12,175
	Finish Hardware	13,500				13,500
	Drywall Gypboard			20,108		20,108
	Vinyl Wall Covering			9,648		9,648
	Vinyl Asbestos Tiling			19,179		19,179
	Acoustical Ceiling			19,257		19,257
	Ceramic Tile			20,851		20,851
	Painting			10,840		10,840
	Specialties	11,274	1,624		14	12,912
	Equipment	15,765	144			15,909
	Drapery Track	856	500			1,356
	Casework			18,440		18,440
	Plumbing			80,774		80,774
	HVAC			93,012		93,012
	Electrical			95,297		95,297
	Direct Job Costs	177,784	58,394	587,247	3,210	826,618
	Job Overhead	4,352	42,154	34,395		80,901
		182,136	100,548	621,643	3,210	907,519
	6% Sales Tax	10,923				10,923
	18.7% Labor Burden		18,800			18,800
	Subtotal "A"					937,249
	10% Overhead and Profit					105,250
	Bond					10,005
	Bid					1,052,504

Figure 8 • Completed Recap Sheet

3.2.1 Uses for the Estimate

The completed estimate serves several purposes, including:

- Bidding the job or negotiating for the work
- Budgeting for the job
- Beginning a work breakdown schedule
- Establishing a cost control program
- Providing legal documentation for use in arbitration or litigation

Instructor's Notes:

3.2.2 Organization of the Estimate

The estimate must be organized in an orderly, systematic fashion so that it can be clearly understood by anyone who may be required to use the information it contains or to step in and complete it if it is incomplete. The measure for determining whether or not an estimate is well organized is the question, "Could someone else pick it up and understand it?"

3.2.3 Errors in Estimating

Regardless of how careful the estimator is, errors occur but by recognizing the most common types of errors, the estimator can locate and correct them before they can adversely affect the company or the project. The most common error is writing down a number incorrectly; the most costly error is leaving an item off the estimate entirely.

The ten most common estimating errors are:

1. Using the wrong scale
2. Misplacing a decimal point
3. Incorrectly adding, subtracting, multiplying, or dividing
4. Omitting items
5. Transposing numbers when copying them from one sheet to another
6. Using incorrect conversion factors
7. Calculating allowances improperly
8. "Mis" errors: mis-reading, mis-measuring, mis-noting, mis-hearing, and mis-understanding
9. Underestimating the length of time required to complete the project
10. Neglecting to double-check all calculations

Any one of these errors can have a tremendous impact on the project and the company.

3.3.0 Example of a Material Estimate

Estimates for different types of materials take somewhat different forms but all are similar in many respects. All are based on determining: total length, such as for curbs, pipe, wiring, or conduit; total area, such as site clearing, concrete finishing, floor covering, wall covering; or total volume such as excavation or concrete materials.

Consider concrete work. In addition to the volume of concrete required, the estimator must also calculate quantities and prices of form work, reinforcing steel, curing compounds, finishing, and related accessories. All of these items make concrete a major cost item in most construction projects, and special care should be given to the takeoff of concrete work.

Concrete itself is taken off in cubic feet and then converted to cubic yards, as is excavation work. Form work is taken off in square feet in contact with the concrete, and therefore is estimated by finding areas. When the form work is priced, the estimator must decide if the forms will be wood or metal and how many times each form can be reused.

Reinforcing steel is taken off in linear feet, according to size of bar, and then converted to pounds. For large quantities, the pounds are converted to tons. Slab finishing and curing compounds are taken off in terms of the square feet of the surface area.

Ensure that you have all the necessary materials to teach the course. Check the Materials and Equipment list at the front of the module.

Have the trainees complete the Participant Activity.

Assign reading of Section 3.3.0 for the next class session.

PARTICIPANT ACTIVITY

Discuss the following:

1. In your own words, write down the purpose of the first three estimates described in this section.
2. What is the most difficult mistake to find when reviewing an estimate? Why?
3. What construction documents are needed before a detailed estimate can be prepared?
4. Where does the estimator obtain material prices?

Go over the Participant Activity with the trainees.

Show Transparency 7 (Figure 9).

PARTICIPANT ACTIVITY

Determine the amount of concrete and form work required by the footing shown in *Figure 9* by following these guidelines.

Task 1

Determine the length of the foundation.

Note: The easiest method of determining any material associated with the perimeter of a building is to first determine the centerline dimension. By using the centerline of the foundation, you do not have to make adjustments for the overlap at the corners. The centerline is determined by taking the total outside dimensions as shown on the foundation plan and deducting four times the wall thickness to account for the four corners.

Centerline = [2 × width] + [2 × length] − [4 × wall thickness]
= [2 × 35.5 ft.] + [2 × 80 ft.] − [4 × 1 ft.]
= 227 ft.

Task 2

Determine the volume of concrete required.

Volume = length × width × height
= 227 ft. × 2 ft. × 1 ft.
= 454 cu. ft.

Task 3

Convert the volume into cubic yards.

Cubic feet ÷ 27 = cubic yards
454 ÷ 27 = 16.8 cu. yd.

Task 4

Add an appropriate figure for a waste allowance (such as 5%) and round the figure to a practical order quantity.

16.8 cu. yd. × 1.05 = 17.7 cu. yd.
= 18 cu. yd. (rounded off)

Task 5

Determine the area of form work required. The area used is the surface area in contact with the concrete. This is called the square feet of contact area (sfca).

Area of contact = length × height
= 227 ft. × 1 ft. × 2
= 454 sq. ft.

Task 6

Add an appropriate figure for a waste allowance (such as 10%).

454 sq. ft. × 1.10 = 499.4 sq. ft.
= 500 sq. ft. (rounded off)

PARTICIPANT ACTIVITY

Instructor's Notes:

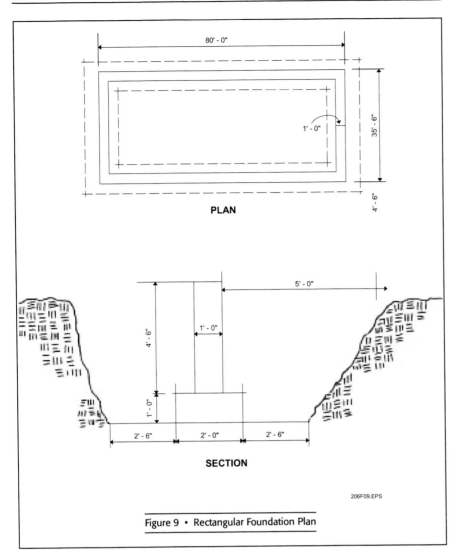

Figure 9 • Rectangular Foundation Plan

Have the trainees complete the Participant Activity. Using a blank transparency, go over the estimates with them. The correct answers are given at the end of this module.

Show Transparency 8 (Participant Activity)

PARTICIPANT ACTIVITY

Refer to *Figure 10* to complete this estimating activity. Follow the same steps as you did in the previous activity.

Task 1

Determine the amount of concrete required for the footing illustrated in *Figure 10*. Include a waste allowance of five percent.

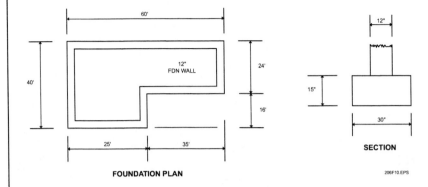

Figure 10 • L-Shaped Foundation Plan

Task 2

Determine the carpet, vinyl base, and paint required for a room 24 feet by 30 feet. The room has a 10 foot ceiling that does not require painting. There are four windows in the room each measuring 3 feet by 6 feet and two doors each measuring 3 feet by 7 feet. Three coats of paint are required (primer plus 2 finish coats). Coverage of paint is estimated at 450 square feet per gallon. Include waste allowances as follows: carpet 7%, vinyl base 5%, and paint 10%.

Alternate Task

Using a room plan available from your employer:

1. Determine the linear feet (LF) of conduit and wiring and the number of fixtures, receptacles, outlets, and switches.

2. Determine the LF of piping required and the number of pipe connections and fixtures.

3. Determine the pounds of duct work required, the number of grilles and registers, and the size and type of equipment required.

PARTICIPANT ACTIVITY

Instructor's Notes:

SUMMARY

Document control in the construction industry is on the rise. With advances in technology from the office to the field, there is no reason for supervisors to be left behind. Supervisors are key players in document control and training is important. For many supervisors, using a computer for document control may seem daunting. Supervisors may need training to understand how to use and maintain construction documents in the system. Take time to learn how document control can decrease rather than increase your workload.

Estimating is an extremely important part of the construction process. Estimates are made to determine the probable costs of a proposed construction project and to determine a budget for the owner. Estimates are also made in the field to determine the cost of changes or the quantities of resources required. All estimates require similar procedures.

While project estimates require each and every part of the project to be finalized, the field estimates may be concentrated on a few major categories, depending on the type of work performed by the contractor.

Instructor's Notes:
_____
_____
_____
_____
_____
_____

Trade Terms Introduced in This Module

Architect's supplemental instruction (ASIs): Additional instructions, interpretations, or minor change orders issued by the architect.

Estimating: Process of determining the cost of work before it is done.

Field orders: Standard form used by a project representative to maintain a concise record of site visits or a daily log of construction activities.

Labor estimate: Number of work-hours or crew-days required to complete the project, and the cost of those hours or days.

Material estimate: Determines the amount of material required for the construction of the project. Also known as a *quantity takeoff*.

Personal digital assistant (PDA): A handheld computer that manages various daily functions, such as contact lists, schedules, things-to-do list, expenses, and e-mail.

Quantity takeoff: See *material estimate*.

Recapitulation (Recap) Sheet: Summary of the estimated costs for material, labor, and equipment.

Requests for Proposal (RFPs): Used in the pre-bid phase by the owner to gather information about how to approach a project and to identify firms that are able to provide the solution.

Submittals: Documents that include information provided by subcontractors and vendors that provide the connecting link between design and construction.

Submittal logs: Documents that track the progress of a submittal.

Summary sheets: Documents used to bring together the quantities and prices for labor, material, and equipment. Also known as *pricing sheets*.

Transmittals: Documents include all information sent about a project.

Worksheets: Used for recording all material requirements during the construction of the project. Also known as *quantity takeoff sheets*.

Ask trainees to review the Glossary for unfamiliar terms.

Review the objectives of the module and then answer any questions the trainees may have.

Have trainees prepare for the Module Examination.

Administer the Module Examination. Be sure to record the results of the Exam on Craft Training Report Form 200 and submit the results to the Training Program Sponsor.

Instructor's Notes:
_____
_____
_____
_____
_____
_____

NATIONAL CENTER FOR CONSTRUCTION EDUCATION AND RESEARCH

MODULE MT206

Answers to Participant Activity

Estimating Activity for L-Shaped Foundation Plan, Section 3.3.0

Task 1
Concrete:
Centerline dimension	=	196 ft.
Volume of concrete	=	196 ft. × (30/12) × (15/12)
	=	196 ft. × 2.5 ft. × 1.25 ft.
	=	612.5 cu. ft /27
	=	22.7 cu. yd.
22.7 cu. yd. × 1.05 (5% waste)	=	23.84 cu. yd.
Estimated amount	=	**24 cu. yd.**

Task 2
Carpet:
24 ft. × 30 ft.	=	720 sq. ft.
720/9	=	80 sq. yd.
80 × 1.07 (7%) waste	=	85.6 sq. yd.
Estimated amount	=	**86 sq. yd.**

Vinyl Base:
(24 + 30) × 2	=	108 ft.
108 – 2(3)	=	102 ft.
102 × 1.05 (5% waste)	=	107.1 ft.
Estimated amount	=	**108 ft.**

Paint:
108 × 10	=	1080 sq. ft.
1080 × 1.10 (10% waste)	=	1188 sq. ft.

Note: Normally no deduction is made for standard door and window openings.

1188/450	=	2.64 gal.
Estimated amount	=	**3 gal. primer**
		6 gal. finish coat

CONTREN™ LEARNING SERIES — USER UPDATES

The NCCER makes every effort to keep these textbooks up-to-date and free of technical errors. We appreciate your help in this process. If you have an idea for improving this textbook, or if you find an error, a typographical mistake, or an inaccuracy in NCCER's Contren™ textbooks, please write us, using this form or a photocopy. Be sure to include the exact module number, page number, a detailed description, and the correction, if applicable. Your input will be brought to the attention of the Technical Review Committee. Thank you for your assistance.

Instructors – If you found that additional materials were necessary in order to teach this module effectively, please let us know so that we may include them in the Equipment/Materials list in the Instructor's Guide.

Write: Curriculum Revision and Development Department
National Center for Construction Education and Research
P.O. Box 141104, Gainesville, FL 32614-1104

Fax: 352-334-0932

E-mail: curriculum@nccer.org

Craft _____ Module Name _____

Copyright Date _____ Module Number _____ Page Number(s) _____

Description
_____
_____
_____
_____

(Optional) Correction
_____
_____
_____

(Optional) Your Name and Address
_____
_____
_____

Project Supervisor

Module MT207-01
Planning and Scheduling

Planning and Scheduling
Instructor's Guide

Module MT207

MODULE OVERVIEW

This module introduces the project supervisor trainee to planning and scheduling in the construction industry. This module teaches the trainee how the planning process works by breaking down a project into easy-to-handle pieces and then setting the schedule for the job site.

PREREQUISITES

There are no prerequisites for this module; however, prior to training with this module, it is recommended that the trainee complete the following modules:
  Project Supervision, Modules MT201 through MT206

LEARNING OBJECTIVES

Upon completion of this module, the trainee will be able to:

1. Describe the link between planning and scheduling.
2. Describe how the planning process is carried out.
3. Write a goal statement and an objective.
4. Break down a project into manageable components that need to be handled before a job begins to its completion.
5. Describe the various resources that need to be considered when planning a job.
6. Assign resources to each activity in a list.
7. Explain how to conduct a job analysis.
8. Develop a bar chart.
9. Explain the purpose of network diagrams.
10. List the benefits of short-interval schedules.

PERFORMANCE OBJECTIVES

This is a knowledge-based module – there is no performance profile examination.

NCCER STANDARDIZED TRAINING PROGRAM

The National Center for Construction Education and Research (NCCER) provides a standardized national program of accredited craft training. Key features of the program include instructor certification, competency-based training, and performance testing. The program provides trainees, instructors, and companies with a standard form of recognition through a National Craft Training Registry. The program is described in full in the Guidelines for Accreditation, published by the NCCER. For more information on standardized craft training, contact the NCCER by writing us at P.O. Box 141104, Gainesville, FL 32614-1104; calling 352-334-0911; or e-mailing info@nccer.org. More information may be found at our Web site, www.nccer.org.

HOW TO USE THIS ANNOTATED INSTRUCTOR'S GUIDE

Each page presents two sections of information. The larger section displays each page exactly as it appears in the Trainee Module. The narrow column ties suggested trainee and instructor actions to each page and provides icons to call your attention to material, safety, audiovisual, or testing requirements. The bottom of each page includes space for your notes.

 If you see the Teaching Tip icon, that means there is a teaching tip associated with this section. Also refer to the suggested teaching tips at the end of the module.

PREPARATION

Before teaching this module, you should review the Module Outline, Learning Objectives, and the Materials and Equipment List. Be sure to allow ample time to prepare your own training or lesson plan and gather all required equipment and materials.

MATERIALS AND EQUIPMENT LIST

Materials:

Transparencies

Markers/chalk

Module Examinations*

Sample company policies regarding tools and equipment and tool inventory programs**

Equipment:

Overhead projector and screen

Whiteboard/chalkboard

*Located in the Test Booklet packaged with this Annotated Instructor's Guide.
**If available on loan from your workplace or other resource.

ADDITIONAL RESOURCES

This module is intended to present thorough resources for task training. The following reference works are suggested for both instructors and motivated trainees interested in further study. These are optional materials for continued education rather than for task training.

Construction Materials and Building Publication No. 91, International Organization for Standardization (ISO), ISO Online Catalog at www.iso.ch.

Professional Construction Management: Including Contracting C M, Design-Construct, and General Contracting, 1991. Donald S. Barrie and Boyd C. Paulson (Contributor). Upper Saddle River, NJ: McGraw-Hill Higher Education.

Construction Management, 1997. Daniel W. Halpin and Ronald W. Woodhead. New York: John Wiley & Sons.

Construction Operations Manual of Policies and Procedures, 2000. Andrew Civitello, Jr. Upper Saddle River, NJ: McGraw-Hill Professional Book Group.

TEACHING TIME FOR THIS MODULE

An outline for use in developing your lesson plan is presented below. Note that each Roman numeral in the outline equates to one session of instruction. Each session has a suggested time of 2 1/2 hours. This includes 10 minutes at the beginning of each session for administrative tasks and one 10-minute break during the session. Approximately 17 1/2 hours are suggested to cover *Planning and Scheduling*.

Topic	Planned Time
Session I. Introduction to Planning	
A. Introduction	_____
B. An Overview of Planning	_____
1. What Is a Plan?	_____
2. Participant Activity	_____
3. Stages of Planning	_____
a. Pre-Construction Planning	_____
b. Construction Planning	_____
c. Participant Activity	_____
d. Implementing the Construction Plan	_____
Session II. The Planning Process and Planning Resources	
A. The Planning Process	_____
1. Establish a Goal	_____
2. Identify the Work to be Done	_____
3. Determine Tasks	_____
4. Participant Activity	_____
5. Communicate Responsibilities	_____
6. Follow-Up	_____
B. Planning Resources	_____
1. Planning Materials	_____
2. Planning Equipment	_____
3. Planning Tools	_____
4. Planning Labor	_____
5. Coordinating With Other Contractors	_____
Session III. Job Analysis and Implementation of the Plan	
A. Job Analysis	_____
1. Participant Activity	_____
B. Implementation of the Plan	_____
1. Job-Site Meetings	_____
2. Participant Activity	_____
3. Planning Checklist	_____
Session IV. Introduction to Scheduling	
A. Types and Benefits of Schedules	_____
1. Types of Schedules	_____
B. Developing a Construction Schedule	_____
1. Formal Planning	_____
2. Participant Activity	_____
3. The Construction Plan	_____
4. The Construction Schedule	_____

Session V. Bar Charts and Network Diagrams
   A. Bar Charts   _____
   B. Network Diagrams   _____
      1. Precedence   _____
      2. Determining the Time Required for Each Activity   _____

Session VI. Short-Term Schedules
   A. Short-Term Schedules   _____
      1. Participant Activity   _____
      2. Participant Activity   _____

Session VII. Schedule Updates
   A. Schedule Updates   _____
      1. Participant Activity   _____
   B. Summary   _____
      1. Summarize module   _____
      2. Answer review questions   _____
   C. Module Examination   _____
      1. Trainees must score 70% or higher to receive recognition from the NCCER.   _____
      2. Record the testing results on Craft Training Report Form 200 and submit the results to the Training Program Sponsor.   _____

Planning and Scheduling

Instructor's Notes:

Instructor's Notes:
_____
_____
_____
_____
_____
_____
_____

ACKNOWLEDGMENTS

The NCCER wishes to acknowledge the dedication and expertise of Roger Liska and Jay Newitt, the original authors and mentors for this module on planning and scheduling.

Roger W. Liska, Ed.D., FAIC, CPC, FCIOB, PE

Clemson University

Chair & Professor

Department of Construction Science & Management

Jay S. Newitt, PhD

Brigham Young University

Professor of Construction Management

We would also like to thank the following reviewers for contributing their time and expertise to this endeavor:

J.R. Blair

Tri-City Electrical Contractors

An Encompass Company

Mike Cornelius

Tri-City Electrical Contractors

An Encompass Company

Dan Faulkner

Wolverine Building Group

David Goodloe

Clemson University

Kevin Kett

The Haskell Company

Danny Parmenter

The Haskell Company

Course Map

This course map shows all of the modules of the *Project Supervision* curriculum. The suggested training order begins at the bottom and proceeds up. Skill levels increase as you advance on the course map. The local Training Program Sponsor may adjust the training order.

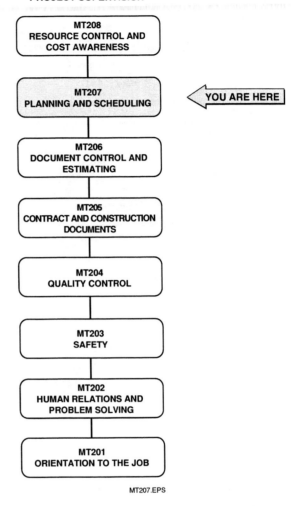

Instructor's Notes:

MODULE MT207

TABLE OF CONTENTS

1.0.0	**INTRODUCTION**	7.1
2.0.0	**AN OVERVIEW OF PLANNING**	7.2
2.1.0	What Is a Plan?	7.2
2.2.0	Stages of Planning	7.3
2.2.1	*Pre-Construction Planning*	7.3
2.2.2	*Construction Planning*	7.3
2.2.3	*Implementing the Construction Plan*	7.5
2.3.0	The Planning Process	7.5
2.3.1	*Establish a Goal*	7.5
2.3.2	*Identify the Work to be Done*	7.6
2.3.3	*Determine Tasks*	7.6
2.3.4	*Communicate Responsibilities*	7.7
2.3.5	*Follow-Up*	7.8
2.4.0	Planning Resources	7.8
2.4.1	*Planning Materials*	7.8
2.4.2	*Planning Equipment*	7.8
2.4.3	*Planning Tools*	7.9
2.4.4	*Planning Labor*	7.9
2.4.5	*Coordinating With Other Contractors*	7.10
2.5.0	Job Analysis	7.10
2.6.0	Implementation of the Plan	7.11
2.6.1	*Job-Site Meetings*	7.12
2.6.2	*Planning Checklist*	7.13
3.0.0	**SCHEDULING**	7.13
3.1.0	Types and Benefits of Schedules	7.13
3.1.1	*Types of Schedules*	7.13
3.2.0	Developing a Construction Schedule	7.16
3.2.1	*Formal Planning*	7.16
3.2.2	*The Construction Plan*	7.16
3.2.3	*The Construction Schedule*	7.17
3.3.0	Bar Charts	7.19
3.4.0	Network Diagrams	7.21
3.4.1	*Precedence*	7.21
3.4.2	*Determining the Time Required for Each Activity*	7.21

3.5.0	Short-Term Schedules	7.23
3.6.0	Schedule Updates	7.26
	SUMMARY	7.27
	REVIEW QUESTIONS	7.29
	GLOSSARY	7.33

LIST OF FIGURES

Figure 1	Planning Checklist	7.13
Figure 2	Daily Work Plan	7.14
Figure 3	Bar Chart	7.15
Figure 4	CPM Network Diagram	7.15
Figure 5	Work Breakdown Structure (WBS) Chart	7.17
Figure 6	List of Work Items for Laboratory Renovation	7.18
Figure 7	Updated Bar Chart	7.20
Figure 8	Resource Profile Developed from Bar Chart	7.20
Figure 9	Precedence Relationship	7.21
Figure 10	Sample Network Diagram for Lab Renovation	7.22
Figure 11	Detail of Network Logic Diagram	7.23

Instructor's Notes:

NATIONAL CENTER FOR CONSTRUCTION EDUCATION AND RESEARCH

MODULE MT207

Planning and Scheduling

Ensure that you have all the necessary materials to teach the course. Check the Materials and Equipment list at the front of the module.

Ask trainees to list additional management responsibilities for the contractor, project manager, and field supervisor other than those mentioned.

OBJECTIVES

Upon the completion of this module, you will be able to do the following:

1. Describe the link between planning and scheduling.
2. Describe how the planning process is carried out.
3. Write a goal statement and an objective.
4. Create a step-by-step list of the tasks that will complete a project.
5. Describe the various resources that need to be considered when planning a job.
6. Assign resources to each activity in a list.
7. Explain how to conduct a job analysis.
8. Develop a bar chart.
9. Explain the purpose of network diagrams.
10. List the benefits of short-interval schedules.

SECTION 1

1.0.0 INTRODUCTION

The contractor, the project manager, and the field supervisor each have management responsibilities for the jobs to which they have been assigned. For example, the contractor's responsibility begins with obtaining the contract, and it does not end until ownership of the project has been transferred to the client. The project manager is generally the person with overall responsibility for coordinating the project. Finally, the field supervisor is responsible for coordinating the installation of the work of one or more workers, one or more crews of workers within the company, and, on occasion, one or more crews of subcontractors. A major portion of these responsibilities includes the tasks of planning and scheduling.

Just as the saw and the hammer are the primary tools of the carpenter, the construction schedule and the estimate are the primary tools of the supervisor. The schedule is a tool used to maintain control over projects under construction. The development of a good schedule will help the supervisor to carefully think the project through in advance, and then use that schedule to communicate the construction plan to everyone involved in the execution of the project. The process of developing the schedule forces the supervisor to mentally build the project and record it on paper. This eliminates many potential errors that could happen later.

Show Transparency 1 (Course Objectives).

Assign reading of Module MT207, Sections 1.0.0 – 2.2.3.

Copyright © 2003 National Center for Construction Education and Research, Gainesville, FL 32614-1104. All rights reserved. No part of this work may be reproduced in any form or by any means, including photocopying, without written permission of the publisher.

PLANNING AND SCHEDULING— INSTRUCTOR'S GUIDE MODULE MT201   7.1

Ask trainees to share examples of situations when they were confused because their supervisors neglected to update the plan with approved changes.

A well thought-out schedule will put the supervisor in a position where *management by objectives* is the rule, rather than *management by crisis*. Do you spend your day running from one problem to another all day long? This may be an indication of a lack of detailed planning and scheduling at the beginning of the project.

SECTION 2

2.0.0 AN OVERVIEW OF PLANNING

Planning is a managerial activity that includes three essential elements: setting objectives, identifying the tasks that need to be done to get the job built, and putting those tasks in the most efficient sequence.

Supervisors play a critical role in planning. Unfortunately, supervisors don't always have the information needed in order to plan well. They are often the first ones to need information, yet supervisors tend to be the last ones to get it. In addition, there may be parts of the plan, such as material delivery, over which supervisors have little control. For all of these reasons, it is very important that supervisors learn to anticipate their needs well ahead of time, clearly communicate their needs to the appropriate person(s), and follow through to be sure that their needs will be met.

Successful planning is largely a matter of remembering that even the most complex project is made up of many small, simple jobs. It also involves knowing that these small, simple jobs remain largely the same from project to project. For example, materials are transported from the storage area to the work area; tools and equipment are brought to the work area; and workers are assigned tasks in every job. True, some of these work activities require special preparation, such as erecting scaffolding. Generally, however, the activities required to complete a job are quite simple. Complexity and confusion come about when no one takes responsibility for organizing these activities in a proper sequence, one that assures that time, energy, and money are used with the maximum efficiency. Neglecting to update the plans as changes occur also contributes to the complexity and confusion of jobs.

2.1.0 What Is a Plan?

A *plan* is an organized sequence of events that identifies all parts of the work to be done. It shows in advance what tasks are required, allowing better control of each task, and it identifies and eliminates (or at least minimizes) obstacles. With the information provided in a plan, a supervisor can direct work efforts efficiently and use resources such as personnel, materials, tools, equipment, and work methods to their full potential.

To a newcomer on a construction site, the variety of activities going on simultaneously seems complex, and indeed it is. There are cranes lifting materials, workers placing concrete, equipment excavating earth, and workers moving about in every direction. Managing a typical construction job is a complex task. A recent study noted that the construction of a ten-story office building comprises more than 85,000 different materials and construction activities. Each of these activities must be coordinated with others in such a way that all tasks can be completed according to the plan.

No two construction projects are alike, making construction a very challenging profession. Even if the same drawings and specifications are used for more than one job, the results will vary. This can be attributed to differences in contractors, locations, suppliers, and costs. These and other variables can make planning difficult.

Faced by such difficulties, some job-site supervisors tend to think it does no good to plan, or they conclude that they cannot possibly know enough about the tools, materials, or workers that will be available for them to make a plan on which they can rely. The fact is, however, that supervisors can be very effective in planning their work, no matter how large or complex the job.

Lacking a single piece of equipment or a single item of material can cause delays that seriously affect productivity and make it difficult to finish the job on time and within the cost estimate. Many managers, including supervisors, are unaware of such problems until after they occur. Although such oversights reflect on the performance of the entire management team, it is usually the supervisors who must deal directly with the effects.

Instructor's Notes:

PARTICIPANT ACTIVITY

Discuss and complete the following:

1. List two activities in your personal life, such as a weekend outing with the family or a fishing trip, that you had to plan during the last month.
2. For each activity you listed above, list the tasks you performed as part of the planning process.
3. If you had *not* planned these activities, what could have happened? Write a brief explanation.
4. Do your job duties include planning? If so, briefly explain how.
5. List two advantages that you see in planning.
6. List any disadvantages that you see in planning.

Have trainees complete the Participant Activity. Discuss their answers.

To avoid delays and the consequences that they cause, the supervisors must be involved in day-to-day planning.

2.2.0 STAGES OF PLANNING

There are various times when planning is done for a construction job. These include the time before a project begins (pre-construction) and throughout the duration of the job (construction).

2.2.1 Pre-Construction Planning

The pre-construction stage of planning occurs prior to the start of construction. Except in a fairly small company or for a relatively small job, the supervisor usually does not get directly involved in the pre-construction planning process.

There are two phases of pre-construction planning. The first is when the proposal, bid, or negotiated price for the job is being developed. This is when the estimator, along with others such as the project manager and the field supervisor, develops a preliminary plan for how the work will be done. This is accomplished by applying experience and knowledge from previous projects. It involves determining what methods, personnel, tools, and equipment will be used and what level of productivity can be expected from them.

The second phase occurs after the contractor is awarded the contract. This phase requires a thorough knowledge of any project documents that pertain to the project scope being planned.

During this stage, the actual work methods and resources needed to perform the work are selected. Here, supervisors might get involved, but as they plan they must adhere to the work methods, production rates, and resources that fit within the estimate that was prepared before the contract was awarded. If the project requires a method of construction different from what is normally followed, the supervisor will usually be informed of what method to use.

2.2.2 Construction Planning

During construction, the supervisor is directly involved in planning on a daily basis. This planning consists of selecting methods of completing tasks before beginning them. Effective planning exposes likely difficulties and enables the supervisor to minimize the unproductive use of personnel and equipment. Effective planning also provides a gauge by which progress can be measured.

All construction jobs consist of several activities. One of the characteristics of an effective supervisor is the ability to reduce each job to its simpler parts and organize a plan for handling each task. Time and cost limits for the project are established by others, and the supervisor's planning must fit within those constraints. Therefore, it is important to understand the following factors that may impact the work:

- Site and local conditions such as soil types, accessibility, or available staging areas
- Climate conditions that should be anticipated during the project

Have trainees complete the Participant Activity. Discuss their answers.

- Timing of all phases of work
- Types of materials to be installed and their availability
- Equipment and tools required and their availability
- Personnel requirements and availability
- Relationships with the other contractors and their representatives on the job

On a simple job, these items can be handled almost automatically. However, larger or more complex jobs force the supervisor to give these factors more formal consideration and study.

The supervisor's primary responsibility is to provide leadership to the crew. This requires the supervisor to anticipate upcoming work instead of worrying about problems resolved last week. In short, it involves *planning*.

It is also necessary to realize that planning and scheduling are closely related. While planning involves determining which activities must be performed and completed and how they should be accomplished, scheduling involves establishing start and finish times or dates for each activity.

The major goals of planning are to:

- Determine the best method for performing the job
- Identify the responsibilities of each person on the work crew
- Determine the duration of each activity
- Identify what tools will be needed to complete a job
- Ensure that the required materials are at the work site when needed
- Make sure that heavy construction equipment is available when required
- Work with other contractors in such a way as to avoid interruptions and delays
- Communicate effectively with everyone involved

Those in a supervisor's crew should be involved in any planning as much as possible. Research indicates that people who are involved in planning are more enthusiastic about putting the plan into action and are therefore more productive and efficient.

When a supervisor fails to understand the factors involved in planning, company profits and employee job security can become threatened. For example, the failure of an electrical contractor to coordinate efforts with the general contractor could result in the work not being started until after the interior of the building has been painted and the general contractor's scaffolding removed. The result: the electrical contractor is faced with the effort and expense of erecting separate scaffolding and possibly the cost of painting the conduit when the job is finished.

The HVAC supervisor who fails to review the drawings and specifications and become familiar with the types of equipment required is likely to install the rough-ins incorrectly. When the equipment is delivered and the error is discovered, the HVAC contractor is faced with the labor and material costs for altering the existing system as well as adjacent work installed by other trades. Profits are reduced, and the schedule of the overall project may be impacted.

PARTICIPANT ACTIVITY

Discuss and complete the following:

1. What are the differences between pre-construction planning and construction planning?
2. What are the similarities between pre-construction planning and construction planning?
3. Describe your role in the pre-construction and the construction planning processes.
4. Describe how you presently get involved in the planning process on your job.
5. Plan the activities that you expect to do tomorrow on your project. Anticipate those steps in the plan that can go haywire and develop a corresponding contingency plan for each.

Instructor's Notes:

This is not to say that even the best supervisor is not sometimes caught by surprise. On every construction job, things happen that are not anticipated in the plan. Unexpected bad weather or the breakdown of equipment can threaten any project with costly delays. What can supervisors do? They can prepare contingency plans. When preparing the original plan, the supervisor should always consider what is to be done if something goes wrong. Answering this question before the unexpected occurs makes emergencies easier to handle and minimizes delays.

2.2.3 Implementing the Construction Plan

Once a plan is set into action, the supervisor must follow up daily to ensure that it is being carried out. If seven working days are allotted to complete a job, the seventh day is too late to determine whether or not the job will be finished on time.

The concept of a short-interval scheduling program should be incorporated into the follow-up program. It involves determining each day's goals and measuring progress according to those goals. For example, a supervisor assigned to place 350 cubic yards (cu. yd.) of concrete in seven days can determine that the crew's daily goal should be 50 cu. yd. (350 divided by 7). This provides a means for measuring daily progress and the progress of the entire task.

Of course, most jobs begin slowly, pick up speed, then slow down again near the end. Therefore, 50 cu. yd. per day should be considered only an estimate. The crew might place only 30 cu. yd. the first day and 40 cu. yd. the second. It might not be until the third day that they reach their average, but then they place 60 cu. yd. on each of the next two days. By averaging out their performance, the supervisor can easily tell whether they will meet their schedule. If at the end of the fifth day, the crew has only placed 240 cu. yd. of concrete, the supervisor can see that the goal may not be reached. Consequently, appropriate action must be taken in an effort to meet the goal, such as increasing the daily estimated output of concrete per day. If the supervisor waits until the afternoon of the seventh day, however, there is nothing that can be done about the problem; it is too late. The advantage of following up on one's plan on a daily basis, then, is obvious.

One word of caution is in order concerning measuring progress by installed quantities. Be aware that it is possible to be ahead of schedule in terms of cubic yards placed, for example, yet be behind in terms of the total project schedule. A supervisor who chooses to install all of the large and/or easy-to-handle components of the job first will appear far ahead of schedule; however, the opposite is actually true because the more difficult tasks remain to be completed. Furthermore, it is important to remember that choosing one's own work sequence can seriously delay other crews and extend the total job schedule.

When following up, the supervisor should use a diary or daily log to document any changes made to the original plan. That way, when faced with a similar situation in the future, reliable information will be available upon which to base a workable plan. For this reason, it is also good for the supervisor to document plans that are successful as originally envisioned. The documentation may also serve as evidence in the case of future claims.

2.3.0 The Planning Process

The planning process can be broken down into five steps:

Step 1 Establish a goal.

Step 2 Identify the work activities that must be completed in order to achieve the goal.

Step 3 Determine what tasks must be done to ensure that those activities will be accomplished.

Step 4 Communicate responsibilities.

Step 5 Follow up to see that the goal is achieved.

2.3.1 Establish a Goal

A *goal* is something that we direct our efforts toward accomplishing. There are many types of goals, including goals that we set within our personal lives and goals that we set at work.

Personal goals may be those that bring a special reward or achievement. Examples of personal goals include improving your golf game, moving to a new area, or taking a vacation with friends or family. These goals are usually self-directed. As individuals, we usually set our own goals and then create specific **objectives** for how we will meet them.

Ensure that you have all the necessary materials to teach the course. Check the Materials and Equipment list at the front of the module.

Refer to end of this module for a teaching tip.

Assign reading of Sections 2.3.0 – 2.4.5 for the next class session.

Show Transparency 2 (Five-Step Planning Process).

At work, the organization or project team within the organization establishes goals in much the same way. Examples of job-related goals include constructing column forms, placing concrete, or setting a water heater. Although supervisors can provide input on how a job goal will be accomplished, they rarely, if ever, establish the goal.

2.3.2 Identify the Work to be Done

The second step in planning is to identify the work to be done in order to achieve the goal. In other words, it is the series of activities that must be done in a certain sequence. The topic of breaking down a job into activities is covered later in the module. At this point, the supervisor should know that, for each activity, one or more objectives must be set.

An objective is a statement of what is desired at a specific time. An objective must mean the same thing to everyone involved, be measurable (so that everyone knows when it has been reached), be achievable, and have everyone's full support.

Examples of practical objectives include:

- By 4:30 p.m. today, my crew will have completed laying two-thirds of the 12-inch interior block wall.
- By closing time Friday, all electrical panels will be installed.

Notice that both examples meet the first three requirements of an objective. In addition, it is assumed that everyone involved in completing the task is committed to achieving the objective, which is the fourth requirement. The advantage in writing objectives for each work activity is that it allows the supervisor to evaluate whether or not the plan is being carried out as developed and the schedule is being followed. In addition, objectives serve as subgoals that are usually under the supervisor's control.

Some construction work activities, such as installing 12"-deep footing forms, are done so often that they require little planning. However, other jobs, such as placing a new type of mechanical equipment, require substantial planning. It is this type of job that requires the supervisor to set specific objectives. Any time supervisors are faced with a new or complex activity, they should take the time to establish objectives that will serve as guides for accomplishing the task at hand. These guides can be used in the current situation as well as in similar situations in the future.

2.3.3 Determine Tasks

To plan effectively, the supervisor must be able to break a work activity down into the smaller tasks of which it is comprised. Large jobs include a greater number of tasks than small ones, but all jobs can be broken down into manageable components.

When breaking down an activity into tasks, the supervisor should first determine how much detail is required. The general rule is that each task should be identifiable and definable. A task is definable if you can assign a specific time to it. It is identifiable when the types and amounts of resources it requires are known. For purposes of efficiency, the supervisor's job breakdown should not be too detailed or complex, unless the job has never been done before or must be performed with strictest efficiency.

For example, a breakdown for the work activity:

Install 12-inch × 12-inch
vinyl tile in a cafeteria

might be the following:

Step 1 Lay out.
Step 2 Clean floor.
Step 3 Spread adhesive.
Step 4 Lay tile.
Step 5 Clean floor.
Step 6 Wax floor.

A more detailed list may be:

Step 1 Lay out.
Step 2 Acquire labor, equipment, and materials.
Step 3 Clean floor.
Step 4 Spread adhesive over part of the floor.
Step 5 Lay tile on that part of floor over which adhesive was spread.
Step 6 Repeat Steps 4 and 5 until the entire floor area has been tiled.
Step 7 Clean floor.
Step 8 Wax floor.

Instructor's Notes:

The supervisor could be even more detailed by breaking down any one of the tasks, such as "lay tile" into subtasks. In this case, however, that much detail is unnecessary and wastes the supervisor's time and the project's money. However, a more detailed breakdown of tasks might be necessary in a case where the job is very complex or the analysis of the job needs to be very detailed.

Every work activity can be divided into three general parts: preparing, performing, and cleaning up.

One of the most frequent mistakes made in the planning process is forgetting the first and third items, preparing and cleaning up. The supervisor must be certain that preparation and cleanup are not overlooked on any job breakdown.

After identifying the various activities that make up the job and developing an objective for each activity, the supervisor must determine what resources the job requires. Resources include labor, equipment, materials, and tools.

In most jobs, these resources are identified in the job estimate, and the supervisor must only make sure that they are on the site when needed. On other jobs, however, the supervisor may have to determine, in part, what is required as well as arrange for timely delivery.

2.3.4 Communicate Responsibilities

A supervisor is unable to complete all of the activities within a job independently and must rely on other people to get everything done. Therefore, construction jobs have a crew of people with various experiences and skill levels available to assist in the work. The supervisor's responsibility is to draw from this expertise to get the job done well and in a timely manner.

Once the various activities that make up the job have been determined, the supervisor must identify the person or persons who will be responsible for completing each activity. This requires that the supervisor be aware of the skills and abilities of the people on the crew.

Have trainees complete the Participant Activity. Go over the scenarios together. Example answers are located at the end of this module.

Ask trainees to list all their crew members on one page and write their skills on another page. If they are missing information, have them take the time to discover their crew's skills in order to enhance their ability to match a worker to a job.

PARTICIPANT ACTIVITY

1. Define *goal* and give two examples, one personal and one professional.

2. Define *objective* and write two examples. Check to see that your examples meet the qualifications of an objective.

Read the following scenarios and complete the activities.

Scenario I

You have been given the job of constructing the formwork for a 5'- high, uniformly thick retaining wall 100 feet long. You have five days in which to complete it. Assume your crew and other resources are on the site and weather will be favorable.

- Write a goal statement for the job.
- Write one or more objectives for the job.

Scenario II

You have been assigned to prepare and paint (a prime and a final coat) the ceiling and walls of six rooms of an office building. All rooms are on the same level and all tools, equipment, materials, and labor are available. You have one crew to do the work. The estimate indicates that you have four eight-hour days to do the work.

- List the identifiable and definable activities for this job.
- Give three reasons why goals and/or objectives might not be met for this job.

PARTICIPANT ACTIVITY

Ask trainees to give examples of what types of tools and equipment can be shared with other contractors on the job site.

Then, the supervisor must put this knowledge to work in matching the crew's skills and abilities to specific tasks that must be accomplished to complete the job.

Upon matching crew members to specific activities, the supervisor must then communicate the assignments to the responsible person(s). Communication of responsibilities is generally handled verbally by the supervisor talking directly to the person that has been assigned the activity. However, there may be times when work is assigned indirectly through written instructions or verbally through someone other than the supervisor. Either way, the crew members should know what it is they are responsible for accomplishing on the job.

2.3.5 Follow-Up

Once the activities have been delegated or assigned to the appropriate crew members, the supervisor must follow up to make sure that they are completed effectively and efficiently. Follow-up involves being present on the job site to make sure that all of the resources are available to complete the work, ensuring that the crew members are working on their assigned activities, answering any questions, or helping to resolve any problems that occur while the work is being done. In short, follow-up means that the supervisor is aware of what's going on at the job site and anticipates whatever is necessary to make sure that the work is completed as scheduled.

2.4.0 Planning Resources

Once a job has been broken down into its tasks or activities, the next step is to assign the various resources needed to perform them.

2.4.1 Planning Materials

The materials required for the job are identified during pre-construction planning and are listed on the job estimate. The materials are then ordered from suppliers who have been identified by the estimating and planning staff as having provided quality materials on schedule and within estimated cost.

A purchase order is normally used to order materials from the supplier. A purchase order lists the quantity, quality, and delivery date of the materials being purchased. Once a purchase order has been submitted, it is usually up to the project manager to ensure that the materials will be delivered to the site or storage area when needed and in the quantity stated. A designated supervisor on the job also gets a copy of the purchase order or other communication so that it will be known when and where the materials will be delivered. Some materials, such as connectors, may be ordered at the time they are needed instead of months in advance. For these items, close communication between field and office management is needed.

The supervisor is usually not involved in the planning and selection of materials, since this is done in the pre-construction phase. However, planning materials for tasks such as job-built formwork and scaffolding is a supervisor's job. Also, the supervisor occasionally may run out of a specific material, such as fasteners, and need to order more. In such cases, the next higher supervisor should be consulted, since most companies have specific purchasing policies and procedures.

The supervisor's major responsibility in regard to materials is to help ensure that necessary items will be available on site as needed. Timely delivery avoids added job expenses due to slack time spent waiting for delivery, shifting workers from one task or one job to another, laying off workers, and rehiring other workers when the materials finally arrive.

The supervisor should follow up at least one week ahead of time on the materials needed for a particular job. This provides the opportunity to prepare alternate plans if the materials will not arrive on time. In most cases all materials should be on site at least one day before they are needed. There are exceptions in cases of special materials or unusual construction processes where the materials are needed on the site sooner.

Before delivery, the supervisor must plan for receiving, storing, and controlling the materials. This plan must include time to unload, inspect and test, store, and retrieve from storage.

2.4.2 Planning Equipment

Planning the use of construction equipment involves identifying the types of equipment needed, the tasks each type must perform, and the time each piece is needed. Much of this is planned during the pre-construction phase;

Instructor's Notes:

Refer to the end of this module for a teaching tip.

however, it is up to the supervisor to work with the home office to make certain that the equipment reaches the job site on time. The supervisor should also coordinate equipment with other contractors on the job, as sharing equipment can save time and money and avoid duplication of effort. In addition, if the equipment breaks down, the supervisor must know who to contact to resolve the problem. The supervisor should also designate some time for routine equipment maintenance on a regular basis in an effort to avoid equipment failure.

Finally, when various pieces of equipment will be working in a relatively small area or one piece will work in conjunction with another, such as a loader and a dump truck or a bulldozer and a scraper, the supervisor must coordinate the use of the equipment. It is a good idea to have an alternate plan ready in case one piece of equipment breaks down, so that the other equipment does not sit idle. This planning should be done in conjunction with the home office or the supervisor's immediate superior.

2.4.3 Planning Tools

Most work activities require the use of hand and/or power tools. The supervisor is responsible for identifying any tools necessary and having the required tools available. Studies show that a lack of well-maintained tools is the cause of many job delays.

The supervisor must know company policies and procedures regarding tools. If the policy is to have craftworkers provide their own tools, then the supervisor must see that the workers bring them to the job and that all tools are in safe working condition. However, if the company provides tools to workers, then the supervisor must do more planning to make sure that all needed tools reach the job site as required. Advance planning is necessary; arrangements should be made at least a week before the job begins, not when the job is about to start. This allows time to collect tools from alternate sources, if needed. The greater the number and types of tools required, the longer the lead time necessary to ensure that they are available.

When the contractor supplies tools, the supervisor must do whatever is necessary to streamline the company's tool control program so that checking tools in and out does not take too much time away from getting the job done. One means of doing this is for workers to check tools in and out through the supervisor, a plan that produces more effective control and also reduces wasted time. When employees are required to provide their own tools, it is wise for the supervisor to have a small inventory of backups for temporary use when employees' tools are lost or damaged. However, when an employee borrows a company tool, the supervisor must follow up to make sure that the employee replaces or repairs missing or broken tools and returns the borrowed tools to the company inventory.

Planning the tools for a job also involves maintaining a tool part inventory and having a tool maintenance program. These items are usually the responsibility of central management, but the supervisor is responsible for informing management when either program is not working.

The supervisor is also responsible for ensuring that workers use the tools properly and safely. This may require some on-the-job training. If on-the-job training is required to teach the crew how to use the tools, the training should be included in the job plan.

2.4.4 Planning Labor

All tasks require some sort of labor because supervisors cannot complete all of the work by themselves. In planning labor, the supervisor must first identify the skills needed to perform the work; determine how many people having those specific skills are needed; and finally, decide who will actually be on the crew.

In many companies, the project manager or job superintendent determines the size of the crew and the make-up of the crew. In this situation, the supervisor is expected to accomplish the goals and objectives with the crew provided. However, in some companies, the supervisor has full authority and responsibility for determining all or part of the make-up of the crew.

If the supervisor is given the responsibility of putting a crew together, it is necessary to know if there are qualified employees available within the company or if they must be hired from outside. Consideration must also be given to how much time should be spent on training, and this time should be included in the job plan. Finally, the labor requirement must be planned

Ensure that you have all the necessary materials to teach the course. Check the Materials and Equipment list at the front of the module.

Show Transparency 3 (Job Analysis Suggestions).

Assign reading of Sections 2.5.0 – 2.6.2 for the next class session.

far enough in advance so that adequate arrangements can be made to have the required personnel on site when the job is scheduled to begin.

Unexpected labor problems, such as absenteeism, turnover, and illness, often occur on construction jobs. Therefore, the supervisor must be ready to deal with them as they happen. This involves anticipating such problems and having contingency plans ready for action.

2.4.5 Coordinating With Other Contractors

No task stands alone. The work performed by one crew must always be coordinated with the work of other contractors and crews on the job site. The most effective way for the supervisor to coordinate the work is to review the work plan of all the contractors on the job each week and then discuss potential or anticipated problems with the next higher supervisor or manager. It is also in everyone's best interest for all supervisors on the job to establish a satisfactory working relationship with one another.

2.5.0 Job Analysis

A *job analysis* is a study of the job aimed at effectively and efficiently integrating all tasks and activities. It is most often used when situations such as inclement weather or late material deliveries threaten to delay work. The supervisor, possibly in conjunction with other managerial or non-managerial personnel on the job, must identify the remaining tasks and, if necessary, alter resources and work methods to complete the job on time and without additional expense. A job analysis can also be conducted during pre-construction to determine the most efficient and effective job plan.

To understand job analysis, consider the tasks involved in taking out the garbage at home. If you use wastepaper baskets, plastic garbage bags, and garbage cans, taking out the garbage may be done as follows:

Step 1 Go outside and carry the garbage can out to the curb.

Step 2 Line the garbage can with a plastic garbage bag.

Step 3 Return to the house.

Step 4 Collect the wastepaper baskets throughout the house.

Step 5 Take each wastepaper basket out to the curb and empty its contents into the garbage can.

Step 6 Put the lid on the garbage can.

The above steps accomplish the task, but a job analysis would show that there is a more efficient way to do the same job:

Step 1 Collect the wastepaper baskets throughout the house.

Step 2 Empty the wastepaper baskets into a garbage bag and secure it.

Step 3 Carry the garbage bag outside, place it in the garbage can, and cover the can.

Step 4 Carry the garbage can to the curb.

The second method is more efficient (assuming that the loaded garbage can is not too heavy to wrestle to the curb) and could have resulted from an analysis of the job in question. There are several ways of doing every job; how you do any job depends on your experience, authority, responsibility, and creativity.

On the job, the supervisor is often instructed on how to do a particular task. Other times, a task is performed in the same manner as in the past. However, if the task is a new one or a particular situation adds an unusual twist to a job, a plan must be established to do it effectively. This involves job analysis. The supervisor should never forget that there is always more than one way to perform any task.

Job analysis begins with a brief written description of the job along with a list of the various identifiable and definable activities involved and the needed resources. Then, the activities and resources are arranged in such a way that they meet the desired goal or objective in the manner that takes the least time and effort and costs the least amount of money.

Some suggestions to consider in conducting a job analysis are:

- Eliminate unnecessary detail.
- Rearrange for optimum sequence.
- Provide better tools, material, and equipment.
- Simplify, where possible, to make things easier to accomplish.
- Keep safety in mind.

Instructor's Notes:

- Use different people, different skills, and/or a different number of people
- Consult with others involved in the job, such as lead men or other crew members.
- Eliminate or minimize unproductive time spent hunting for materials, tools, or equipment.

A job analysis that takes these factors into account produces a plan that will more than pay for the time spent in preparing it. Of course, the amount of time spent in performing a job analysis should be in proportion to the size and complexity of the job. However, even small jobs benefit from job analysis.

2.6.0 Implementation of the Plan

After the supervisor has developed a plan, the plan must be implemented. The first step in the implementation process is to make sure that all of the required resources are on the job so work can proceed as scheduled.

It cannot be overemphasized that the key to effective planning lies in taking the time to be sure that resources are available when needed. This involves knowing who to contact to obtain tools, equipment, materials, and manpower, as well as knowing the various company policies related to these areas.

Have trainees complete the Participant Activity. Discuss their answers and the scenarios together. Suggestions can be found at the end of this module.

PARTICIPANT ACTIVITY

1. Write a brief description of the last project you did in your home or apartment. Then, break down the project into activities. List the tools, equipment, labor, and materials you used to do each activity.

2. In your company, who determines the following?
   - Size and make-up of the crew
   - The type and quantity of materials
   - The type, size, and quantity of equipment
   - The type and number of tools

3. If you run out of material on your job, what procedure do you follow to get more?

4. For the job you are now working on:
   - List the types of materials being used.
   - List the types of labor and the quantity being used.
   - List the types of tools being used and how they are provided.
   - List the types of equipment being used.

Read the following scenario and complete the activities.

You are in charge of constructing a 12' × 24' storage building having an 8' eave height and a standard hip roof. The foundation slab is already in place. Your job is to erect the wall framing, install wall and roof plywood sheathing, install all the windows and doors, and apply the felt paper and asphalt shingles to the roof. In addition, you are to paint all of the exposed exterior wood. No interior finishing is to be done. Assume that all materials are available, and you can get all the necessary tools and equipment. In addition, the required supply of skilled and unskilled labor is also available.

- Write a brief description of the work to be done.
- Write one general goal statement for the entire job.
- Break down the job into tasks or activities.
- Assign an estimated time to complete each task, and write an objective for each.
- Identify the tools, equipment, and labor needed for each task, and determine how many of each (except for hand tools) is required.
- State the estimated time needed to do the entire job, assuming that there will be no weather delays.

PARTICIPANT ACTIVITY

Have trainees complete the Participant Activity. Discuss their answers. Suggestions can be found at the end of this module.

The next step in implementation is to make sure that the work flows smoothly. The supervisor should spend as much time as possible with the crew. When problems arise, the supervisor should be available to immediately notify the appropriate person(s) and obtain instructions, if necessary.

Implementation also involves coordinating the supervisor's efforts with those of other crews and contractors. For this reason, it is vital that a spirit of communication and cooperation be developed and maintained with everyone on site, to promote smooth workflow and timely completion of tasks.

2.6.1 Job-Site Meetings

Responsibility for satisfying the needs of the crew begins with the supervisor. The supervisor is responsible for determining what these needs are and communicating them through the proper channels. The supervisor must accomplish planning for the crew, as it is unlikely that anyone else will do it. Communication is the key.

One of the most important places for communication between the supervisor and others on the job is the job-site meeting. There are many types of job-site meetings: craft coordination meetings, safety meetings, and crew meetings for planning and assigning work. In the case of crew meetings, in particular, the supervisor should follow these guidelines:

- Start the meeting on time.
- State the purpose of the meeting.
- Present all information pertinent to the purpose of the meeting.
- Ask questions throughout the meeting to make sure everyone understands what is being discussed.

PARTICIPANT ACTIVITY

Discuss and answer the following questions:

1. What does it mean to implement a plan? Give two examples.
2. Briefly describe the plan of the job you are now working on, and state how it is being implemented.
3. What are the purposes of job-site meetings? Give three examples.
4. Describe the last job-site meeting you attended.
   - What was the purpose of the meeting?
   - What did you do at the meeting?
   - Do you feel the meeting was successful? Why or why not?
   - What was decided at the meeting? Did it affect you?

Read the following scenario and complete the activity.

On a specific project, you are in charge of placing the concrete for 36 column footings and piers. You are under a tight schedule, but you have all of the equipment, materials, tools, and experienced personnel needed. You call a short crew meeting at 8:00 a.m. the day before the job. Prepare for the meeting by doing the following:

- List the items you will talk about in the order that you will present them.
- Decide on the time you will devote to each topic discussed.
- List what you hope to accomplish for each item on the agenda and what decisions you hope to reach.
- List what you must bring with you to the meeting.
- Identify topics that might lead to disagreements or disputes. How will you handle them?
- Determine what will you do after the meeting to ensure that your goals are reached.

PARTICIPANT ACTIVITY

Instructor's Notes:

- Take time to answer questions and clear up confusion.
- Before adjourning, ask for questions and provide answers.
- If a complex problem needs to be solved, use good problem-solving techniques.
- If a plan must be established, use the guidelines presented in this chapter.

2.6.2 Planning Checklist

The supervisor can make daily planning more efficient and less time consuming by using the planning checklist in *Figure 1*.

It is recommended that the supervisor carry a small notepad, microcassette recorder, or PDA. These can be used to plan and make notes that will help the supervisor remember details. In addition, it may be helpful to put the daily work plan in writing using a form such as the one shown in *Figure 2*.

As the job proceeds, the supervisor should, from time to time, refer to the plan that was established to ensure that all tasks are being done in the sequence and up to the standard planned. This is usually referred to as *analyzing the job program*. Experience shows jobs that are not built according to work plans usually end up costing more than estimated and take more time.

SECTION 3

3.0.0 SCHEDULING

Scheduling begins with the planning phase, which is when the project is broken down into a number of activities or work items. The scheduling phase then determines the relationships among all of the work items: when they should be done and what resources are needed to complete them.

To be effective, the construction schedule should be developed by those who will use it, usually the project manager or the project superintendent. With some companies, the company scheduler develops all the schedules company-wide. However, input from the job foremen, superintendents, project managers, estimators, and other key managers is essential if it's expected that the schedule will actually be used to mon-

PLANNING CHECKLIST

For each task involved in the job, identify the following:

- Work to be accomplished
- People who will perform the tasks
- Start and finish dates
- Location
- How the work is to be performed, including:
  - Sequence of activities
  - Delivery of materials
  - Delivery time of equipment and tools
  - Safety considerations
  - Coordination with other crews and trades
  - Quality control considerations

Figure 1 • Planning Checklist

itor and control the project. Only a schedule developed as a team effort will prove useful to the entire management team.

3.1.0 Types and Benefits of Schedules

As a project supervisor, you will be exposed to a number of scheduling types and techniques. There are many tools, including software programs, available to help you with scheduling. Whether you use a manual or a computerized system, understanding and performing scheduling is one of the keys to your success. It is important to be aware of types of schedules. Learning how to schedule your projects will make you an efficient, productive project supervisor.

3.1.1 Types of Schedules

There are a number of different scheduling techniques available, but effective scheduling requires only two basic methods to be learned: the **bar chart** and the **network diagram.**

The bar chart is a timeline showing the day, week, or month when certain project activities should be accomplished. From this information,

Ensure you that have all the necessary materials to teach the course. Check the Materials and Equipment list at the front of the module.

Ask trainees to give examples of how they as crew members or supervisors have benefited from scheduling.

Assign reading of Sections 3.0.0 – 3.2.3 for the next class session.

Ask trainees how they distinguish between priorities, and what is the success rate with their system.

"PLAN YOUR WORK, AND WORK YOUR PLAN = EFFICIENCY"

Plan of _____ Date _____

PRIORITY	DESCRIPTION	✓ When Completed ✗ Carried Forward

207F02.EPS

Figure 2 • Daily Work Plan

---

the supervisor arranges to have on hand the resources needed to perform the work. The bar chart is an excellent scheduling tool, and each supervisor should learn how to use it to maintain better control of projects. For an example, look at the bar chart in *Figure 3*.

The format of the chart makes it clear that the work item, *Excavate,* must be accomplished during the first four days of the project—July 1, 2, 3, and 7.

The network diagram, *Figure 4,* is a flow chart or road map of how the project should proceed. It traces progress from the beginning of the project to its end, just as a map would show a complete route. Notice that each box represents an activity. The arrows or lines show the relationships between the activities. An arrow from a prior activity indicates that the prior activity must be finished before the following activity can begin—a ruling principle referred to as *network logic* or *precedence*.

For example, the supervisor cannot begin the activity *Pour Concrete* until the activity *Set Reinforced Steel* is completed. Likewise, *Set Forms* cannot be completed until the activities *Build Forms* and *Drive Piles* are completed.

To realize the benefits of project scheduling, try to imagine building a project without construction drawings. Just as the architect's or engineer's drawings are your step-by-step guide to *what* to build, the project schedule is a step-by-step guide of *when* to build.

Instructor's Notes:

Module MT207 • Planning and Scheduling

Show Transparencies 4 and 5 (Figures 3 and 4).

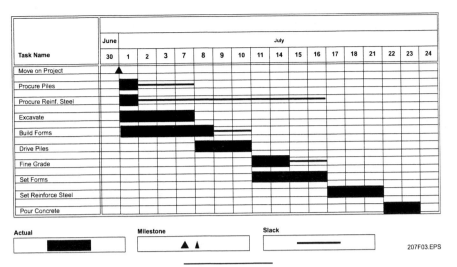

Figure 3 • Bar Chart

Explain how the Figure 3 bar chart works. If you have examples of other bar charts or network diagrams, distribute them to the trainees and review the differences.

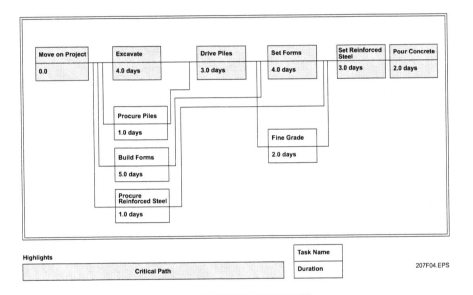

Figure 4 • CPM Network Diagram

PLANNING AND SCHEDULING— INSTRUCTOR'S GUIDE MODULE MT201

Have trainees complete the Participant Activity. Discuss their answers.

In general, the purpose of a schedule is to:

- Force detailed thinking to solve construction process problems on paper before they become actual problems
- Bring order to the building process
- Balance the use of labor, equipment, and materials that go into a project
- Serve as a communication tool among everyone involved in the construction process
- Provide information for controlling the construction process
- Put management in charge of the project

The schedule shows step-by-step procedures to guide the supervisor. The supervisor must remember a schedule is not a set of rules that can never change. It is a tool for anticipating changes and adapting to them, so that work can progress smoothly and efficiently. For any scheduling process to be effective, those using it must have input into it and believe in it.

3.2.0 Developing a Construction Schedule

There is no one best way to develop a construction schedule. Various techniques are used, from a simple listing of things to be done to very detailed computations that require the use of a computer. Midway between these extremes are the bar chart and the network diagram. For the supervisor's purposes, these tools lend themselves very well to construction scheduling.

As different as one scheduling technique is from another, they all require one basic element: knowing what is to be done. To obtain that basic element, the scheduler must break down the construction process into its various steps.

3.2.1 Formal Planning

In formal planning, the important thing is for the scheduler to go through the complete process. There is always a temptation to take shortcuts in scheduling, but each part of the process has value. One of the major benefits of developing a formal plan is that it forces management to think through every construction step in detail and, perhaps, in new and different ways.

Formal planning forces the scheduler to identify the things to be done, figure out when they have to be done, and, in relationship to other activities, determine how long it will take to do them. There are three major aspects of formal planning:

- It requires that the planner know the work to be performed and be able to visualize each activity.
- It provides a schedule that can be used for making necessary arrangements well in advance of actually performing the work.
- It provides a method for controlling the flow of work in the sequence necessary to accomplish the required activities.

PARTICIPANT ACTIVITY

Discuss and answer the following:

1. List projects that you have worked on that have had formal project schedules.
2. Describe the types of schedules, how they were used, and who developed them.
3. Did the various schedules help or hinder the progress of the jobs?

PARTICIPANT ACTIVITY

3.2.2 The Construction Plan

A formal plan is a system that helps the project team complete a project. Regardless of how simple the project is, that system must include:

- An objective that must be reached within a given time
- Manageable project components
- A set of activities that must be performed to reach the objective
- Time and resource requirements for each activity
- A sequence that identifies which activities must be completed before other activities can begin
- A schedule that combines activities, resources, and the sequencing of activities
- A control and monitoring system for enforcing the schedule

Instructor's Notes:

For construction projects, the objective is to complete a quality project on time and within budget. These three elements — time, quality, and cost — are directly related to each other.

- When *time* becomes the driving force to the project, quality suffers and costs increase.
- When *quality* becomes the driving force, costs increase and time typically increases.
- When costs become the driving force, quality decreases and time may also adjust.

The ideal schedule is a balance of time, cost, and quality. A schedule will help to provide a quality project on time and under budget.

The activities are the steps to follow in reaching the objective. Time and resource requirements include the duration of each activity and the tools, materials, manpower, and equipment needed to complete it. The sequence is based on each activity's relationship to, and dependence on, other activities. The schedule integrates the activities into a practical and efficient sequence. The control and monitoring system is a means for tracking progress and comparing it to the scheduled objectives.

3.2.3 The Construction Schedule

The construction schedule combines all of the elements above into the three basic parts of every formal plan.

The **Work Breakdown Structure (WBS)** provides a hierarchy of what has to be done and is a listing of organized phases of the project. The Work Breakdown Structure, depicted in *Figure 5*, is the basis for all project planning. The WBS should identify every task on the project. This is done using three steps:

Step 1 Focus on major deliverables.

Step 2 Create tasks for every deliverable.

Step 3 Evaluate summary tasks for oversight value.

The WBS chart is useful because it graphically breaks the project down. The items in Tier 1 will be completed first, with subsequent tasks following. However, this type of diagram can quickly become unwieldy, even for small projects. That's when the WBS in outline form makes sense.

Show Transparency 6 (Figure 5).

Refer to the end of this module for a teaching tip.

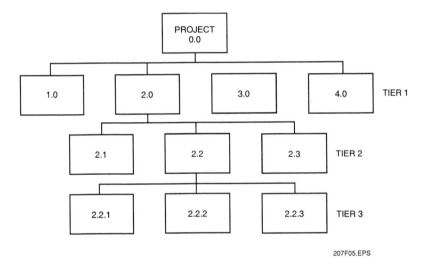

Figure 5 • Work Breakdown Structure (WBS) Chart

Ask trainees to give examples of how they would resolve a difference in the project detail of a WBS as given by planners and as needed by supervisors.

The first step in the work breakdown is to create a list of the activities that must be performed to meet the objectives of the project. This step does not take into consideration how the activities will be accomplished or who will do them. It does not mention how long the activities will take or the exact sequence in which they must be done. The work breakdown focuses only on the major activities themselves. *Figure 6* is a typical work breakdown for renovating a laboratory.

As many schedulers break down the project activities, they make sure the activity descriptions consist of an *action verb,* a *noun,* and a *location.* This ensures that the schedule activities are communicated to everyone reading the schedule. This is particularly important if the schedule is to be computerized.

It is common with computer-generated schedules to have the computer select only the activities that will be under construction for the next few weeks that a particular subcontractor is responsible for. In this case, a description such as *Footings* alone, without the full schedule, would not provide the necessary information. Is this *Excavate, Form, Reinforce, Pour, Strip the Footings,* or all of the footing activities? And which footings? For example, *Form Footings, North Side* gives anyone reading the schedule the precise information necessary to carry out the scheduled objectives.

For construction projects, the work breakdown structure can include inspections, ordering materials, and receiving materials, as well as the construction activities. Planners have considerable leeway as far as the level of detail is concerned. Their own experience and the policies of the company should serve as a guide in deciding the level of detail needed.

A relatively inexperienced supervisor may need a very detailed work breakdown structure to make it possible to control progress at a fine level of detail. A more experienced supervisor, on the other hand, may be able to operate with fewer, broader activities with less detail because of greater familiarity with the process and a lesser need to control individual pieces of the operation as it progresses.

WBS Outline for Laboratory Renovation

1.0 **Project Kick-Off**
2.0 **Obtain Parts**
    2.1 Obtain Paint
    2.2 Obtain Vinyl Floor Coverings
    2.3 Obtain Fume Hood
    2.4 Obtain Chemical Sink
3.0 **Preliminary Construction**
    3.1 Strip Room
    3.2 Replace Existing Fume Duct
    3.3 Install New Fume Hood
    3.4 Rough Plumbing
    3.5 Rough Electrical
    3.6 Repair Floor
    3.7 Repair Walls and Ceiling
4.0 **Installation**
    4.1 Wall Cabinets
    4.2 Install 1/3 Base Cabinets
    4.3 Install Chemical Sink
    4.4 Install 2/3 Base Cabinets
5.0 **Finishing Work**
    5.1 Stain and Lacquer Cabinets
    5.2 Paint Walls and Ceiling
    5.3 Finish Plumbing
    5.4 Finish Electrical
    5.5 Lay Vinyl Floor
6.0 **Closeout**

Figure 6 • List of Work Items for Laboratory Renovation

The difference between the amount of detail given by the planner in the WBS and the amount of detail needed by the supervisor from the WBS can lead to problems. The planner may be extremely experienced and able to operate without much detail in a work breakdown. However, the supervisor actually using the breakdown may need more detail than the planner provides. For this reason, a work breakdown should be structured for the level of experience of the person using it, not the experience level of the individual drawing it up.

Instructor's Notes:

With a large, complex project, the scheduler may want to start the work breakdown with milestones or major phases of the project only. Then, after the major phases are developed, the scheduler fills in the details of each phase with subphases or individual activities. On large projects, the activity level may not be reached until a few months prior to that phase or sub-phase. Once the work breakdown structure has been completed, the bar chart or network diagram can be started.

3.3.0 Bar Charts

Because of its ease of development and simplicity, the bar chart is the most widely accepted type of schedule used for construction projects. It is easy to understand and the supervisor can use it to see not only when an activity is scheduled to begin and end, but also exactly when workers, subcontractors, and materials must be available. In effect, it is a two-dimensional picture of a process conducted over time. The activities that make up the process are listed down the left-hand side of the chart in the order in which they will be performed (refer to *Figure 3*). Across the top of the chart is a line indicating the time involved in the project, usually in calendar or workdays, or both. Each square on the bar chart grid indicates a specific time unit—a working hour, a day, a week, or a month, depending on the time scale used.

A scheduled activity is shown by a horizontal bar drawn on the same grid line as the activity's listing on the left and directly under the appropriate time unit(s). The bar begins under the working day on which the activity begins, and its length reflects the duration of the activity. For example, the activity *Build Forms* in *Figure 3* has a duration of five days, since it takes up five time units (days) on the grid. The solid part of the bar indicates the number of days required for the activity; the light line extended past the solid bar indicates the duration from the earliest possible starting date to the latest possible finishing date for the activity. This is called *activity float*, or *slack time*. With most bar charts, this line is not shown. The typical bar chart only shows the earliest an activity can start and finish.

A bar chart provides an easy way to quickly check and record the progress of a job. Weekly or monthly, the supervisor can review the as-planned chart and mark on it the as-built progress of each activity, giving a visual representation of each activity progress. The as-built bar is generally drawn below the as-planned bar. Anyone reading this type of updated bar chart can easily compare the planned schedule with the actual or as-built schedule as shown in *Figure 7*.

The solid vertical line between July 14 and 15 is the update date. Everything to the left should be finished. This particular schedule shows started and finished activities with the solid triangle. Activities which have started but not finished have a solid starting triangle and a clear ending triangle indicating a new scheduled finish date, as seen in the activities *Fine Grade* and *Set Forms*. This could be done by hand with a colored pen or marker.

You will also note that in some activities, such as *Procure Piles*, the activity was not completed as early as possible, but it was completed within its float time. Also, notice that some activities had longer or shorter durations than originally estimated. This information is noted and used for future schedules.

Another benefit of a bar chart is that it gives the supervisor a simple tool for developing **resource profiles.** A resource profile is a projection of the resource requirements for a particular project. This can be done for any type of resource but is usually done for labor and equipment.

The resource profile takes the bar chart and notes on the bars how many resource units are required for each activity. For example, in the simple bar chart shown in *Figure 8,* say that Activities A, B, and C each require four laborers per day. Activities D and E each require three laborers per day. Activities F and G each require five laborers per day. To determine the daily labor requirements, the supervisor would simply total the number of laborers required during each time unit and plot them on a resource profile (also shown in *Figure 8*).

Ensure that you have all the necessary materials to teach the course. Check the Materials and Equipment list at the front of the module.

Have trainees discuss the advantages and disadvantages of working with a schedule. If time permits, have trainees draw a bar chart or network diagram for the project they are currently supervising or working on.

Assign reading of Sections 3.3.0 – 3.4.2 for the next class session.

Show Transparencies 7 and 8 (Figures 7 and 8).

Explain how a resource profile is developed from a bar chart using Figure 8.

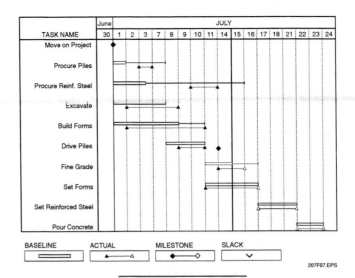

Figure 7 • Updated Bar Chart

Figure 8 • Resource Profile Developed from Bar Chart

Instructor's Notes:

3.4.0 Network Diagrams

Network diagrams, or CPM networks, are becoming more and more common, due to the increased use of computers in many construction companies. To develop a network diagram, the activities in the work breakdown structure are arranged in a logical sequence from beginning to end in the form of a flow chart. The resulting network diagram is not a schedule in itself; in fact, some schedulers consider the network preparation to be part of the planning phase. It is one of the steps in developing a schedule. At this point, it still needs the consideration of resources and durations and have the activity start and finish dates calculated in order for it to become a schedule.

3.4.1 Precedence

The fundamental concept in a network diagram is that the supervisor cannot start the next activity without first completing the prior activity. This is the concept of *precedence*, and it is illustrated simply and clearly in *Figure 9*.

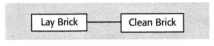

Figure 9 • Precedence Relationship

Precedence is dictated by the logical requirements of the activities themselves. For example, a crew cannot clean brick until it has finished laying the brick.

Ask the following questions when developing a network diagram that places each work item in the proper sequence:

- What activities must be done immediately before this activity can start?
- Which activities must immediately follow this activity?
- Which activities can be done at the same time?

Figure 10 is the network diagram for the laboratory renovation broken down in *Figure 6*.

The arrows indicate precedence. For example, *Strip Room* (activity 2) must be completed before activities 4, 5, or 6 *Replace Existing Fume Duct, Rough Plumbing,* and *Rough Electric* can begin.

Remember that the network diagram is only one part of the total planning, scheduling, and control process. Its real importance becomes apparent as the supervisor uses it to make the plan a reality. For monitoring and supervising the job, the supervisor may find it helpful to transfer the information on the network diagram onto a bar chart, a form better suited to everyday management.

The major advantage of network schedules is that they show the relationships among all the activities. Because of this, they are more effective in forecasting the effects of changes and delays. They are ideal for computer applications, and they have more credibility in legal cases. They also predict float or slack time and critical activities with greater accuracy than a bar chart. Their main disadvantage is that they can become complex, typically require training to become an effective tool, and are not well accepted nor understood by many field supervisors.

Many schedulers who have used networks for some time wouldn't think of starting a project without first drawing the network diagram, because it causes them to think through the project in detail, forcing them to mentally build it and solve many construction-related problems before they even set foot on the project.

3.4.2 Determining the Time Required for Each Activity

This step is accomplished by dividing the quantity of work by the production rate of the workers, as shown in the following equation:

$$\text{Activity Duration} = \frac{\text{Materials Quantity}}{\text{Production Rate}} + \text{Modifier}$$

Materials quantity refers to the quantity of work. *Production rates* are based on past performance. The rates may be modified as necessary to fit current job conditions, such as a work area where the maneuverability is either extremely easy or extremely difficult. The *Modifier* is a factor you add to reflect these conditions.

Ask trainees who have used network diagrams to discuss their experiences with the group.

Have trainees relate Figures 6, 10, and 11 to review how network diagrams are developed and used.

Show Transparencies 9A and 9B (Figure 10) and 10 (Figure 11)

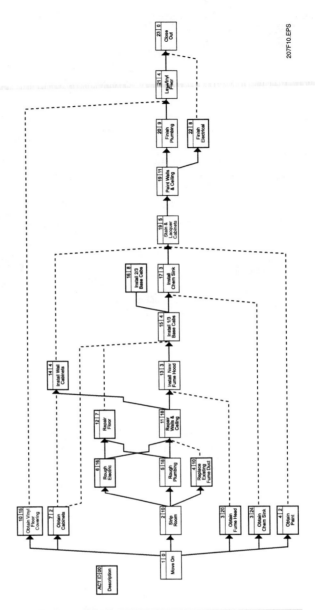

Figure 10 • Sample Network Logic Diagram for Lab Renovation

Instructor's Notes:

Module MT207 ◆ Planning and Scheduling

For example, if you have eight rest room stalls to build (materials quantity) and you can build two per day (production rate), your activity duration would be eight divided by two, or four days. If, for this job, materials have to be carried up three flights of stairs, you might add an additional day to cover the extra work caused by the installation being on the third floor.

Reasonable estimates of how long specific activities will take can be obtained from:

- Subcontractors
- Records of past jobs
- Manufacturers
- Equipment suppliers
- Cost-estimating manuals
- The supervisor's personal experience

These estimates, however, must be refined to allow for unusual aspects of the current project.

See *Figure 11* and refer back to *Figure 10* to see the network logic diagram for the laboratory remodel with the durations for each task. The individual task durations can be added up along the longest path to determine the time required to complete the entire project. They can also be used to predict resource requirements, since the timing of each activity determines when personnel, materials, and equipment are needed.

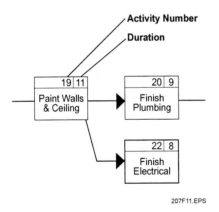

Figure 11 • Network Logic Diagram

The durations on this network are printed above the activity descriptions and to the right of the activity number. The longest path and the critical path are indicated on this network by double lines around the activity box.

The exercise of going through the three steps of formal planning produces information on the sequence in which each activity should start and finish and the number of days the total project requires. This information is well worth the tedium of doing the exercise. This is especially true for large or complex projects involving a large number of work activities, because it increases the supervisor's control of project activities.

3.5.0 Short-Term Schedules

One of the most important control techniques available to the supervisor is the **short-term schedule,** sometimes referred to as a *short-interval schedule,* a *short-interval production schedule (SIPS),* or a *short-term construction activity plan (CAP).* By whatever name, the goal of short-term scheduling is to extract a detailed production schedule from the overall construction schedule covering no more than three months.

Three months is the maximum, but a short-term schedule is normally much shorter. The three-month schedule allows time for coordinating material acquisitions and deliveries and to meet milestones set by the overall project schedule. Shorter schedules, covering only one or two weeks, are used by the supervisor to plan the use of resources and to provide manageable measuring sticks for monitoring production.

Short-term schedules always consider the relationship between production and cost. When production is controlled, so is cost; therefore, the purpose of production control is to reduce costs or at least keep costs within the estimate. As a result, short-term scheduling is often combined with a production control reporting system that monitors production by reporting resources used and units of work completed.

Where an overall project schedule may consider only time, a short-term schedule always considers time and manpower and often considers materials, equipment, and other factors as well.

Ensure that you have all the necessary materials to teach the course. Check the Materials and Equipment list at the front of the module.

Assign reading of Section 3.5.0 for the next class session.

Teaching Tip

Use a blank transparency and have the class walk you through item three in the Participant Activity. Some of the answers are found at the end of this module.

PARTICIPANT ACTIVITY

1. List the major steps involved in creating a construction schedule.
2. Discuss which member(s) of the project team should develop the construction schedule.
3. Draw a network diagram that illustrates the following logic conditions:
   - Activity A must be done before Activities B and C.
   - Activity B must be done before Activity D.
   - Activity C must be done before Activity E.
   - Activities D and E must be done before Activity F.

PARTICIPANT ACTIVITY

The information used in the short-term schedule comes from the estimate and the work breakdown structure. The short-term schedule makes the estimate operational by applying estimate data and the various job plans into a day-to-day schedule of events.

For example, assume that a 12-inch masonry block wall has been estimated at 4,250 sq. ft. Historical records show a daily production rate of 280 sq. ft. per crew. Consequently, the supervisor's short-term schedule should include using 15 crew days for this activity (4,250 sq.ft. ÷ 280 sq. ft. per day = 15.2 crew days). If 15 crew days exceed the schedule, it may be necessary to use more than one crew, if available.

An important purpose of a short-term schedule is to make it easier for the supervisor to compare actual production with estimated production so that if actual production begins to slip behind the estimate, the problem can be recognized and steps can be taken to correct it — before it gets out of hand.

Sample Short-Term Schedule

A carpentry crew on a retaining wall project is about to form and pour catch basins and put up wall forms at the same time. The crew has already installed a number of catch basins, so the supervisor is sure that they can put the basins in within the estimated time. However, the supervisor is concerned about their production of wall forms.

The supervisor refers to the estimate or estimate breakdown and finds the following:

- Production factor for wall forms: 16 hours per 100 sq. ft.
- Work to be done by measurement: 800 sq. ft.
- Budget (800 × 16 ÷ 100): 128 work hours

The carpenter crew consists of:
- 1 carpenter supervisor
- 4 carpenters
- 1 laborer

If the crew remains the same (six employees) the work should be completed in about 21 crew hours (128 work hours ÷ 6 employees). This information gives the supervisor and the crew a definite time frame for the work, and provides a measuring stick for monitoring whether or not the crew meets its production goals.

The principle of the short-term schedule is to translate production into work hours or crew hours and to schedule work so that it can be accomplished within the estimate. In addition, establishing a production target or goal often motivates a crew to produce more than the estimate requires.

Short-term schedules can also be used when a quantity of work is to be the measure of production. For example, a construction company has a contract to build a flat-roofed warehouse 60 feet × 400 feet. The height of the building to the fascia is 20 feet. Plans call for plywood siding. The total area of the siding, with deductions for door and window openings, is 17,000 sq. ft. The production factor used in the estimate is 200 sq. ft. per work hour.

Instructor's Notes:

Module MT207 ❖ Planning and Scheduling

The supervisor has assigned two employees to do the work, so the short-term schedule for the job can be determined by answering two questions:

1. When should the employees finish the job in order to be within the estimate?
2. How much area should be covered at the end of the first full day?

PARTICIPANT ACTIVITY

Discuss and answer the following:

1. Discuss which member(s) of the project team should develop short-term schedules.
2. How often are short-term schedules used on your jobs?
3. What is the effect when they *are* used? When they *aren't* used?
4. Develop a two-week schedule for the lab renovation project described in *Figure 10*.
5. Briefly list the advantages gained through bar chart scheduling.
6. Complete a bar chart for the following project:

   (*Note:* A blank bar chart form is provided for your use.)

A small one-story commercial building is to be constructed on the site of an existing small frame structure. It is 30 ft. × 60 ft. in plan. The exterior and interior walls are of concrete block. The roof is comprised of bar joists on long-span bar joists covered with a steel roof deck, rigid insulation, and built-up roofing. The ceiling is suspended acoustical tile. The floor is a concrete slab on grade with an asphalt tile finish. The interior finish on all walls is paint. The project has been broken down into 14 steps (not in any particular order) and a construction time estimate has been made for each as follows:

- *Demolition* – includes the demolition of the present structure, removal of debris, and rough grading of the site. **2 days**
- *Foundations* – the construction of the foundation for the new structure. **3 days**
- *Underground services* – the installation of water and sewer service from mains in the street. **1 day**
- *Exterior walls* – the construction of the exterior block walls. **6 days**
- *Interior walls* – the construction of the interior block walls. **3 days**
- *Roof steel* – includes the installation of the long-span joists, the bar joists, and the steel roof deck. **2 days**
- *Floor slab* – assumed to include fine grading, sand fill and compaction, membrane waterproofing, casting, and finishing of the floor slab. **3 days**
- *Floor finish* – includes the installation of asphalt tile and base mold. **2 days**
- *Rough plumbing and heating* – includes setting the heating unit and rough ductwork, the installation of rough plumbing above the floor slab, and vents. **3 days**
- *Finish plumbing and heating* – includes the final installation of the heating radiators, controls, and the installation of conduit, service inlet, and meter box. **3 days**
- *Finish electrical* – the installation of wire and fixtures. **3 days**
- *Rough carpentry* – the installation of rough door frames and display window framing. **2 days**
- *Finish carpentry* – the installation of door trim, hanging of doors, and so on. **4 days**
- *Ceiling* – the installation of the suspended acoustic ceiling. **3 days**

PARTICIPANT ACTIVITY

Classroom

Divide trainees into groups of four to five to complete the Participant Activity. At the end of 30 minutes, bring the class back together. Compare and discuss their bar charts and scheduling order. Some of the answers are found at the end of this module.

Audiovisual

Use Transparency 11 (blank bar chart grid) to sketch out the bar chart in the Participant Activity.

PLANNING AND SCHEDULING— INSTRUCTOR'S GUIDE MODULE MT201

Ensure that you have all the necessary materials to teach the course. Check the Materials and Equipment list at the front of the module.

Assign reading of Section 3.6.0 for the next class session.

PARTICIPANT ACTIVITY
(continued)

Activity	Time

PARTICIPANT ACTIVITY

In this case, the production target is in square feet rather than in work hours. The answer to the first question gives the supervisor and the crew a goal for the entire task. The answer to the second question gives them a way of monitoring their progress as they go along.

3.6.0 Schedule Updates

To gain the most from a schedule, it must be updated at established intervals. Updating allows the supervisor to realign a schedule to compensate for changes that occur. The schedule becomes the primary tool for monitoring job progress.

Schedule updates should be performed weekly or monthly, depending on the size and complexity of the job. In general, if the project duration is less than a year, the updates should be done on a weekly or biweekly basis. If the project duration is more than a year, updates can be performed monthly unless the complexity of the job dictates otherwise. Each job should be evaluated individually to determine the frequency of the updates.

The supervisor updates the schedule by simply taking the information gathered from the short-term schedules and posting it to the master schedule. The master schedule is likely to be in the form of a bar chart. Activities that have been completed are indicated on the master schedule with a colored marker. Activities in progress are indicated with a different-colored marker along with notes on when they were started and what percentage has been completed.

Instructor's Notes:

By comparing the completion or anticipated completion dates of each activity with those on the master schedule, the supervisor knows how the overall project is proceeding and how much it is ahead or behind schedule.

PARTICIPANT ACTIVITY

Discuss and answer the following:

1. List jobs you have worked on that used schedule updates.
2. How often were the updates performed?
3. What are the advantages to frequent updates?

◄ PARTICIPANT ACTIVITY ►

SUMMARY

The key to planning is defining goals and determining how to accomplish them. Obviously, the further in advance the supervisor plans, the more time there will be to eliminate or work around any obstacles that arise. The supervisor must learn methods of developing, reviewing, and revising a job plan if jobs are to be built on time and within budget.

The questions that should arise during planning are essentially:

- What is the job, and what tasks make up the job?
- What tools, equipment, and materials are needed and in what quantity?
- In what order should the tasks be scheduled?
- What needs to be planned for the next day or week?
- What alternatives are there if bad weather or other unexpected events occur and the plan cannot be carried out as intended?

The purpose of project scheduling is to help the supervisor think out the project in detail using a plan. The plan will complete the project in the shortest time with the least cost and highest quality. The plan then becomes a communication and monitoring tool that can be used throughout the project to show the relationships between the estimate and actual production. A schedule does not tell how to build a project. Rather, it provides an orderly, organized guide for the construction and a means by which everyone involved can know whether job goals are being met.

An effective project schedule is made up of a number of related parts:

- Work breakdown structure
- Network diagram
- Bar chart
- Short-term schedules
- Periodic updates

Each performs a separate but interrelated function.

Smaller construction projects can be controlled with very simple schedules, work breakdowns, and simple bar charts. Large and complex projects require extensive scheduling devices.

The development of the schedule is a team effort. Involve all members of the team so that the final product reflects the group's combined experience and expertise.

Have trainees complete the Participant Activity. Discuss their answers.

Instructor's Notes:

Have trainees complete the Review Questions. Discuss the correct answers, located at the end of this module.

Review Questions

1. The planning process involves ____.
   a. listing the tasks that need to be done, because the supervisor is usually the first one to receive information about all aspects of a project
   b. sequencing the various tasks that are needed to complete a project, because the individual activities required to complete even large projects are rather simple
   c. reusing previous plans, sketches, and specifications, because projects often involve similar results
   d. waiting until enough information is known about the available tools, materials, or workers in order for supervisors to make a reliable plan

2. The stages of planning include all of the following *except* ____.
   a. setting goals and putting those tasks in the most efficient sequence
   b. determining the proper order of events
   c. breaking the project into smaller tasks
   d. analyzing past mistakes

3. The pre-construction stage of planning entails ____.
   a. the leveraging of any existing project documents that are relevant to the project being planned
   b. the supervisor getting involved in the project
   c. selecting the actual work methods and resources needed to perform the work
   d. the estimator determining what personnel, tools, and equipment will be used

4. The construction stage of planning entails ____.
   a. the supervisor's involvement on a weekly basis
   b. the supervisor determining time and cost limits for the project
   c. considering such local factors as climate and soil conditions
   d. concentrating on past problems that have been resolved in order to avoid these problems in the future

5. All the following are objectives of planning *except* ____.
   a. getting the job done with the needed resources and within the budget
   b. determining which equipment to use and assessing its availability
   c. assigning responsibilities to each employee involved in the project
   d. making all decisions for employees so that they can concentrate on their specific tasks, thus saving time

6. The planning process includes ____.
   a. the supervisor setting goals
   b. establishing objectives for large-scale activities
   c. developing a detailed plan for tasks that are performed frequently
   d. making sure crew members know their individual responsibilities

PLANNING AND SCHEDULING— INSTRUCTOR'S GUIDE MODULE MT201

7. An objective should ideally ____.
   a. be something that most, though not necessarily all, employees will interpret in the same manner
   b. have specific, measurable subgoals
   c. be enforced by the supervisor even if crew members do not fully approve
   d. provide a long-term view of the tasks involved in a project

8. The reasons for writing objectives include all of the following *except* ____.
   a. enabling supervisors to see if work is being performed as planned
   b. eliminating delays
   c. helping keep a project on schedule
   d. giving the supervisors extra control over a project

9. When a supervisor breaks a project down into tasks, it is important to have ____.
   a. only a detailed list of tasks and sub-tasks if needed
   b. one individual responsible for each task
   c. tasks that are definable, meaning that there is a clear idea of what resources are needed
   d. tasks that are identifiable, meaning that there is a clear, established timeframe for the activities

10. The statement "By 12:00 p.m. on Tuesday, all windows will be installed" is an example of ____.
    a. a goal
    b. a job breakdown
    c. an objective
    d. a job estimate

11. If supervisors run out of materials, they should ____.
    a. order more themselves
    b. appoint a crew member to order more
    c. notify their own supervisor
    d. notify their suppliers

12. The required amounts and types of material for a job are usually determined ____.
    a. by the supervisor
    b. from a job estimate
    c. by suppliers
    d. from a construction-phase job analysis

13. The supervisor's main responsibility in planning material is to ____.
    a. monitor the delivery and usage of materials
    b. list needed items
    c. order from trusted suppliers
    d. select appropriate materials

14. When planning equipment, the supervisor must do the following *except* ____.
    a. find a contact person if equipment is not working properly
    b. specify what tasks the equipment will perform
    c. avoid sharing equipment with other contractors, because this often results in confusion and logistic problems
    d. be responsible for the development of contingency plans if equipment malfunctions

Instructor's Notes:

Module MT207 ◆ Review Questions

15. Job analysis is used primarily in order to _____.
    a. identify the personnel and skill set needed for a project
    b. provide an estimate of how many resources are needed for a project
    c. determine the most efficient way to perform a task
    d. determine the amount of time needed to do a job

16. Job analysis is often performed _____.
    a. during the construction phase as a means of planning
    b. when materials are not delivered on time
    c. at the beginning of large-scale projects
    d. after the job is completed, in order to evaluate what did and did not work

17. Job analysis is particularly useful for the following types of jobs *except* _____.
    a. jobs involving a set of instructions that the supervisor must follow
    b. jobs involving new activities
    c. jobs requiring strict adherence to deadlines
    d. jobs involving events that were not anticipated

18. One of the steps involved in conducting a job analysis is to _____.
    a. determine how to meet a goal even if it takes more time and money than anticipated
    b. avoid spending time in consultation with crew members, because time is of the essence
    c. make it a priority to find the least expensive tools, materials, and equipment
    d. get rid of excess detail

19. The following are advantages gained through using bar chart scheduling *except* _____.
    a. it is easier to use than a network diagram
    b. it indicates start and finish times for activities
    c. it offers a comparison between as-built and as-planned activities
    d. it is the most accurate way of indicating float or slack time

20. One difference between bar charts and network diagrams is that _____.
    a. bar charts are not really a schedule, but rather a step involved in creating a schedule
    b. bar charts are difficult to use and generally require some degree of training
    c. network diagrams involve the concept of precedence
    d. network diagrams do not show the relationships among activities as effectively as bar charts do

PLANNING AND SCHEDULING— INSTRUCTOR'S GUIDE MODULE MT201

Instructor's Notes:
_____
_____
_____
_____
_____
_____
_____

NATIONAL CENTER FOR CONSTRUCTION EDUCATION AND RESEARCH

GLOSSARY

Trade Terms Introduced in This Module

Bar chart: A time line showing the day, week, or month when certain project activities should be accomplished.

Network diagram (CPM Network): A flow chart or road map of how the project should proceed.

Objective: A statement of what is desired at a specific time. It must be measurable, it must be achievable, and it must have everyone's full support.

Resource profile: A projection of the resource requirements for a particular project, usually done for labor and equipment.

Short-term schedule: Extracts a detailed production schedule covering no more than three months from the overall construction schedule. Also called a *short-internal schedule* and *short-term construction activity plan (CAP)*.

Work Breakdown Structure (WBS): A hierarchy of what has to be done to complete a project, listed by organized phases of the project.

Ask trainees to review the Glossary for unfamiliar terms.

Review the objectives of the module and then answer any questions the trainees may have.

Have trainees prepare for the Module Examination.

Administer the Module Examination. Be sure to record the results of the Exam on Craft Training Report Form 200 and submit the results to the Training Program Sponsor.

MOD MT207-01—TEACHING TIPS

The following are suggested activities or instructional methods to help you teach the material in this AIG.

Section 2.3.0 The Planning Process

Read and discuss each of the five steps in detail and then ask the trainees to give examples from their experiences of each step in the planning process as it relates to their craft.

Ask trainees for examples of plans gone awry and what contingency plans were implemented.

Section 2.4.3 Planning Tools

1. If available, distribute samples of your company's policy regarding tools and equipment and ask trainees for other rules they have had to follow.
2. Distribute completed and blank samples of a tool and tool part inventory and a tool maintenance program. Discuss how each sample works.
3. Ask trainees to complete the blank sample tool and tool part inventory and the tool maintenance sheets for their particular job sites.

Section 3.2.3 The Construction Schedule

Create deliverables for the four major deliverables in a typical project. Step through the Work Breakdown Structure (Figure 5) using simple project deliverables that trainees can easily follow. Ask trainees to give you examples for Tiers 2 and 3. Refer to Figure 6 for ideas.

NATIONAL
CENTER FOR
CONSTRUCTION
EDUCATION AND
RESEARCH

MODULE MT207

Answers to Review Questions

	Answer	Section Reference
1.	b	2.0.0
2.	d	2.1.0
3.	c	2.2.1
4.	c	2.2.2
5.	c	2.2.2
6.	c	2.3.0
7.	a	2.3.2
8.	b	2.3.2
9.	d	2.3.3
10.	c	2.3.2
11.	c	2.4.1
12.	b	2.4.1
13.	a	2.4.1
14.	b	2.4.2
15.	c	2.5.0
16.	b	2.5.0
17.	a	2.5.0
18.	d	2.5.0
19.	d	3.3.0
20.	d	3.4.0

National Center for Construction Education and Research

MODULE MT207

Answers to Participant Activities

Participant Activity Section 2.3.3

1. *Goal:* Something to accomplish, such as going on vacation or getting a job built.
2. *Objective:* A statement, that notes what is desired at a specific time, such as those examples provided in the trainee manual.

Scenario 1 *Goal Statement (example):* Constructing the formwork for a retaining wall is the goal of this job. Objective (example): Given the needed resources, by the end of the fifth work day, a retaining wall will be constructed according to the drawings and specifications.

Scenario 2 An example of a listing:

 a. Gather needed tools, equipment, and materials
 b. Paint ceilings in all 6 rooms with primer
 c. Paint ceilings in all 6 rooms with final coat
 d. Paint walls in all 6 rooms with primer
 e. Paint walls in all 6 rooms with final coat
 f. Clean up and put away tools and equipment

Some reasons might be interruptions, financial problems, and lack of resources.

Participant Activity Section 2.5.0

Scenario

Work to be done: Construct a 12' × 24' storage building having an 8' eave height and a standard hip roof. Erect the wall framing, install wall and roof plywood sheathing, install all the windows and doors, and apply the felt paper and asphalt shingles to the roof. Paint all of the exposed exterior wood.

General Goal Statement: Construct a storage building to specification and complete exterior painting and roofing.

Job Activity Breakdown: Have participants work through the various activities needed, reminding them to organize the activities according to the three phases of a job. Have them write objectives for each activity and assign a resource and timeline for completion of each. Remember to check on the objectives to see if they are specific, attainable, and measurable and that they support the overall goal and time frame.

Planning
Construction
Cleanup

Participant Activity Section 2.6.1

1. Implementing a plan requires making sure that all the required resources are ready and available. It requires spending time with the crew to make sure the work flows smoothly. It also requires coordinating with other supervisors and other crews.

   Examples:

   • Researching what vendors were used on similar jobs in the past can save time.

   • Spending time initially with the crew and observing them and giving them help as needed will help to make sure that the project is up and running.

   • Taking time to establish relationships and meeting with other contractors, supervisors, and crews that you will be working with will help to implement your plan.

2. Individual responses
3. The purpose of meetings is to keep good communication going, identify risks to completing the project as planned, solve problems, and distribute information.
4. Individual responses

Participant Activity Section 3.5.0

1. A construction schedule includes identifying activities, resources needed, and sequencing of tasks or activities.

   The ideal schedule is a balance of time, cost, and quality. A schedule will help to provide a quality project on time and under budget. The activities are the steps to follow in reaching the objective. Time and resource requirements include the duration of each activity and the tools, materials, manpower, and equipment needed to complete it. The sequence is based on each activity's relationship to, and dependence on, other activities. The schedule integrates the activities into a practical and efficient sequence. The control and monitoring system is a means for tracking progress and comparing it to the scheduled objectives.

2. Those who will use it, usually the project manager or the project superintendent should develop the construction schedule. With some companies, the company scheduler develops all the schedules company-wide. Input from the job foremen, superintendents, project managers, estimators, and other key managers is essential if it's expected that the schedule will actually be used to monitor the project.

Participant Activity Section 3.5.0

1. Project supervisors should develop short-terms schedules to plan the use of resources and to provide manageable measuring sticks for monitoring production.
2. Individual answers
3. Individual answers

CONTREN™ LEARNING SERIES — USER UPDATES

The NCCER makes every effort to keep these textbooks up-to-date and free of technical errors. We appreciate your help in this process. If you have an idea for improving this textbook, or if you find an error, a typographical mistake, or an inaccuracy in NCCER's Contren™ textbooks, please write us, using this form or a photocopy. Be sure to include the exact module number, page number, a detailed description, and the correction, if applicable. Your input will be brought to the attention of the Technical Review Committee. Thank you for your assistance.

Instructors – If you found that additional materials were necessary in order to teach this module effectively, please let us know so that we may include them in the Equipment/Materials list in the Instructor's Guide.

Write: Curriculum Revision and Development Department
National Center for Construction Education and Research
P.O. Box 141104, Gainesville, FL 32614-1104

Fax: 352-334-0932

E-mail: curriculum@nccer.org

Craft _____ Module Name _____

Copyright Date _____ Module Number _____ Page Number(s) _____

Description _____

_____

_____

_____

_____

(Optional) Correction _____

_____

_____

_____

(Optional) Your Name and Address _____

_____

_____

Project Supervisor

Module MT208-01

Resource Control and Cost Awareness

Resource Control and Cost Awareness
Instructor's Guide

Module MT208

MODULE OVERVIEW

This module introduces the project supervisor trainee to resource control and cost awareness. This module teaches the trainee to control resources on the job, solve problems affecting productivity, maintain cost reporting, and perform a simple production analysis.

PREREQUISITES

There are no prerequisites for this module; however, prior to training with this module, it is recommended that the trainee complete the following modules:
   Project Supervision, Modules MT201 through MT207.

LEARNING OBJECTIVES

Upon completion of this module, the trainee will be able to:

1. State why it is important to control resources on the job.
2. Define productivity and explain how it differs from production.
3. List several factors that affect productivity and determine how to solve problems relative to productivity.
4. Describe how job-site productivity can be improved.
5. Describe how a five-minute rating is developed and how it can be used to improve productivity.
6. Explain how a crew balance chart is developed and how it can be used to improve productivity.
7. Describe how to complete a supervisor's delay survey and how it can be used to improve productivity.
8. Describe how to control the various job resources.
9. Explain the importance of being aware of the costs on the job.
10. Define estimated cost, actual cost, and projected cost.
11. Describe the different parts of a job reporting system.
12. Define production analysis and perform a simple production analysis.
13. Discuss the impact of improper cost reporting.

PERFORMANCE OBJECTIVES

This is a knowledge-based module – there is no performance profile examination.

NCCER STANDARDIZED TRAINING PROGRAM

The National Center for Construction Education and Research (NCCER) provides a standardized national program of accredited craft training. Key features of the program include instructor certification, competency-based training, and performance testing. The program provides trainees, instructors, and companies with a standard form of recognition through a National Craft Training Registry. The program is described in full in the Guidelines for Accreditation, published by the NCCER. For more information on standardized craft training, contact the NCCER by writing us at P.O. Box 141104, Gainesville, FL 32614-1104; calling 352-334-0911; or e-mailing info@nccer.org. More information may be found at our Web site, www.nccer.org.

HOW TO USE THIS ANNOTATED INSTRUCTOR'S GUIDE

Each page presents two sections of information. The larger section displays each page exactly as it appears in the Trainee Module. The narrow column ties suggested trainee and instructor actions to each page and provides icons to call your attention to material, safety, audiovisual, or testing requirements. The bottom of each page includes space for your notes.

 If you see the Teaching Tip icon, that means there is a teaching tip associated with this section. Also refer to the suggested teaching tips at the end of the module.

PREPARATION

Before teaching this module, you should review the Module Outline, Learning Objectives, and the Materials and Equipment List. Be sure to allow ample time to prepare your own training or lesson plan and gather all required equipment and materials.

MATERIALS AND EQUIPMENT LIST

Materials:

Transparencies

Markers/chalk

Module Examinations*

Sample quality control programs**

Sample production analysis**

Equipment:

Overhead projector and screen

Whiteboard/chalkboard

*Located in the Test Booklet packaged with this Annotated Instructor's Guide.
**If available on loan from your workplace or other resource.

ADDITIONAL RESOURCES

This module is intended to present thorough resources for task training. The following reference works are suggested for both instructors and motivated trainees interested in further study. These are optional materials for continued education rather than for task training.

Construction Materials and Building Publication No. 91, International Organization for Standardization (ISO), ISO Online Catalog at www.iso.ch.

Professional Construction Management: Including Contracting C M, Design-Construct, and General Contracting, 1991. Donald S. Barrie and Boyd C. Paulson (Contributor). Upper Saddle River, NJ: McGraw-Hill Higher Education.

Construction Management, 1997. Daniel W. Halpin and Ronald W. Woodhead. New York: John Wiley & Sons.

Construction Operations Manual of Policies and Procedures, 2000. Andrew Civitello, Jr. Upper Saddle River, NJ: McGraw-Hill Professional Book Group.

TEACHING TIME FOR THIS MODULE

An outline for use in developing your lesson plan is presented below. Note that each Roman numeral in the outline equates to one session of instruction. Each session has a suggested time of 2 1/2 hours. This includes 10 minutes at the beginning of each session for administrative tasks and one 10-minute break during the session. Approximately 15 hours are suggested to cover *Resource Control and Cost Awareness*.

Topic	Planned Time
Session I. Introduction to Productivity Analysis	
A. Factors Affecting Productivity	_____
B. Participant Activity	_____
C. Measuring Job-Site Productivity	_____
1. The Five-Minute Rating	_____
2. Crew Balance Chart	_____
3. Delay Survey	_____
D. Participant Activity	_____
Session II. Productivity Improvement and Resource Control	
A. Productivity Improvement	_____
B. Resource Control	_____
1. Material Control	_____
2. Tool Control	_____
3. Equipment Control	_____
4. Labor Control	_____
5. Participant Activity	_____
Session III. Quality, Cost, and Resource Control	
A. Quality Control	
B. Cost and Resource Control	_____
1. Control Standards	_____
2. Cost Awareness	_____
3. Participant Activity	_____
Session IV. The Reporting System	
A. Cost Coding	_____
B. Types of Reports	_____
C. Participant Activity	_____
Session V. Production Analysis	
A. Production Analysis	_____
B. Participant Activity	_____
Session VI. Impact of Improper Reporting	
A. Impact of Improper Reporting	_____
B. Participant Activity	_____
C. Summary	
1. Summarize Module	_____
2. Answer Review Questions	_____
D. Module Examination	_____
1. Trainees must score 70% or higher to receive recognition from the NCCER.	_____
2. Record the testing results on Craft Training Report Form 200 and submit the results to the Training Program Sponsor.	_____

Resource Control and Cost Awareness

Instructor's Notes:

Instructor's Notes:

ACKNOWLEDGMENTS

The NCCER wishes to acknowledge the dedication and expertise of Roger Liska, the original author and mentor for this module on resource control and cost awareness.

Roger W. Liska, Ed.D., FAIC, CPC, FCIOB, PE

Clemson University

Chair & Professor

Department of Construction Science & Management

We would also like to thank the following reviewers for contributing their time and expertise to this endeavor:

J.R. Blair

Tri-City Electrical Contractors

An Encompass Company

Mike Cornelius

Tri-City Electrical Contractors

An Encompass Company

Dan Faulkner

Wolverine Building Group

David Goodloe

Clemson University

Kevin Kett

The Haskell Company

Danny Parmenter

The Haskell Company

Course Map

This course map shows all of the modules of the *Project Supervision* curriculum. The suggested training order begins at the bottom and proceeds up. Skill levels increase as you advance on the course map. The local Training Program Sponsor may adjust the training order.

PROJECT SUPERVISION

- **MT208 — RESOURCE CONTROL AND COST AWARENESS** ⬅ YOU ARE HERE
- **MT207 — PLANNING AND SCHEDULING**
- **MT206 — DOCUMENT CONTROL AND ESTIMATING**
- **MT205 — CONTRACT AND CONSTRUCTION DOCUMENTS**
- **MT204 — QUALITY CONTROL**
- **MT203 — SAFETY**
- **MT202 — HUMAN RELATIONS AND PROBLEM SOLVING**
- **MT201 — ORIENTATION TO THE JOB**

Instructor's Notes:

MODULE MT208

TABLE OF CONTENTS

1.0.0	**INTRODUCTION**	8.1
2.0.0	**PRODUCTIVITY ANALYSIS**	8.1
2.1.0	Factors Affecting Productivity	8.2
2.2.0	Measuring Job-Site Productivity	8.3
2.2.1	*The Five-Minute Rating*	8.3
2.2.2	*Crew Balance Chart*	8.6
2.2.3	*Delay Survey*	8.7
3.0.0	**PRODUCTIVITY IMPROVEMENT**	8.7
4.0.0	**RESOURCE CONTROL**	8.10
4.1.0	Material Control	8.10
4.2.0	Tool Control	8.11
4.3.0	Equipment Control	8.11
4.4.0	Labor Control	8.12
5.0.0	**QUALITY CONTROL**	8.13
6.0.0	**COST AND RESOURCE CONTROL**	8.14
6.1.0	Control Standards	8.14
6.2.0	Cost Awareness	8.14
6.2.1	*Estimated Cost*	8.15
6.2.2	*Actual Cost*	8.15
6.2.3	*Projected Cost*	8.16
7.0.0	**THE REPORTING SYSTEM**	8.16
7.1.0	Cost Coding	8.16
7.2.0	Types of Reports	8.16
7.2.1	*Daily Time Report*	8.17
7.2.2	*Material Installation Report*	8.18
7.2.3	*Equipment Utilization Report*	8.18
7.2.4	*Summary Report*	8.18
7.2.5	*Cost Summary Report*	8.19
8.0.0	**PRODUCTION ANALYSIS**	8.23
9.0.0	**IMPACT OF IMPROPER REPORTING**	8.25
	SUMMARY	8.25
	REVIEW QUESTIONS	8.27
	GLOSSARY	8.31

LIST OF FIGURES

Figure 1	•	Five-Minute Rating	8.5
Figure 2	•	Crew Balance Chart	8.6
Figure 3	•	Delay Survey	8.8
Figure 4	•	Typical Daily Time Report	8.17
Figure 5	•	Quantity Report – Stripping and Cleaning Forms	8.18
Figure 6	•	Supervisor's Quantity Report	8.19
Figure 7	•	Daily Equipment Report – Excavating Soil	8.20
Figure 8	•	Weekly Summary – Set Anchor Bolts	8.21
Figure 9	•	Weekly Summary – Excavation	8.22
Figure 10	•	Weekly Summary – Strip and Clean Forms	8.22
Figure 11	•	Weekly Cost Summary	8.23

Instructor's Notes:

NATIONAL CENTER FOR CONSTRUCTION EDUCATION AND RESEARCH

MODULE MT208

Resource Control and Cost Awareness

OBJECTIVES

Upon the completion of this module, you will be able to do the following:

1. State why it is important to control resources on the job.
2. Define productivity and explain how it differs from production.
3. List several factors which affect productivity and determine how to solve problems relative to productivity.
4. Describe how job-site productivity can be improved.
5. Describe how a five-minute rating is developed and how it can be used to improve productivity.
6. Explain how a crew balance chart is developed and how it can be used to improve productivity.
7. Describe how to complete a supervisor's delay survey and how it can be used to improve productivity.
8. Describe how to control the various job resources.
9. Explain the importance of being aware of costs on the job.
10. Define estimated cost, actual cost, and projected cost.
11. Describe the different parts of a job reporting system.
12. Define production analysis and perform a simple production analysis.
13. Discuss the impact of improper cost reporting.

SECTION 1

1.0.0 INTRODUCTION

Every job uses many resources, the major ones being time, money, materials, labor, tools, and equipment. How these resources are used on the job determines whether or not the project is built on time and within the estimate.

The supervisor is responsible for controlling the use of resources and knowing how they affect job cost. Specifically, the supervisor first needs to understand how the costs in the job estimate are derived. This understanding is called **cost awareness**. Once work begins, the supervisor must know how to control the resources available. This control is called **resource control**. How resources are used directly affects the cost of doing the work; therefore, if supervisors do an effective job of controlling the use of resources, they will also effectively control costs.

SECTION 2

2.0.0 PRODUCTIVITY ANALYSIS

Productivity is a term heard a lot in the construction industry today. It is usually measured in terms of the amount of work accomplished per work hour and, as a result, many believe that the deciding factor in productivity is how fast employees work. This leads to supervisors blaming the lack of productivity on workers' attitudes and workers thinking of productivity as a speed-up effort that they should resist.

Copyright © 2003 National Center for Construction Education and Research, Gainesville, FL 32614-1104. All rights reserved. No part of this work may be reproduced in any form or by any means, including photocopying, without written permission of the publisher.

Ensure that you have all the necessary materials to teach the course. Check the Materials and Equipment list at the front of the module.

Ask trainees for examples of cost awareness and resource control programs at their workplace and why they are or are not effective.

Show Transparency 1 (Course Objectives).

Assign reading of Module MT208, Sections 1.0.0 – 2.2.3.

Ask trainees for additional factors that have affected productivity on the job and what they did or what was done to change it.

It is true that when **production** is low, workers are accomplishing less in the time they are on the job. However, workers cannot work efficiently and quickly if they do not have the materials, tools, equipment, and instructions they need to proceed. They cannot work well if they do not have the skills and knowledge to perform the work they are assigned or lack the supervision needed to coach and train them. Some will not work unless motivated by competent supervisors who demonstrate enthusiasm.

In order for productivity to improve, managing for productivity must be a policy at all levels of the company. The supervisor must be the first and most persistent productivity manager in the company. If the supervisor continually applies principles of good productivity management on the job, there will be a significant improvement in the productivity and morale of the crew. In addition, the company will improve its competitive position, and others in the company will recognize the need to improve productivity management in their own areas of responsibility. This, in turn, enhances everyone's job security.

Defining productivity is difficult because it is influenced by so many things. Too often, production is confused with productivity.

Production is simply output and is measured by the number of units produced or the quantity of material installed.

Productivity, on the other hand, relates to the efficiency with which the workers' work is being performed. It is measured in units placed in a designated time period such as 100 bricks per hour. The greater the efficiency, the lower the cost. Thus, production and productivity are related concepts, yet their applications are different.

Field supervisors evaluate the performance of their crews by their level of production. They assume that because everyone appears busy and the crew installs X amount of material or places Y number of fixtures, labor productivity is satisfactory. This attitude must be overcome! The supervisor must continually seek ways to improve productivity of the workforce, especially if production analyses show that work is behind schedule or over estimate.

Because it is the supervisor's objective to complete the job within a given time and within the estimated cost, the supervisor is the one who must establish and maintain the production rate necessary to meet—and, if possible, exceed—that objective. The supervisor must also do whatever possible to increase productivity and, therefore, the company's long-range performance in the field. This cannot be done by simply comparing production with the established estimate. Such comparisons help bring the job in at the estimated profit, but it does nothing to improve company performance. This is because estimates are usually based on past experience, and if poor productivity occurred in the past, it is part of those estimates. Consequently, using the estimate as the only measuring stick of productivity locks the company into its past mistakes—mistakes that could hinder its competitiveness in the market.

2.1.0 Factors Affecting Productivity

Many factors on a typical construction job affect productivity, but the major factors are:

- Excessive overtime
- Errors and omissions in plans and specifications
- Numerous change orders
- Complex designs
- Incomplete designs
- Poor coordination of trades
- Ineffective supervision
- Excessive reassignment of personnel from job to job on the same project
- Material shortages
- Adverse temperature and humidity conditions
- Inadequate lighting
- Uncontrollable water in the ground
- Government regulation
- High rate of absenteeism
- High job turnover rate
- High accident rate
- Lack of trained personnel
- Poor worker attitude
- Inadequate crew size and composition

Instructor's Notes:

- Size and duration of project
- Timeliness of decisions
- Uncontrolled coffee and meal breaks
- Impractical quality control tolerances
- Time of day and day of week.
- Inadequate temporary facilities, such as parking, change rooms, and rest rooms
- Job cleanliness
- Available work space

The supervisor must be aware of these factors, anticipate them, and take action whenever possible to prevent them from having a negative effect on production. If they do become problems, the supervisor must take action to correct them within the policies and procedures of the company.

2.2.0 Measuring Job-Site Productivity

As noted earlier, productivity is the output of the effort expended on the job. It is measured in such terms as dollars per cubic yard and hours per unit or square foot. There are many methods available to measure the efficiency (productivity) of getting the work done. Some are rather simple, and others are quite complicated.

Among the most common methods is production analysis. Production analysis compares actual cost to estimated cost or compares actual labor and equipment production rates to the original estimates of these factors. Production analysis is a warning device that lets the supervisor know during the job whether or not the job is progressing on schedule and within budget. It warns of problems, but it does not tell how to bring the job back in line without increasing costs. In addition, the information from the production analysis covers work that has already been completed. It does not provide an accurate assessment of the impact of the problem on work yet to be completed.

The supervisor must have other production analysis techniques that will both spot problems quickly and accurately and also point toward practical, cost-effective solutions. Three such methods are discussed below:

- The five-minute rating
- The crew balance chart
- The supervisor's delay survey

There are more formal techniques than these that provide better information, but they require support and resources from upper management.

These three activities are based on activity sampling techniques. **Activity sampling** is the practice of observing and classifying a small percentage of a total activity – similar to opinion polling, where a small percentage of the total population is surveyed to get an estimate of how the public feels about a particular issue. Activity sampling measures the work on a particular activity at a specific time in order to estimate the effectiveness of activity performance. Since activity sampling can provide answers within minutes, the supervisor can give immediate attention to production that appears to be below standard.

This section focuses on activity sampling techniques for work being done by a crew. However, there are also techniques for analyzing the efficiency of the entire labor force on a project.

2.2.1 The Five-Minute Rating

The purpose of the five-minute rating is to:

- Indicate delays on the job and the extent of their importance
- Measure the effectiveness of a crew
- Indicate where more thorough, detailed planning could result in cost savings

---

PARTICIPANT ACTIVITY

1. Define *production* and give an example.
2. Define *productivity* and give an example.
3. List three things on your job that you feel affect productivity. Be specific and provide as much detail as possible.
4. Explain how the following affect productivity:
   - Excessive use of overtime
   - Poor coordination of trades
   - Adverse temperature and humidity conditions
   - Poor worker attitude

PARTICIPANT ACTIVITY

Ask trainees to complete the Participant Activity. Discuss their answers.

The five-minute rating uncovers construction delays such as crews or crafts interfering with each other, poor work methods, and shortages of equipment and materials. In addition, it can be used to measure the efficiency of a work method.

For example, a five-minute rating shows where using two trucks instead of three would be adequate or where a twelve-person crew can be reduced to ten. To conduct a five-minute rating, the supervisor must watch the crew in such a way that they do not know they are being observed. Using a form such as the one in *Figure 1*, the supervisor notes the activity or inactivity of the crew members at intervals of 30 seconds, 1 minute, or 2 to 3 minutes from the time they begin a particular task to the time they complete it. The supervisor should watch at least one cycle of the activity in order to obtain the most complete information; however, this method can also be used to assess a portion of an activity. The supervisor develops an effectiveness rating, expressing the time worked as a percentage of the total time spent.

Figure 1 shows a five-minute rating analysis for erecting a precast concrete panel. The observations are made every minute and are listed on the left. Across the top of the form are the individual skill classifications of the members of the crew. The right-hand side, opposite each time designation, describes the activity or task going on during that specific time. A check mark indicates that the crew member was working on the particular task when the observation was made. A blank space indicates that the crew member was not working or otherwise being productive at the time the observation was made.

The example shows that at 2:14 p.m., ironworker number one was involved in loading the panel but at 2:18 p.m., that ironworker was not involved in aligning the panel. In fact, at 2:18 p.m. that ironworker was doing no productive work.

For the purpose of the five-minute rating, *working* is defined as:

- Carrying, holding, or supporting

- Participating in such active physical work as:
  — Measuring
  — Laying out
  — Reading drawings
  — Filling in time cards
  — Writing orders
  — Giving instruction
  — Operating a machine or piece of equipment
  — Discussing the work, provided it can be determined that such is the case

Non-working personnel are those:
  — Waiting for another person to finish work
  — Talking while not actively working
  — Attending self-operating machines, unless engaged in useful task
  — Walking empty-handed

After the supervisor's observations are completed, the total number of check marks and the total number of boxes are each added up. In *Figure 1*, the total number of check marks is 36, and the total number of boxes, whether they contain check marks are not, is 78. The check mark total (36) is divided by the box total (78), resulting in a .46, or 46 percent, five-minute rating. A five-minute rating of less than 60 percent is generally considered unsatisfactory.

The five-minute rating is so named because no member of the crew should be observed for less than five minutes. In fact, the minimum observation should be the number of minutes equivalent to the number of workers. Five-minute ratings are quick, economical, and relatively easy to use. They give an overview of the work situation and guide the supervisor in solving problems. However, the results are only indicators, not conclusive evidence, of poor performance and should not be used as a basis for discipline or discharge.

The way to increase crew effectiveness is to have as many members of the crew working as much of the time as possible. *Figure 1* shows that, during erection of precast concrete panels,

Instructor's Notes:

Module MT208 • Resource Control and Cost Awareness

	Ironworker #1	Ironworker #2	Carpenter #1	Carpenter #2	Carpenter #3	Welder	
Start	1	2	3	4	5	6	
2:13	✓						Crew waiting for panel to be placed
:14	✓	✓	✓	✓	✓		Installing panel / welding waiting
:15	✓	✓	✓	✓	✓		Tack rebar
:16		✓	✓	✓	✓		Install upper brace bolts
:17		✓		✓	✓		Install braces
:18			✓	✓	✓		Align panels
:19			✓	✓	✓		Ditto
:20			✓	✓	✓		Ditto
:21	✓	✓	✓				Unhook crane rigging
:22	✓	✓	✓				Ditto
:23						✓	Welder tacks rebar/crew waits for next panel
:24						✓	
:25						✓	

Job Name: XYZ Fast Food
Supervisor: S. Lord

Total Time Units: 78   Effective Time Units: 36   Five-Minute Rating: 46%

208F01.EPS

Figure 1 • Five-Minute Rating

some crew members are busy only for relatively short periods. This should suggest to the supervisor that perhaps not all of the people in the crew are needed or that the work method could be changed to result in more of the crew working more of the time. In this particular case, perhaps the supervisor could increase productivity by removing one ironworker from the crew and having the welder perform the ironworker's tasks — assuming, of course, that the welder has been trained in ironwork. Making such a change in the crew would change the five-minute rating results: 36 effective time units divided by 65 (78 – 13) total time units equals .55 or 55 percent effectiveness — a 19 percent improvement!

Show Transparency 2 (Figure 1).

Explain to the trainees how the five-minute rating works, using Figure 1. Alternatively, take a field trip and complete one or more of the three types of production analyses on site.

Show Transparency 3 (Figure 2).

Explain to the trainees how the crew balance chart works, using Figures 2.

2.2.2 Crew Balance Chart

Determination of proper **crew balance** is critical. Crew balance is defined as the proper deployment and efficient use of personnel involved in the work process. In other words, it's having the right number and type of personnel assigned to each job. The interrelationships among members of the crew and their equipment can be displayed graphically on a crew balance chart.

A crew balance chart is a bar chart that summarizes the results of the five-minute rating. It can also be used by itself to graphically depict the efficiency of each person in the crew. As *Figure 2* shows, a vertical bar is drawn for each worker on the crew. The shaded areas show the times that the crew member was actually working; the blank areas show the unproductive or nonworking times. The crew balance chart tells the supervisor at a glance which workers were most productive (those with the most shaded areas) and which were less productive (those with the most blank areas).

The left side of *Figure 2* shows an example of a three-person mechanical crew preassembling sections of an air supply duct. The process involves bringing raw materials to the fabrication site, fastening them up, fitting and aligning them, and then carrying the assembled sections away. This cycle is repeated continuously.

A close look at the chart shows that one of the journeymen was not needed and could have been assigned to a different task. Properly balancing the crew and the work would produce a crew balance chart like that on the right side of *Figure 2*. Of course, each task must be analyzed separately. If, for instance, removing completed sections of duct work and bringing raw materials to the work area requires 30 minutes instead of 15, the supervisor might consider keeping a third person on the crew to reduce the waiting time of the worker assembling the duct. Unless time data can be accurately predicted before an operation begins, the formulation of a crew balance chart is usually a result of close observation of several cycles of the same work process.

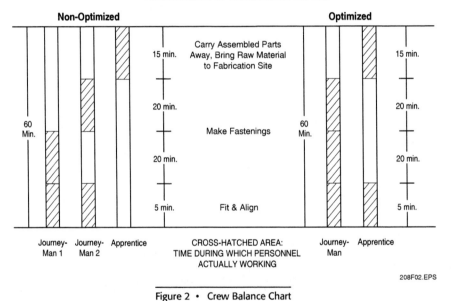

Figure 2 • Crew Balance Chart

Instructor's Notes:

2.2.3 Delay Survey

The **delay survey**, shown in *Figure 3*, is a tool for identifying delay problems in the field. The survey form lists probable causes of job delays and gives the supervisor space to record the number of hours involved in the delay and the number of workers affected.

Once the supervisor has completed the form with information from observations, the number of delay hours are multiplied by the number of delayed workers to determine the number of worker hours lost in connection with each type of delay. The columns are then totalled and the sums entered at the bottom. This form can be used by supervisors to compare the whole crew, or supervisors may elect to have the crew fill it out themselves.

A delay survey is not an exact method for determining the impact of delays on job progress, but it does provide some helpful information on delays and the magnitude of their effect on productivity.

SECTION 3

3.0.0 PRODUCTIVITY IMPROVEMENT

How can a supervisor improve productivity? There are two basic ways: increase production without increasing labor costs and decrease labor costs without reducing production.

Consider this example: A crew of three electricians is hanging lighting fixtures at a rate of three per hour. The supervisor adds a new person to the crew, and the production goes up to three and a half fixtures per hour. What is the result? Production has increased, but productivity has decreased, because the 33 percent increase in labor has increased production by only a half of a fixture per hour — 17 percent. If, however, the supervisor changes the work method so that the three electricians can install four fixtures per hour, their productivity has increased by 33 percent without increasing costs. The result: an overall decrease in unit cost.

Improving productivity, then, is a matter of finding ways to use labor, material, and equipment as efficiently as possible.

Studies show that major increases in productivity can be realized by:

- Improving supervision
- Using innovative materials, equipment, and tools
- Prefabricating as many components as possible and installing prefabricated units
- Motivating workers
- Standardizing as many processes as possible
- Pre-planning as many activities as possible
- Assuring adequate tools, equipment, and materials are available when needed
- Implementing job-site safety programs
- Providing on-the-job training
- Studying job activities and altering work methods as needed
- Changing the crew make-up

Some of the above can be done wholly or in part by the supervisor. The balance requires support and action from others in the company. It is critical for the supervisor to understand what productivity is and be able to supervise the crew efficiently and make changes in work methods when necessary.

There are work improvement methods which the supervisor can use to improve productivity. These methods are not intended to be used by themselves to locate areas within the company that need to be improved from an operational or work method standpoint. Such improvements are the responsibility of upper management.

Employees often feel they are judged by how hard rather than how smart they work. Unfortunately, this opinion often reflects the way they are evaluated, even by supervisors who know better. For example, skilled craftworkers know from experience that time spent preparing the workplace and gathering materials often saves installation time. But preparatory activities rarely appear productive, and there is often a fear of being criticized for wasting time. As a result, the task takes longer than it should, due to improper preparation.

Ensure that you have all the necessary materials to teach the course. Check the Materials and Equipment list at the front of the module.

Ask trainees to describe how they used or would like to use human relations skills, fact gathering, and work improvement methods to increase production.

Explain to trainees how the delay survey works using Figure 3.

Assign reading of Sections 3.0.0 – 4.4.0 for the next class session.

Show Transparency 4 (Figure 3).

DELAY SURVEY

Supervisor's Name

Craft/Crew Designation

Date of Evaluation

Number in Crew(s)

PROBLEM CAUSING DELAYS

Work Hours Lost

	Number of Hours	×	Number of Workers	=	Work Hours
1. Waiting for materials at warehouse					
2. Waiting for materials not received or not ordered					
3. Waiting for tools or tools not available					
4. Waiting for equipment					
5. Equipment breakdowns					
6. Changes/rework–design errors					
7. Changes/rework–prefabrication errors					
8. Changes/rework–field errors					
9. Move to other work					
10. Waiting for information					
11. Interference with other crews					
12. Overcrowded working areas					
13. Plant coordination/authorizations					
14. Other					
TOTALS					

Figure 3 • Delay Survey

Instructor's Notes:

PARTICIPANT ACTIVITY

This five-minute rating shown below was made on a concrete placement activity. Based on the data indicated:

1. How do you evaluate the crew's productivity? Why?
2. Analyze the operation and suggest ways to increase productivity without increasing cost.
3. Prepare a crew balance chart.
4. Analyze the crew balance chart and determine if the crew makeup can be changed to make the operation more efficient.
5. Could the work method be altered to result in a more efficient operation?

Job Name: XYZ Fast Food
Supervisor: S. Lord

Time	Laborer #1	Laborer #2	Laborer #3	Laborer #4	Finisher #1	Finisher #2	Activity
11:03	✓						Truck arrives
:04	✓						Chute attached
:05	✓						Placing begins
:06	✓	✓	✓	✓			Placing
:07	✓	✓	✓	✓	✓		Finishing begins
:08	✓	✓			✓	✓	Finishing
:09	✓	✓		✓	✓	✓	Ditto
:10	✓	✓		✓	✓	✓	Ditto
:11		✓		✓	✓		Ditto
:12			✓	✓	✓		Ditto
:13			✓	✓	✓		Ditto
:14	✓				✓		Ditto
:15		✓	✓		✓	✓	Chute removed / truck leaves
Totals							

208F04.EPS

PARTICIPANT ACTIVITY

Have trainees complete the Participant Activity. Discuss their answers. Some suggestions for answers are located at the end of this module.

Use a blank transparency to sketch a crew balance chart.

Ask trainees to give examples of their best methods for receiving, inspecting, and storing materials.

In such cases, the supervisor must use human relations skills to exchange information and suggestions with the crew, to analyze and improve work methods, and to involve them in the decision-making process.

All work improvement begins with gathering the facts related to the situation. Changes in methods cannot be based on hunches or gut reactions. Here, the productivity-measuring methods described earlier come into play. It is also helpful for the supervisor to involve other supervisors who may be affected by the resulting changes. They should be informed about what is happening, how they will be involved, and how the outcome of the process will be used.

Once facts are gathered, the supervisor must analyze all relevant factors using job analysis methods. How can the job be done differently? How can operations be simplified, unnecessary activities eliminated, task sequences improved, and new tools, equipment, and materials used? This part of the process should involve crew members in defining and establishing the improved work method — one which will increase the production rate without increasing the cost of the job. Finally, the supervisor must get a commitment from the workers that they will implement and follow the plan. A follow-up procedure should be implemented to see if, in fact, productivity improves; if it doesn't, the plan should be revised accordingly.

SECTION 4

4.0.0 RESOURCE CONTROL

The best way to control cost on the job is to control the use of:

- Materials
- Tools
- Equipment
- Labor

This section discusses the methods for controlling each of these resources.

4.1.0 Material Control

The control of material begins with purchasing and receiving. This means that the supervisor must be involved in project planning as well as job-site receiving, storage, and installation. It is the supervisor's responsibility to make certain that necessary materials are on site when work begins.

Once the material arrives at the site, the supervisor may be responsible for seeing that it is unloaded, inspected, stored, handled, and installed correctly. In addition, material must be protected from damage or loss while in storage.

When receiving material, the supervisor should follow these guidelines:

Step 1 Inspect the material for damage before unloading it. Note all physical damage on the delivery ticket. Refuse to unload severely damaged material until you check with your supervisor. Do not sign for or accept damaged material without prior approval.

Step 2 Check the material received against the packing slip and the purchase order to verify that the correct type(s) and quantity have been shipped.

Step 3 If time does not permit careful inspection at the time of delivery, write "Received ___ (number) boxes/crates subject to inspection at a later time." Keep in mind that each state has its own laws governing how many days the supervisor has to complete the inspection, and the supervisor should obtain this information from the company.

Step 4 Return the receiving ticket to the designated person for processing. Check with your superior to see who the ticket should go to.

If the material is to be stored, it must be protected. Store material where it is protected from weather and other elements that could harm it.

Store material where it can be found later. On big jobs, always record storage locations on an inventory form. Be sure to protect material from theft.

Instructor's Notes:

In addition to correct receipt, inspection, and storage, the supervisor must know how material is to be used; that is, how the material is to be worked, altered, and installed. It is also the supervisor's responsibility to determine work methods that make the most efficient use of the material as well as the labor to install it.

To ensure efficiency:

- Get the material to the job on time.
- Check the condition of all materials to make sure they are not damaged and that they meet specifications.
- Protect material from damage during on-site transportation and installation.
- Ensure that all craftworkers are trained in the handling, use, and installation of the material.
- Install material correctly.
- Return any unused material (overage) to the appropriate place.

The supervisor must know and follow all company policies and procedures related to the purchase, receipt, inspection, and storage of material, including the completion of necessary paperwork. If the supervisor is unfamiliar with company policies or procedures or needs assistance in completing forms properly, the next higher supervisor should be consulted.

4.2.0 Tool Control

Controlling small tools and supplies such as first-aid kits and fire extinguishers is not an easy task, due largely to the fact that these items can be used off the job site.

Small tools must be located where they can be found and must be protected from weather, theft, and misuse. It is the supervisor's responsibility, in most cases, to obtain the tools and ensure they are used for the purpose for which they are intended and to see they are repaired when necessary.

Studies show that the key to an effective tool control program lies in limiting:

- The number of people allowed access to stored tools
- The number of people held responsible for tools
- The way in which tools are returned to storage

As part of a tool control program, the supervisor should consider the following:

- Who is responsible for providing all or part of the needed tools—the worker or the company? Refer to the company's policies and procedures for this information and make it known to all workers in the crew.
- Make it clear that craftworkers are personally responsible for the care of the tools they use. Make unannounced safety and maintenance checks on each tool.
- Provide training in the safe and correct use of all tools.
- Lock up all company-owned tools when they are not in use.
- Keep accurate check-in and check-out records. Refer to the list above for some of the applicable methods for doing this.
- Stamp, paint, or etch all company tools with permanent identification marks.
- Have spare parts available for all tools.
- Replace all tools that are in disrepair.

4.3.0 Equipment Control

Proper planning is the key to equipment control. All work should be planned to determine exactly what is needed, when, and for how long. Obtaining equipment may be the supervisor's responsibility; if not, company policies and procedures regarding equipment requisition should be incorporated into the job-site practices.

For the supervisor, controlling equipment also means making sure the equipment is being used according to the job plan and ensuring that required maintenance is performed on schedule. Lack of maintenance leads to overly expensive repairs and costly delays. If a piece of equipment breaks down or arrives at the job site in poor condition, the supervisor should contact the appropriate person(s) immediately so that workable equipment can be obtained without delay. Good planning gives the supervisor sufficient lead time to inspect equipment before it is needed and replace faulty equipment quickly.

Ensure that you have all the necessary materials to teach the course. Check the Materials and Equipment list at the front of the module.

Have trainees complete the Participant Activity. Discuss their answers.

Assign reading of Sections 5.0.0 – 6.2.3 for the next class session.

4.4.0 Labor Control

Labor is the most difficult item to control and has the greatest impact on every job's bottom line. Improving labor productivity reduces costs and increases profits, yet many studies show that construction personnel are only 35-40 percent efficient. That is, during any typical construction hour only 21-24 minutes are spent in productive work! The balance is wasted on nonproductive or indirect tasks and is thus a waste of dollars.

Work sampling studies indicate that a typical worker's day is broken down into the following:

- Receiving instructions
- Reaching for materials, tools, and equipment
- Searching for materials, tools, and equipment
- Carrying or otherwise moving materials, tools, and equipment
- Working at the trade
- Working at trades other than the one they know
- Waiting for other tradesmen
- Waiting for materials, tools, and equipment
- Walking and loading
- Being idle for no apparent reason
- Talking about topics other than work
- Personal delays
- Planning, discussing, and laying out

Studies also show the reasons that workers often have unproductive days. These reasons include a lack of:

- Information
- Work to do or interference by others
- Tools
- Motivation
- Ability
- Desire

PARTICIPANT ACTIVITY

Discuss and answer the following:

1. What does controlling resources on the job mean? Give an example.
2. One of the supervisor's responsibilities is to control the use of materials. Discuss what you would do to ensure that materials such as lumber or connectors are not wasted on the job site.
3. Who provides what tools on your job?
4. When you are in charge of a job requiring heavy equipment, what do you do to control its use?
5. Controlling labor takes a large part of the supervisor's time. Outline a program you would use to effectively control labor.

PARTICIPANT ACTIVITY

Lack of information and tools can be corrected by effective planning and supervision. Lack of motivation and ability can be corrected by effective utilization, leadership, and training. Lack of desire can be reversed by the supervisor who must instill in the worker a sense of integrity, initiative, and loyalty.

In summary, labor and its costs can be controlled by:

- Careful planning that includes letting workers know what is expected of them.
- Proper scheduling and informing each worker when assigned tasks are to be done and how much time has been allowed to complete them
- Making sure that workers have all the resources they need to work efficiently
- Supervising effectively

Instructor's Notes:

SECTION 5

5.0.0 QUALITY CONTROL

Quality is the one standard that applies to every job. It is specified in the drawings and specifications in terms varying from the makes and models of air handlers to the strength of concrete used in footings and the number of coats of paint on walls and ceilings. Typical contract documents contain hundreds of such quality standards.

Another important aspect of quality is the manner in which the work is performed. The supervisor is responsible for performing all work effectively and efficiently. Prior to starting a job activity and while performing the activity, the job supervisor must make certain that tasks are completed in accordance with the job plan and schedule. This helps to assure conformance to quality standards. For example, before and during placement of a concrete slab, the supervisor should perform the following quality control activities:

- Check that all concrete decking is clean.
- Verify that any concrete bonding agent is applied in accordance with the manufacturer's recommendations.
- Verify the grade installation of all screeds.
- Verify that tolerances are within project specifications.
- Make sure that installation procedures meet project specifications.
- Obtain concrete test cylinder specifications, if required as part of the supervisor's job, and deliver them to the testing laboratory.

This is not necessarily a complete list of quality control activities for the job of placing a concrete slab. Each company must develop its own series of checklists, based on the type of work it does, the manner in which it performs job tasks, and the details of the contract documents (especially the drawings and specifications).

The supervisor and crew must perform all work in strict accordance with the drawings and specifications. This requires that the supervisor be familiar with the specifications and drawings prior to starting the job and forbids changing these documents without prior approval of the project's designated authority.

To support supervisors' efforts, the contractor should develop and maintain an effective quality management (control/assurance) program. Many companies have developed written quality control/assurance policies and programs; supervisors should determine if such programs exist in their company and become familiar with their role in carrying out the programs in the field. In addition, if the company has quality-related teams or committees, the supervisor (if a member) should participate positively to help solve quality-related problems.

Whether or not a company has a formal quality management program, the supervisor is still responsible for building a quality job. This demands not only adhering to drawings and specifications, but adequate training of those workers who do the construction. For example, a carpenter must know how to measure, cut, and fit a *quality* joint that will hold up under specified loads. This knowledge is not addressed directly by the contract documents, but it is certainly implied, and it is up to the supervisor to make sure that all work is done by trained and experienced craftworkers using up-to-date and correct methods, tools, and equipment.

Supervisors should never attempt to build the job their own way, especially if construction methods are described in the drawings or specifications. On jobs in which supervisors have little or no experience, they should check with their supervisors prior to starting the work to discuss the methods they plan to use. If they find the drawings and/or specifications to be unclear, they should not start any work until their supervisor answers questions and clarifies instructions.

If work is not performed correctly the first time, it will have to be redone. This is referred to as *rework*. Research has shown that 12 percent of the total in-place cost of the average project is due to the need to perform rework. This has a negative impact on the company's profit margin and its ability to remain competitive. This, in turn, affects the ability of the company to retain all of its employees. It is every supervisor's job to plan and perform work so as to minimize the need for rework.

Ask trainees to give examples of how quality control can impact cost control.

Ask trainees for examples of successes they have had or have seen by controlling resources on the job.

A supervisor should never fall into the trap of assuming that the next job can be done exactly like the last one. Too often, this assumption leads to unnecessary rework that drastically delays job progress and dramatically increases project costs. Quality control must be part of every decision the supervisor makes to control resources and to complete the work on time and within the estimate. See the *Quality Control* module of this book for more detailed information.

SECTION 6

6.0.0 COST AND RESOURCE CONTROL

What is control? Control is a means of measuring actual performance, comparing it to estimated performance, and directing work to accommodate any changes in the plans. The control process can be broken down into the following steps:

- Establish standards in specific measurable terms or units.
- Measure performance against the standards.
- Correct deviations from the standards and the plan.

The supervisor's success is based on how effectively the job resources are controlled, specifically:

- Personnel and their performance
- Material and its use
- Equipment and its performance
- Tools and their use

In order to control resources, the supervisor must be given certain specific information at the beginning of the job, such as the number of labor hours, the amounts of material, and the dollar estimates of particular tasks. This information is usually part of the job schedule. If it is not, the supervisor should ask for it, because without this information there are no standards against which to measure the performance of the crew and its crew leader.

6.1.0 Control Standards

Cost, labor hours, material estimates, and task duration are some of the main control standards that supervisors use to control resources. On the job, the supervisor is continually measuring the efficiency of the crew based on historical numbers and actual results.

In most cases, cost is the major control standard. For example, consider a crew placing concrete for a reinforced footing foundation. The estimate shows that 100 cubic yards (cu. yd.) of concrete must be placed. The cost of placing the concrete was estimated to be $12,500 or $125/cu. yd. ($12,500 divided by 100). As the job progresses, the supervisor calculates that the job is actually costing $160/cu. yd. Obviously, this is over budget and something must be done immediately or the job will wind up costing almost 30 percent more than estimated.

Labor hours can also be used as a control standard. A company estimates that installing all the plumbing fixtures in an office building will take 130 crew hours. At the point where half the fixtures are installed, however, the supervisor notices that 80 crew hours have been spent. Immediately, it can be seen that at the current rate of production, the job will take another 80 hours, for a total of 160 crew hours — 30 hours over budget. Here again, some action must be taken to improve the productivity of the crew in the second half of the job.

Obviously, control standards are essential for identifying and correcting situations that threaten to lose money for the project. Consequently, control standards are important for the entire company, because no company can continue to do jobs and employ workers if it is consistently losing money. Faced with steady losses, the company would end up terminating employees and perhaps even going out of business — actions that would certainly affect the supervisor's job. So the supervisor has personal reasons for being continually aware of job costs and resources.

6.2.0 Cost Awareness

There are three types of costs associated with a construction project: **estimated cost**, **actual cost**, and **projected cost**. The supervisor must consider all three in controlling costs and resources on the job.

Instructor's Notes:

6.2.1 Estimated Cost

The contract agreement between the project owner and the contractor states the total estimated cost of construction. This estimated cost is a combination of direct costs, indirect costs, and expected profit, as determined by the contractor. Direct costs are those directly associated with job activities including materials, labor, equipment, and tools. Indirect costs (sometimes referred to as *overhead*) consist of the contractor's expenses for maintaining business operations such as the job office, the company office, bonds, insurance, utilities, payroll checks, safety, and similar items. Direct costs are related directly to building the job; indirect costs are incurred regardless of whether or not the job is built.

There are many uses for estimated cost, and the most important of these is as a control standard. In this context, the estimated cost can be considered a budget for the project, since it defines the maximum amount the supervisor can spend without endangering the company's profit. If job costs exceed this budget, the contractor's profit shrinks, perhaps to the point where company funds must be used to complete the work. It is a part of the supervisor's responsibility to ensure that jobs are built for the estimated cost.

In addition to standard cost information, the estimate provides standard production rates for labor and equipment. This information is valuable for evaluating the efficiency of construction operations.

6.2.2 Actual Cost

Actual costs are those incurred in doing the work. They must be accurately tracked and recorded if a true financial record of the project is to be kept. Actual material costs are tracked through bills received from material suppliers and, in most cases, should agree with the estimated material costs.

Actual labor costs are derived from time cards. Time cards tell management how many hours each craft worker worked on a specific activity. Gathering this information is the role of the job reporting system described later in this module. Actual equipment costs are obtained in the same way as actual labor costs.

> **Classroom**
>
> Divide trainees into small groups of 4-5 and have them complete the Participant Activity. Allow 15 minutes. Discuss their answers. Correct answers are located at the end of this module.

PARTICIPANT ACTIVITY

The following information is known about a job which involves placing 12-inch concrete block.

Quantity of block:	10,000	Wages:	Mason: $15.00/hr.
Cost per block:	$0.80		Helper: $11.50/hr.
Crew:	2 masons and 1 helper	No equipment, scaffolding,	
Production Rate:	400 blocks/day (8-hr day)	or other items are needed	

1. Find the total job cost and the unit direct job cost.

After two weeks (10 working days) the following information is known:
  Total actual block cost: $8,000.00 ($0.80/block); Crew has placed 5,000 blocks.

2. If the same production rate is maintained, what will be the projected cost of the job?

After the job is complete, a review of the field reports and material invoices indicates:
  Actual block cost: $8,000.00   Actual labor production: 350 blocks/day

3. What was the actual total direct job cost and the actual unit direct cost?

4. Relative to the data presented after 10 working days, how are you doing on the job?

5. How has the information given for the completed job above changed your approach to your job?

Ensure that you have all the necessary materials to teach the course. Check the Materials and Equipment list at the front of the module.

Assign reading of Sections 7.0.0 – 7.2.5 for the next class session.

By tracking these actual costs, management knows the direct costs of the work being performed and can compare them to the estimate. The goal, of course, is to keep actual direct costs less than or equal to the estimated direct costs. In addition, by comparing production rates in the estimate with those reported from the field, management can analyze actual production rates of labor and equipment on the job.

6.2.3 Projected Cost

Projected cost is an estimate of the expenses that will be incurred in completing the job. It differs from the original estimated cost in that it is calculated when the job is in progress and takes into account actual job conditions.

For example, consider the concrete job presented earlier in this chapter, where 100 cu. yd. of concrete are to be placed for an estimated cost of $12,500. This estimate is based on the job taking four days to complete. At the end of the second day of work, the supervisor determines that 40 cu. yd. of concrete have been placed at a direct cost of $6,000.00 or $150/cu. yd. There are 60 cu. yd. yet to be placed, and if the current unit cost remains at $150/cu. yd., it will cost the company $150/cu. yd. × 60, or $9,000.00, to finish the work. In this case, the projected cost of the job is $6,000.00 (work completed) + $9,000.00 (work remaining), or $15,000.00. This is $2,500 over estimate.

Projected costs are critical for determining the financial status of any project. By comparing projected costs to the original estimated costs, management can tell if there is enough money available in the budget to finish the work.

SECTION 7

7.0.0 THE REPORTING SYSTEM

Before a supervisor can decide whether any job is being done within the allocated time and estimate, information must be gathered. There must be a system that reports on:

- Labor time spent performing the work
- Equipment time spent performing the work
- Quantity of material installed within a specific amount of time

7.1.0 Cost Coding

The basis of any field reporting system is a cost coding system. The system is based on the codes the estimator assigns to each work item as the estimate is prepared. This code can be a number and/or letter designation, and each item's code is unique.

The code designation works like a personal social security number. All labor, material, and equipment used on a work item is charged to an account corresponding to its code, just as all social security amounts withheld from an employee's paycheck are credited to that individual's social security account. For example, consider the code designation 103.02, where 103 is the activity *earth excavation* and .02 is *by machine*. At the beginning of the project, a project account is set up for this activity, based on the estimate. Then, each hour of machine excavation on the project is reported using the code 103.02 and those hours are charged against the account.

There are various types of coding systems in use today. The best ones are tailored to the individual company's needs, although many companies use the Construction Specifications Institute Uniform Index as a basis for their coding systems. Whatever system is used, the supervisor must keep in mind that every hour of construction time and every dollar spent for construction must be charged against a valid, active account number. If the entire system is not taken seriously or is used inappropriately, the information it produces will be useless, and the supervisor will lose financial control of the project.

7.2.0 Types of Reports

A reporting system can be defined as the various reports that provide information on the hours and costs incurred in the construction of a job. In a small construction company, the reporting system may consist of only a few simple reports; in a larger company, the reporting system may include many reports, each requiring considerable information and demanding considerable time to complete. The supervisor should be familiar with and follow the company's policies and procedures regarding reporting.

Although the size of the reporting system and the nature of the reports vary from company to com-

Instructor's Notes:

7.2.1 Daily Time Report

The daily time report, usually completed by the supervisor, indicates how many hours each worker spends performing specific work items. The work items are usually indicated by code numbers, and in some companies the code numbers are preprinted on the time cards, which makes completing the form easier.

The daily time report must be accurate. The supervisor must never estimate the number of hours spent on an activity; how much time is spent must be recorded exactly, to the hour. If the supervisor gets information from the workers on the crew, it is a good idea to verify this information from time to time, to ensure that it is being reported accurately.

When the supervisor sends the daily time report to the company office, the accounting department uses it to write payroll checks, and the project manager uses it to determine actual production rates and direct costs.

Figure 4 is an example of a typical daily time report. Notice that spaces are provided for recording who worked on what job and how many hours, regular and overtime, were spent on each activity that has its own code designation.

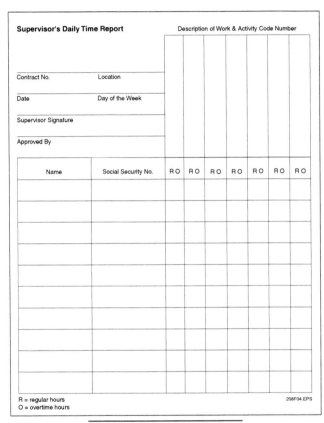

Figure 4 • Typical Daily Time Report

Show Transparency 5 (Figure 6).

Explain to trainees how the Quantity Report works, using Figures 5 and 6.

7.2.2 Material Installation Report

The material installation report records how much material is installed in connection with each work activity within a specified amount of time. The form, sometimes called a *quantity report*, is usually completed once a week and sent to the project manager for use in determining actual production rates and whether or not the job is on schedule. Material installation reports are known by various titles throughout the industry.

Figures 5 and *6* are examples of typical material installation report forms. *Figure 5* is a quantity report for stripping and cleaning forms. It requests complete information on how much work is done for each day of the week, the total for the week, and the total to date. *Figure 6* is a supervisor's quantity report for the activity of placing anchor bolts. It requests information on the quantity of bolts placed to date.

7.2.3 Equipment Utilization Report

The equipment utilization report records what equipment was used during a certain interval, how many hours it was used, and what it accomplished. The completed form is used by the project manager to determine whether the equipment production rate is adequate and whether the schedule is being met.

Figure 7 is an example of an equipment utilization report. It provides spaces for indicating what pieces of equipment worked on what activity (designated by account number) and for how many hours. The form also allows people other than the supervisor to enter information, such as the hourly rate of owning and operating, leasing, or renting the equipment and the total equipment cost of performing the specific activity.

7.2.4 Summary Report

Data from all the daily field reports is summarized, usually weekly, and recorded on summary report forms. A summary form gives management a complete picture of each activity, including current crew production rate, equipment production rate, and actual costs by activity of labor, material, and equipment. In addition, it compares each of these factors to the original estimate.

QUANTITY REPORT

Day or Week Ending ___October 17___ , 20_01_   Reported by ___J. Smith___

Account No.	Total Last Report	QUANTITY FOR						Total to Week	Total to Date	Item No.
		Mon.	Tues.	Wed.	Thur.	Fri.	Sat.			
4260	500 sq ft					1000		1000 sq ft	1500 sq ft	

Figure 5 • Quantity Report – Stripping and Cleaning Forms

```
                    SUPERVISOR'S QUANTITY REPORT


Job No.:  1           Unit:  Anchor Bolt      Craft:  Laborer, Carpenter     Account No.:  2834

Description of Work:  Set Anchor Bolts


a)   Original Budget Quantity                                                    100

b)   Budget Incl. Auth. Change No's -- to -- -- Quantity                         ----

c)   Current Estimated Completion Quantity                                       100

d)   Units Previously Reported                                                    90

e)   Units Completed Today                                                        10

f)   Units Completed To Date


g)   Supervisor's Estimated Units to Complete                                    ----

h)   Total Work Hours Worked Today       Signed   Supervisor      Gen. Supervisor



Field Engineer's Progress Measurement:

Field Engineer's Remarks:      All Bolts Set

                               Field Engineer                        Date: 10/18
```

Figure 6 • Supervisor's Quantity Report

Summary report forms vary from company to company, but typical examples are shown in *Figures 8, 9,* and *10*. *Figure 8* is a weekly summary form for the activity of setting anchor bolts. Notice that it provides space for recording how many hours each day the individuals on the crew worked on this activity, as well as the total number of hours for the week, the wage rate for each worker, and a total labor cost. At the bottom of the form is space for summarizing what has been done to date on this activity, the payroll costs for the work performed to date, the average unit cost, and the actual production rate. *Figure 9* shows the same type of information summary for excavation and *Figure 10* for stripping and cleaning forms.

7.2.5 Cost Summary Report

All the information from the activity summary reports is brought together on a cost summary report such as that shown in *Figure 11*. This type of report gives management a complete picture of all the activities of a project and sufficient information to conduct a production analysis.

Explain to trainees how the Daily Equipment Report works, using Figure 7.

SUPERVISOR'S DAILY EQUIPMENT REPORT

Date __10/13/__ Job __Account No. 3010__ Shift __1st__ Supervisor _____
Materials Handled __Topsoil__
Sandy Soil ____ Clay Loam ____ __X__ Clay Loam ____ Gravelly or Stony Soil ____ Rock ____ Other ____

Condition of Haul Road: Rough ____ Smooth ____ Dry ____ Damp ____ Muddy ____ Soft ____ Sticky or Slippery ____
Grade __None__ Operator _____

Unit	Unit No.	Attachment	Cost Code	Hours Worked	Pay Hours	L.T. Mech.	L.T. Serv.	Hour Rate	Earning
Air Compressor									
Air Tool									
Air Tool									
Bit, Plant									
Boiler									
Bucket									
Batcher, Concrete									
Bin, Aggregate									
Cart, Concrete									
Cement Gun									
Cement Plant									
Concrete Mixer									
Concrete Paver									
Crane, Crawler									
Crane, Truck									
Crusher									
Drag Line									
Engine, Gas									
Grader									
Grader									
Hoist									
Pump, Gas									
Pusher									
Pusher									
Pusher									
Roller, Road									
Roller, Compact									
Saw, Chain									
Saw, Table									
Saw, Radial									
Scraper									
Scraper									
Shovel, Gas									
Shovel, Attach.									
Tractor, Crawler	1	Blade	8	8	8	--	--	25.00	200.00
Tractor, Rubber									
Tractor, Dozer									
Tractor, Dump									
Tractor, Flat									
Tractor, Service									
Tower, Hoist									
Wagon, Dump									

Figure 7 • Daily Equipment Report – Excavating Soil

Instructor's Notes:

Have trainees complete the Participant Activity. Discuss their answers.

Account No. _____ Sheet No. _____

Class of Work **Set Anchor Bolts**   Week Ending **10-17**   Job No. **1**

Classification			13th	14th	15th	16th	17th	Total Hours	Rate	Amount	
Labor			---	---	8	8	4	20	5.00	100	00
Carpenter			---	---	8	8	4	20	7.00	140	00
TOTAL			---	---	16	16	8	28	---	240	00
	Quantity Work in Place	Payroll Costs	Labor Average Unit Cost	Average Quantity Per 8 Hr Day	Quantity Work in Place	Payroll Costs	Labor Average Unit Cost	Average Quantity Per 8 Hr Day			
Previous											
This Week	100 ea.	240.00	2.4	40							
Total	100 ea.	240.00	2.4	40							

208F08.EPS

Figure 8 • Weekly Summary – Set Anchor Bolts

PARTICIPANT ACTIVITY

Discuss and answer the following:

1. Define *field reporting system* and explain its purpose.
2. List two advantages and two disadvantages of a field reporting system.
3. Does your company have a field reporting system? If so, what types of forms do you fill out and what kinds of information do you record on them?
4. How is the actual amount of material installed each day or week determined on your jobs? Who makes the determination? Do you feel the method gives accurate information? Why or why not?
5. Who keeps track of the equipment time used on your job?

PARTICIPANT ACTIVITY

Show Transparency 6 (Figure 9), Transparency 7 (Figure 10), and Transparency 8 (Figure 11).

Explain to trainees how the two weekly summaries (Figures 9 and 10) combine to complete a weekly cost summary (Figure 11).

Account No. __636__ Sheet No. __2__
Class of Work __Excavation__ Week Ending __10-17__ Job No. __1__

Classification		13th	14th	15th	16th	17th	Total Hours	Rate	Amount
Tractor 1		8	—	—	—	—	8	9.00	72 00
TOTAL		8	—	—	—	—	8	9.00	72 00

	Quantity Work in Place	Payroll Costs	Labor Average Unit Cost	Average Quantity Per 8 Hr Day	Quantity Work in Place	Payroll Costs	Labor Average Unit Cost	Average Quantity Per 8 Hr Day
Previous								
This Week	400 cu yd	72.00	.18	400 cu yd				
Total	400 cu yd	72.00	.18	400 cu yd				

208F09.EPS

Figure 9 • Weekly Summary – Excavation

Account No. _____ Sheet No. _____
Class of Work __Strip and Clean Forms__ Week Ending _____ Job No. _____

Classification		13th	14th	15th	16th	17th	Total Hours	Rate	Amount
Laborer		—	—	—	6	8	14	5.00	70 00
Laborer		—	—	—	6	8	14	5.00	70 00
TOTAL		—	—	—	12	16	28	—	140 00

	Quantity Work in Place	Payroll Costs	Labor Average Unit Cost	Average Quantity Per 8 Hr Day	Quantity Work in Place	Payroll Costs	Labor Average Unit Cost	Average Quantity Per 8 Hr Day
Previous	500 sq. ft.	100.00	.20	400 sq. ft.				
This Week	1000 sq. ft.	140.00	.14	572 sq. ft.				
Total	1500 sq. ft.	240.00	.16	500 sq. ft.				

208F10.EPS

Figure 10 • Weekly Summary – Strip and Clean Forms

Instructor's Notes:

Weekly Cost Summary

Cost Code	Work Description	Quant.	Estimated			To Date			This Week			To Date		Projected	
			Total Cost	Unit Cost	Quant.	Total Cost	Unit Cost	Quant.	Total Cost	Unit Cost	Quant.	Saving	Loss	Saving	Loss
4026	Set anchor bolts	EA	100	900	0.00	—	—	—	100	840	8.40	00.00		—	—
4260	Excavate	cu. yd.	370.4	209.81	.57	—	—	—	400	322	.81	112.19		—	—
3010	Strip & clean forms	sq. ft.	2000	160.00	.08	300	40	.13	1000	140	.14				98

Account No. 4026
JOB Haley's Store
Week Ending 10-17
Page 1 of 1
Job No. 1
Prepared by Jones

Figure 11 • Weekly Cost Summary

SECTION 8

8.0.0 PRODUCTION ANALYSIS

Once information from the various field reporting forms has been summarized, it can be analyzed to determine actual unit costs and production rates. This analysis involves nothing more than comparing actual costs and/or production figures to the estimated ones.

The supervisor should know that production analysis is an ongoing process and that performance is evaluated wholly or in part on the results of the analysis. Consequently, it is helpful for the supervisor to understand the mechanics of such an analysis and the types of decisions which may result from it.

Two sources of information are needed to do a production analysis: the report forms listed in the previous sections, and the project estimate. As noted earlier, the estimate provides the standards of material costs, labor and equipment production rates, labor and equipment costs, and the total unit costs for performing a specific job—the standards against which the crew's and its crew leader's performances are measured.

As an example of a production analysis, consider the information on the activity *setting anchor bolts,* shown in the report forms (*Figures 8* and *11*) in the previous section. A summary of that information reveals the following:

ITEM	ESTIMATED	ACTUAL
Quantity of Bolts	100	100
Cost per Bolt	$6.00	$6.00
Crew Production Rate	4 bolts/hr	5 bolts/hr
Unit Cost	$9.00/bolt	$8.40/bolt

The analysis shows that the estimator accurately predicted the material quantity and the cost per bolt. It also shows that crew production is better than estimated, resulting in a total unit cost lower than what was estimated—an excellent reflection on the crew supervisor. This *5 bolts/hour* figure now becomes part of the company's historical data for estimating similar jobs in the future.

Materials

Ensure that you have all the necessary materials to teach the course. Check the Materials and Equipment list at the front of the module.

Teaching Tip

Review the sample production analysis with the trainees and any other samples not included in the module.

Homework

Assign reading of Section 8.0.0 for the next class session.

Ensure that you have all the necessary materials to teach the course. Check the Materials and Equipment list at the front of the module. Answers are located at the end of this module.

Have trainees complete the Participant Activity. Discuss their answers. Answers are located at the end of this module.

Assign reading of Section 9.0.0 for the next class session.

PARTICIPANT ACTIVITY

The following information is for the partially completed activity *stripping and cleaning forms* — a labor-intensive task with no material costs.

ITEM	ESTIMATED	ACTUAL
Quantity of Form Work	2000 sq. ft	1500 sq. ft
Crew Production Rate	125 sq. ft./hour	115 sq. ft./hour
Unit Cost	$ 0.08/sq. ft.	$0.12/sq. ft.

1. Analyze this data. What conclusions can you reach about the crew's production? Are you satisfied with it? If it continues at the same pace, will you complete the job on time?
2. What kinds of things could you do, without increasing the total cost of the activity, to get the job done on schedule?
3. If the job continued at the current actual unit cost, would it save or lose money at the end? How much would be saved or lost?
4. If you had to get the job done on schedule, as determined by the estimate, what would your production rate have to be to strip and clean the last 500 square feet?

PARTICIPANT ACTIVITY

Next, consider the activity *excavate and stockpile soil* as presented in the reporting forms (*Figures 9* and *11*) in the previous section. An analysis of this information shows the following:

ITEM	ESTIMATED	ACTUAL
Quantity of Soil	370.4 cu. yd.	400.0 cu. yd.
Equipment Production Rate	60 cu. yd./hr	50 cu. yd./hr
Unit Cost	$0.57/cu. yd.	$0.81/cu. yd.

Notice that there is no material cost connected with this activity. Therefore, only the amount of material and the equipment production rate is analyzed. Obviously, more material was actually handled than was originally estimated. This could be due to inaccurate estimating or unanticipated field conditions, and the specific reason must be determined to prevent such errors in the future.

The production rate of the equipment also varies from the estimate. Reasons why this occurred — such as estimating error, inclement weather, difficult site conditions, or unmotivated equipment operator — must be determined, and appropriate action taken to prevent repeating the error on another job.

Once the job is done, the supervisor cannot do much about correcting deviations from the estimate. However, if the job is still in progress when the supervisor is given the results of the analysis, the information can be used to identify problems and make corrections so that the balance of the job can be completed on time and within budget. Among the steps that might be taken are putting more labor on the job and working overtime — both of which would increase the cost of doing the work. However, three things can be done that may not increase the cost of the work:

- Change crew make-up
- Improve crew productivity
- Change the work method

These steps would allow the work to be accomplished in less time, although the second item is difficult to implement when the job is already partially completed.

Another option is to bring additional resources onto the job to get the work done on time. However, the decision to do this normally lies with upper management, not the supervisor.

Instructor's Notes:

SECTION 9

9.0.0 IMPACT OF IMPROPER REPORTING

Faulty reporting due to honest mistakes can happen, and it is occasionally excusable. Faulty reporting due to carelessness or indifference is inexcusable and must be avoided. The effort required to report with care is well worth the supervisor's time.

One type of faulty reporting — **spreading** — should always be avoided. Spreading is the practice of reporting labor hours from activities which are over budget as labor hours from activities that are under budget. Spreading occurs when a supervisor has one or more job activities which are running over estimate and one or more that are well within the estimate. In order to report all activities within estimate, the decision is made to "spread" the losses into the activities that are within budget when reporting on the project.

Spreading has several disastrous effects. The most crucial is that it hides potential flaws in the estimating process. Without being aware of it, management continues to use the faulty information in estimating future jobs and thereby unwittingly places other supervisors in the position of having to meet impossible production standards. The results are frustration, cost overruns, and unprofitable jobs that endanger everyone's job security.

Spreading will always have negative results. No matter what the actual hours are on each activity, the supervisor must report them accurately.

SUMMARY

To construct a job within the estimated cost and schedule, the supervisor must control the use of materials, tools, equipment, and labor at each step. This requires an awareness of effective productivity measuring and analytical techniques. Further, the individual must know how to improve work methods, when necessary, through the application of these techniques.

Supervisors should be familiar with the company's quality control program and fully understand their responsibilities in carrying out the program in the field.

It is vital that the supervisor understand the various costs connected with a job and why it is necessary to be constantly aware of them. It is important to also realize how production rates affect costs and what role the field reporting system plays in controlling those costs. Finally, it must be understood that reporting information incorrectly can drastically affect the jobs of the crew and the supervisor.

Classroom

Ask trainees to complete the Participant Activity. Discuss their answers. Answers are located at the end of this module.

PARTICIPANT ACTIVITY

Discuss and answer the following:

1. Describe how improperly reporting labor hours could affect how you are evaluated as a supervisor.

2. Read the following scenario and then answer the questions.

   It is estimated that 140,000 bricks are required for a particular task. The estimate also shows that a production rate for the crew is 10 hours per 1000 bricks. After the job begins, the supervisor discovers that the actual crew production rate is 13 hours per 1000 bricks.

   1. If the supervisor spreads 2 hours per 1000 bricks to a different code so that the inaccurately estimated production rate continues to appear accurate, what is the total number of hours mistakenly reported to upper management?

   2. If the average hourly rate is $12.00 per hour and the crew is comprised of four people, what is the total cost error?

   3. No matter who provides the information for the field reporting system, the information should be:
      a._____
      b._____

Instructor's Notes:
_____
_____
_____
_____
_____
_____
_____

Review Questions

Have trainees answer the Review Questions. Discuss the correct answers, located at the end of this module.

1. Productivity is defined as _____.
   a. the skill level of workers on a project
   b. the percentage of the project completed
   c. the number of workers on a project over a specific period of time
   d. the amount of units installed in a specific time period

2. Production is defined as _____.
   a. the number of workers on the clock
   b. the amount of material installed on a project
   c. the budgeted amount of labor and materials
   d. the rate of unit installation

3. In order to increase productivity, the supervisor should do the following except _____.
   a. provide training
   b. supply motivation
   c. require overtime
   d. limit breaks

4. A common error made in productivity analysis is to _____.
   a. observe workers too closely
   b. look for ways to improve productivity even if there are no problems
   c. examine the productivity of the company as a whole rather than each work group
   d. evaluate production rates by comparing them with those established in the estimate

5. Production analysis is used to _____.
   a. identify problems that can affect an upcoming project
   b. examine how identified problems can affect an upcoming project
   c. compare production rates of equipment to rates determined in the estimate
   d. determine how to get a project on track again once problems have occurred

6. Activity sampling is used in order to _____.
   a. observe and evaluate a large percentage of work being done
   b. determine, within a matter of days, the areas in which production needs to improve
   c. gauge the effectiveness of a specific task at a particular time
   d. evaluate how efficiently crew members feel a task is being performed

7. The five-minute rating technique is used in order to do the following except _____.
   a. show the interactions among crew members and their equipment over a five-minute time span
   b. help construct more detailed plans in the future, if needed
   c. discover shortages of materials
   d. show which crew members are not working efficiently

8. In order to conduct a five-minute rating, the supervisor should _____.
   a. observe crew members every five minutes
   b. observe at least several cycles of a task
   c. note only periods of crew member inactivity, not activity
   d. observe crew members at a set interval from the beginning of a task until its completion

9. Activities that are considered productive include all of the following except _____.
   a. completing time cards
   b. waiting until other crew members finish their work
   c. giving instructions to other crew members
   d. carrying material

10. One major way productivity can be improved is to _____.
    a. employ additional workers for a task
    b. attempt to lower the costs of labor as long as production is not adversely affected
    c. avoid wasting time motivating workers
    d. avoid using unfamiliar materials

11. A supervisor should evaluate workers primarily on the basis of _____.
    a. how efficiently they perform a task
    b. how hard they work
    c. how fast they perform a task
    d. how few rest breaks they take

12. The following involve steps for improving a work method except _____.
    a. involving other supervisors in the process
    b. depending on intuition for solutions to problems that need to be addressed
    c. using job analysis
    d. using the crew balance chart in order to gain insight

13. Another way to improve work methods is to _____.
    a. eliminate time spent gathering up materials before the project begins
    b. have the supervisor establish a plan and then instruct crew members about what the proper work method will be for all projects
    c. stick with the plan that was agreed upon before the project until productivity improves
    d. get worker consent that they will indeed follow the plan

14. In order to control materials in an effective manner, the supervisor should do the following except _____.
    a. ask upper management what the most effective work method is for the materials and project at hand
    b. record storage locations for large projects on an inventory form
    c. have the materials delivered to the job site on time
    d. be familiar with the method of installation for the materials

15. An effective way to control tools is to _____.
    a. allow all workers to have open access to tools
    b. have a check-in/check-out system
    c. ask the workers if they know the safe operation of the tools that they will be using
    d. make regularly announced tool safety checks

16. The hardest item to control on a job is the resource of _____.
    a. equipment
    b. labor
    c. material
    d. tools

Instructor's Notes:

17. It is important that the supervisor be aware of the costs of doing a project for the following reasons *except* _____.
    a. it allows the contractor to determine if a project is measuring up to its estimated cost, determined before the project began
    b. it will aid in quality control
    c. the contractor will receive the profit that was estimated before the project began
    d. it will help the project be completed on time

18. The supervisor must control how material, tools, equipment, and labor are used in the construction process. This is referred to as _____.
    a. cost awareness
    b. cost control
    c. resource awareness
    d. resource control

19. One of the steps involved in the control process is _____.
    a. making a list of needed skill sets
    b. making a list of needed personnel
    c. having measurable standards
    d. allowing for changes from the standards

20. In order to control job resources in an effective manner, the supervisor must know the following things at the beginning of the project *except* _____.
    a. the actual cost
    b. the number of labor hours involved
    c. the cost estimates of various tasks
    d. the amounts of material involved

21. In order to have standards against which to measure the performance of the crew, the supervisor should obtain the necessary information from the _____.
    a. inventory form
    b. job schedule
    c. job diary
    d. job analysis

22. Estimated job costs are derived from _____.
    a. the projected cost
    b. a contract, in which the project owner determines the expenses
    c. direct costs, indirect costs, and the profit that is anticipated
    d. the job breakdown

23. Indirect costs consist of _____.
    a. expenses for obtaining resources
    b. expenses for business operations independent of a project's actual construction
    c. the actual cost minus the estimated cost
    d. the actual cost minus the projected cost

24. The estimated cost serves the following purposes *except* _____.
    a. to indicate the maximum amount of money that can be spent on a project
    b. to provide valuable information about job conditions
    c. to indicate production rates for items such as labor
    d. to help determine a project's efficiency

25. The estimated cost of a project while the job is in progress is called _____.
    a. the projected cost
    b. the direct cost
    c. the actual cost
    d. the indirect cost

26. Actual costs are determined from all of the following *except* _____.
    a. time cards
    b. the job reporting system
    c. bills from suppliers
    d. a job analysis

27. A reporting system is used to _____.
    a. determine whether a job is progressing within the estimate
    b. determine the skill level of crew members on a particular task
    c. estimate the costs and amount of hours spent in a project
    d. supply information about a job by completing numerous complex reports about the tasks involved

28. A cost coding system involves _____.
    a. the supervisor giving a unique code to such items as labor and material
    b. an estimator assigning a number/letter to work items during the preparation of the estimate
    c. identifying the project as a whole with one number/letter code
    d. estimating the costs of each work item in a project

29. One of the purposes of the daily time report is _____.
    a. to provide an estimate of hours involved in a task
    b. to assess indirect costs associated with a project
    c. to indicate how long various pieces of equipment were used
    d. to calculate a project's actual production rates

30. After the field data is obtained, it is _____.
    a. distributed to all crew members
    b. summarized in a final report, usually once every month or two
    c. utilized in order to measure production
    d. used in order to make an initial project schedule

Instructor's Notes:

NATIONAL CENTER FOR CONSTRUCTION EDUCATION AND RESEARCH

GLOSSARY

Trade Terms Introduced in This Module

Activity sampling: The practice of observing and classifying a small percentage of a total activity.

Actual cost: Costs incurred in doing the work. These must be accurately tracked and recorded if a true financial record of the project is to be kept.

Cost awareness: Understanding how the costs in the job estimate are derived.

Crew balance: The proper deployment and efficient use of personnel involved in the work process; having the right number and type of personnel assigned to each job.

Crew balance chart: Bar chart that summarizes the results of the five-minute rating. Also can be used by itself to graphically depict the efficiency of each person in the crew.

Delay survey: Identifies delay problems in the field. The survey form lists probable causes of job delays on the job and gives the supervisor space to record the number of hours involved in the delay and the number of workers affected.

Estimated cost: Combination of direct costs, indirect costs, and expected profit, as determined by the contractor.

Five-minute rating: Uncovers construction delays such as crews or crafts interfering with each other, poor work methods, and shortages of equipment and materials. Can be used to measure the efficiency of a work method.

Production: Output that is measured by the number of units produced or the quantity of material installed.

Production analysis: Compares actual cost to estimated cost or compares actual labor and equipment production rates to the original estimates.

Productivity: Relates to the efficiency with which the workers' work is being performed.

Projected cost: An estimate of the expenses that will be incurred in completing the job. It differs from the original estimated cost in that it is calculated when the job is in progress and takes into account actual job conditions.

Resource control: Controlling time, money, materials, labor, tools, and equipment. How these resources are used on the job determines whether or not the project is built on time and within the estimate.

Spreading: The practice of reporting labor hours from activities which are over the estimate as labor hours from activities that are under budget.

Ask trainees to review the Glossary for unfamiliar terms.

Review the objectives of the module and then answer any questions the trainees may have.

Have trainees prepare for the Module Examination.

Administer the Module Examination. Be sure to record the results of the Exam on Craft Training Report Form 200 and submit the results to the Training Program Sponsor.

NATIONAL
CENTER FOR
CONSTRUCTION
EDUCATION AND
RESEARCH

MODULE MT208

Answers to Review Questions

	Answer	Section Reference
1.	d	2.0.0
2.	b	2.0.0
3.	c	2.1.0
4.	d	2.1.0
5.	c	2.2.0
6.	c	2.2.0
7.	a	2.2.1
8.	d	2.2.1
9.	b	2.2.1
10.	b	3.0.0
11.	a	3.0.0
12.	b	3.0.0
13.	b	3.0.0
14.	a	4.1.0
15.	b	4.2.0
16.	b	4.4.0
17.	b	6.0.0
18.	b	6.0.0
19.	c	6.0.0
20.	a	6.1.0
21.	d	6.1.0
22.	c	6.2.1
23.	b	6.2.1
24.	c	6.2.1
25.	a	6.2.3
26.	b	6.2.2
27.	a	7.0.0
28.	b	7.1.0
29.	d	7.2.1
30.	c	7.2.1

Answers to Participant Activities

Participant Activity Section 3.0.0

1. The efficiency for the Five-Minute Rating is about 51%, which is below 60%; the crew's productivity could be improved.

2. There are many correct answers to this question. One might suggest eliminating laborer #3, if his or her work could be done by finisher #1 and laborer #4 or laborer #2.

3. The instructor can obtain this using the procedure presented in the Trainee Guide.

4. & 5. There are many correct answers to these two questions. The instructor must be sure they are consistent with the material previously presented in this exercise. An example for question 4 may be to replace laborer #4. An example for question 5 may include removing the chute during one of the finishing tasks or using more efficient finishing tools to decrease the total amount of time.

Particpant Activity, Section 6.2.2

1. Material=$8,000.00; Labor=$8,300.00; Total=$16,300.00. Unit cost=$1.63 per block.

2. Material=$8,000.00; Labor=$9503.50 (based on 229 hours); Total=$17,503.50 or about $1.75 per block.

3. Material=$4000.00; Labor=$3320.00; Total=$7320.00

4. Response should indicate a negative outcome.

5. Response should indicate good job performance.

Participant Activity Section 8.0.0

1. The crew is not being as productive as it could be. The quantity of work being produced is off by 25%. The crew's productivity is off by 10 sq. ft./hour. The unit cost overrun is .04/sq. ft. If things continue this way, the job will not be completed on time.

2. You might observe the crew and determine what is slowing them down. See if their efficiency can be improved. Replace crewmembers with more experiences, faster workers.

3. Money ($80) would be lost if things continue this way.

4. Your production rate would need to be slightly more than 166 sq. ft./hour.

Participant Activity Section 9.0.0

1. The supervisor's work (production) evaluation would be based on incorrect figures and thus the evaluation would be invalid.

2. A. 280 hours

   B. $13,440.00

3. Accurate, complete

CONTREN™ LEARNING SERIES — USER UPDATES

The NCCER makes every effort to keep these textbooks up-to-date and free of technical errors. We appreciate your help in this process. If you have an idea for improving this textbook, or if you find an error, a typographical mistake, or an inaccuracy in NCCER's Contren™ textbooks, please write us, using this form or a photocopy. Be sure to include the exact module number, page number, a detailed description, and the correction, if applicable. Your input will be brought to the attention of the Technical Review Committee. Thank you for your assistance.

Instructors – If you found that additional materials were necessary in order to teach this module effectively, please let us know so that we may include them in the Equipment/Materials list in the Instructor's Guide.

Write: Curriculum Revision and Development Department
National Center for Construction Education and Research
P.O. Box 141104, Gainesville, FL 32614-1104

Fax: 352-334-0932

E-mail: curriculum@nccer.org

Craft _____ Module Name _____

Copyright Date _____ Module Number _____ Page Number(s) _____

Description
_____
_____
_____
_____

(Optional) Correction
_____
_____
_____

(Optional) Your Name and Address
_____
_____
_____